Anatomy and Physiology for Nurses

and Students of Human Biology

Anatomy and Physiology for Nurses

and Students of Human Biology

W. GORDON SEARS
M.D. (Lond.), M.R.C.P. (Lond.)
Hon. Consultant Physician, Mile End Hospital, London

and

R. S. WINWOOD
M.B., M.R.C.P.
Consultant Physician, Whipps Cross Hospital, London

FIFTH EDITION

EDWARD ARNOLD

First Published 1941
by Edward Arnold (Publishers) Ltd
41 Bedford Square, London WC1B 3DP

Reprinted 1942, 1943 (twice), 1944 (twice), 1945, 1947, 1948, 1950
Second Edition 1951
Reprinted 1952, 1953, 1954, 1955, 1956, 1957
Third Edition 1958
Reprinted 1959, 1960, 1961, 1962, 1964
Fourth Edition 1965
Reprinted 1967, 1969, 1971, 1972
Fifth Edition 1974
Revised Reprint 1976, 1978

Boards Edition ISBN 0 7131 4218 9
Paper Edition ISBN 0 7131 4219 7

Printed in Great Britain by
Butler & Tanner Ltd, Frome and London

Preface to the Fifth Edition

This book was first written over thirty years ago and every endeavour has been made to keep it up to date with each new edition. It was intended to be a basic text book on the subject for nurses and other students of Human Biology and to include brief references to clinical subjects which could be supplemented by consulting appropriate works when necessary.

The fact that over half a million copies have been sold has encouraged us to believe that it serves a useful purpose.

It has been our endeavour to keep the contents as simple and readable as possible, but at the same time to maintain scientific accuracy compatible with modern advances in knowledge and technical application.

For examination purposes a book of this type is intended to be used under the direction of a competent teacher who will indicate the various portions and details which the candidate is expected to know. On the whole, therefore, more detail than is essential is probably included for purposes of reference.

The type has been reset, the diagrams redrawn and increased in number.

We earnestly hope that the book will continue to be useful and we are very grateful to the nursing profession and others for the support they have given to us in this and our other works for so many years.

W. Gordon Sears and R. S. Winwood
London 1974

Contents

	Preface to the Fifth Edition	v
1	Introduction	1
2	The Tissues	8
3	The Body as a Whole	33
4	The Skeleton	40
5	The Joints or Articulations	90
6	The Muscular System	108
7	The Circulatory System	127
8	The Blood (Haemopoietic system)	164
9	The Lymphatic System	179
10	The Anatomy of the Digestive System	187
11	The Physiology of the Digestive System	215
12	Accessory Organs of Digestion	223
13	Metabolism, Diet and Nutrition	233
14	The Skin. Regulation of Body Temperature	254
15	The Respiratory System	259
16	The Urinary System	275
17	The Ductless Glands or Endocrine System	289
18	The Nervous System	306
19	The Autonomic Nervous System	336
20	The Special Sense Organs	340
21	The Reproductive System	363
	Appendix	380
	Miscellaneous Questions	393
	Weights and Measures	395
	Index	396

1 Introduction

Life is, perhaps, the most mysterious fact in the Universe and it is not unreasonable that man has devoted much study to this phenomenon. The results of his labours have produced the science of Biology. In its broad sense, this subject embraces all living matter, both animal and vegetable, in all its forms, both visible and microscopic.

The study of the simplest forms of life contributes to the better understanding of those which have attained a more complicated and advanced degree of development in the scale of Nature.

From the earliest concepts of the subject, many of which were grossly inaccurate, careful study and the application of logical thinking, backed by evidence supported by ever-growing scientific techniques, has provided an enormous amount of knowledge. Much of this knowledge is so advanced and so specialized that it can only be appreciated by the few and even they would be the first to admit that such knowledge is incomplete and always capable of further expansion.

Human Biology may be studied as a pure science. On the other hand, for doctors, nurses and many other workers it is the practical application of this knowledge to the understanding of disease and the general well-being of the human race that is of major importance.

However, in order to attain this understanding some familiarity with science in general is essential, and in order to apply it to full advantage there must be further appreciation of the workings of the human mind and the development of one of the greatest of human attributes, namely sympathy.

To return to the basic aspects of Human Biology, this has numerous subdivisions which include Anatomy, Physiology and Biochemistry. However, 'the divisions of the sciences are like the branches of a tree that join in one trunk' (Francis Bacon) and they are, therefore, more or less closely related to one another.

Anatomy is the study of the parts of the body, their form, position and relationship to each other. This knowledge has been obtained by careful dissection and further expanded by the detailed study of the structure of the various tissues under the microscope (Histology).

A greater understanding of the subject has been obtained by studying the anatomy of other members of the animal kingdom, the development of the adult creature from its conception in the ovum or egg (Embryology) and a general consideration of the known facts of evolution.

Physiology is the study of the functions of the body as a whole and of the individual structures and organs contained therein. Some of this is reasonably simple; some involves complicated chemical, physical and electrical details.

Every living structure, whether it be animal or vegetable, is derived, so far as we know, from another living structure. It has the power of growth and reproduction, and its life is dependent upon its ability to absorb non-living material which it builds up into the framework of its own body.

Before considering living matter, it is necessary to go back a step further and ascertain the nature of the **chemical substances** of which it is composed, and which are, therefore, found in the human body as a whole.

Broadly speaking, there are two types of matter: elements and compounds. The latter may be divided into inorganic and organic.

An **element** is a substance which contains only one kind of matter. The following are the most important elements found in the human body: carbon, hydrogen, nitrogen, oxygen, sulphur, phosphorus, chlorine, iodine, sodium, potassium, magnesium, calcium and iron. Of these, oxygen and nitrogen sometimes occur in their uncombined natural state. The others are found combined with one another in the form of compounds.

A **chemical compound** is a combination of two or more elements in fixed proportions forming an entirely new substance in which the individual elements apparently lose their identity, thereby differing from a simple mixture. Every part of a compound has exactly the same composition and properties as every other part.

Inorganic compounds are relatively simple combinations of the elements found especially in non-living matter such as minerals, water and salts.

The essential feature of organic compounds is the presence of the element carbon, usually combined with hydrogen and oxygen.

In addition, nitrogen and other elements may be also included and form compounds of a highly complicated nature which are found specially in living matter.

The main organic compounds found in the body are:

Carbohydrates, Fats	containing carbon, hydrogen and oxygen.
Proteins	containing nitrogen and other elements, in addition to carbon, hydrogen and oxygen.

Going back one stage further in the structure of matter and in order to understand some of the principles which must be considered in Physiology, it is necessary to have some knowledge of the atomic theory which was propounded by Dalton about one hundred and eighty years ago. This has been the basis of chemical science ever since.

Stated simply this implies:

1. The basis of all matter is the atom.

2. If further subdivided, an atom consists of protons, neutrons and electrons. The protons and electrons each carry a unit charge of electricity. That of the former is positive and that of the latter negative. The neutrons, as their name implies, are electrically inactive.

3. Every atom consists of a central particle or nucleus around which constantly revolve in their own orbit one, a few or many smaller electrons. On an astronomical scale these are rather like planets revolving round the sun.

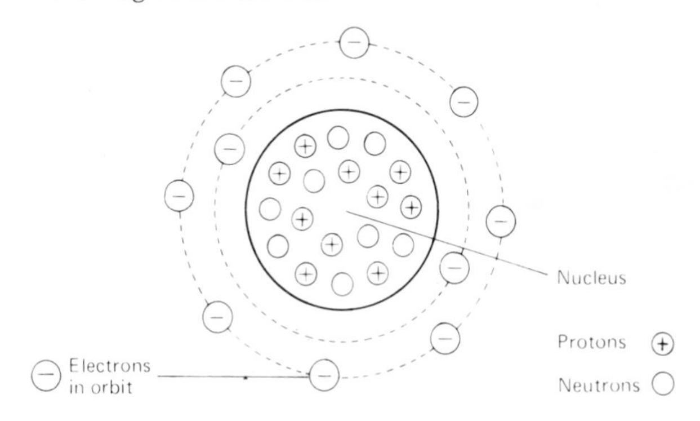

Fig. 1 Diagram of an atom with ten positive protons in the nucleus and ten negative electrons in orbit.

4. The nucleus consists of a compressed mass of protons and neutrons. Because there is a preponderance of protons over neutrons it is positively charged with electricity. In order to render the atom

as a whole electrically neutral it has an appropriate number of circulating negative electrons.

5. For example, the atom of hydrogen carries one electron; that of carbon, six; nitrogen, seven; oxygen, eight; sodium, eleven; chlorine, seventeen, and so on.

6. Under ordinary circumstances the atoms of each element are stable (with the exception of radium, etc.). In 1919 Rutherford succeeded in splitting the atom and his work has led step by step to the modern science of Atomic Fission.

7. The atoms of most elements have the property of combining with atoms of other elements to form the molecules of new compounds.

8. This power can best be visualized by imagining that the atom of each element has one or more hooks or bonds which can link up with a similar hook or hooks of another atom or atoms:

e.g. hydrogen, sodium and chlorine have one hook;
oxygen, calcium and sulphur, two;
nitrogen, three;
carbon, four, and so on.

Thus, one atom of sodium can combine with one atom of chlorine to form one molecule of the compound, sodium chloride or common salt. Using standard chemical symbols this might be expressed thus:

Na — Cl = Sodium Chloride

and one atom of oxygen can combine with two of hydrogen:

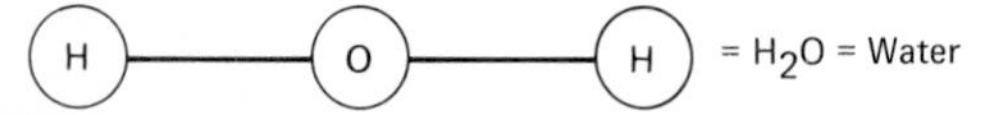

Nitrogen having three 'hooks' can link up with three atoms of hydrogen to make a molecule of the compound ammonia.

9. In other words, the atoms of elements by union with those of other elements form the molecules of chemical compounds which have a larger mass than the individual elements which compose them. Therefore in scientific terms an atom is said to have its own atomic weight and a compound it molecular weight.

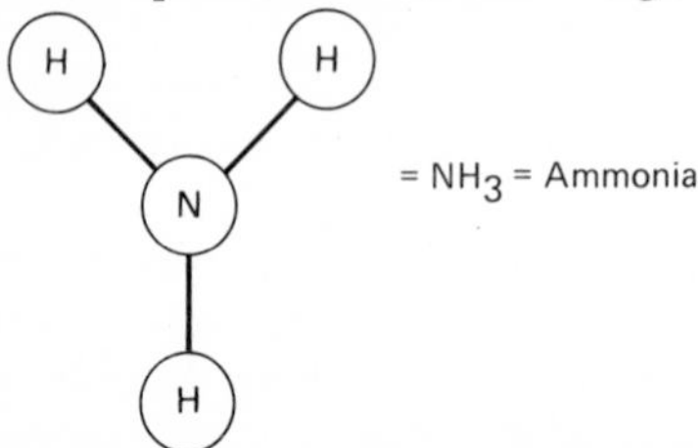

The atomic weight of an element is the average weight of an atom of that element in relation to the weight of an atom of hydrogen, which is taken as 1 (e.g. carbon—12, iron—56, lead—207).

Because the nucleus of some elements does not always have the same number of neutrons but still has the identical number of protons and electrons, there is a slight variation in the weights of individual atoms of a particular element. Those atoms which vary in weight from the standard are called **isotopes**. Many of these are unstable and are radioactive, discharging their nuclei (alpha particles) at high velocity.

Such radioactive isotopes introduced into the body can be detected and traced by a Geiger counter. Thus radioactive iodine given to an individual may be traced to the thyroid gland. In larger doses radioactive isotopes, e.g. radium, may be used to destroy the abnormal cells which occur in cancer.

The following table lists the elements present in the body and their respective symbols:

carbon (C)	chlorine (Cl)
hydrogen (H)	iodine (I)
nitrogen (N)	sodium (Na)
oxygen (O)	potassium (K)
sulphur (S)	magnesium (Mg)
phosphorus (P)	calcium (Ca)
	iron (Fe)

Many other elements are present in only minute amounts and are known as **trace elements.** Some of these are essential to life; examples are cobalt, copper, manganese, molybdenum and possibly selenium. Chromium may possibly protect arteries from atherosclerosis. Some trace elements serve no useful purpose in the body but are contaminants.

10. Returning to the example of sodium chloride, it will be recalled that the atom of sodium has eleven electrons and that of chlorine seventeen. In effecting the combination to form a compound there is a rearrangement of the electrons in such a way that the sodium atoms become positively charged with electricity and the chlorine atoms become negatively charged. Such electrically charged atoms are called **ions**.

11. When such compounds are dissolved in water some, but not all, of the molecules of the compound become ionized into their individual electrically charged atoms. Some of these atoms will carry a positive charge and the others a negative charge. In the case of sodium chloride, the sodium ions are positive (+) and the chlorine ions negative (−)

Atoms in this state of ionization, because they carry an electric charge, are referred to as electrolytes, and the solution containing them, an electrolyte solution. It is in this form that many salts circulate in the water contained in the blood and tissue fluids of the body.

12. Another aspect of this subject is the 'acid/alkaline' reaction of body fluids. The hydrogen ions (H^+) have a positive charge and are acid, while the hydroxyl (HO^-) are negative and cause alkalinity. If the H ions and the HO ions are equally balanced the reaction of the fluid will be neutral:

$$H^+ + HO^- = H_2O \text{ (or water which is neutral)}$$

For practical purposes a scale has been devised and numbered from 1 to 14. The central figure of 7 is taken to represent neutrality and the hydrogen ion concentration is indicated by the symbol pH.

Fluids having a pH of 1 to 7 are acid and those with a pH exceeding pH 7 are alkaline.

It must be understood that this is a scale only used to measure very small degrees of acidity and alkalinity. For example, the pH of the blood is kept constant at 7.4, i.e. very slightly alkaline but never acid. That of urine is usually slightly acid with a pH of 5 to 6.

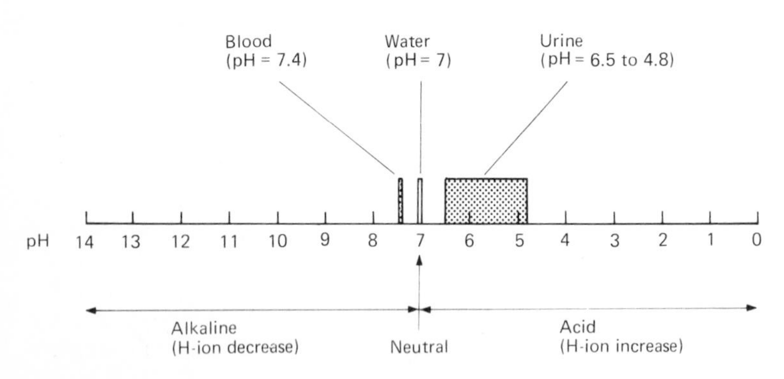

Fig. 2 Diagram illustrating hydrogen ion concentration (pH) in the body.

One of the most important functions of the various salts present in the body is to keep the pH of the blood constant and there is a constant interchange of various positive and negative ions in the tissues to maintain this equilibrium. Any excess of acid hydrogen ions is excreted in the urine by the kidney cells.

On the other hand, one of the most important end-products of carbohydrate and fat metabolism is carbon dioxide (CO_2). When this is dissolved in water it forms a weak acid (carbonic acid):

$$\underset{\text{carbon dioxide}}{CO_2} + \underset{\text{water}}{H_2O} \rightleftharpoons \underset{\text{carbonic acid}}{H_2CO_3} \rightleftharpoons \underset{\text{hydrogen ion}}{H^{+ve}} + \underset{\text{bicarbonate ion}}{HCO_3^{-ve}}$$

This, when ionized, also liberates acid H ions in considerable quantity which would tend to lower the pH of the blood towards the neutral figure of 7. However, the respiratory centre in the medulla of the brain is particularly sensitive to any change in the pH of the blood and immediately causes an increase in the rate and depth of breathing which is maintained until the excess of carbon dioxide (and at the same time the excess of hydrogen ions in the blood) is removed.

There are also certain salts together with the blood plasma which themselves are neutral but which have the power of reacting with hydrogen ions without becoming acid. These are called buffer substances which also help to maintain the pH of the blood at a constant level.

2 The Tissues

(General Histology)

The cell

The unit of the animal kingdom is the living cell. In fact, the most primitive form of animal life is a single-celled organism, the amoeba, which is found in pools of stagnant water and can be seen when a drop is examined under the microscope.

The cell is the structural unit or brick with which the human body is built. Just as many kinds of brick may be used in the construction of a large building, so many different types of cell are

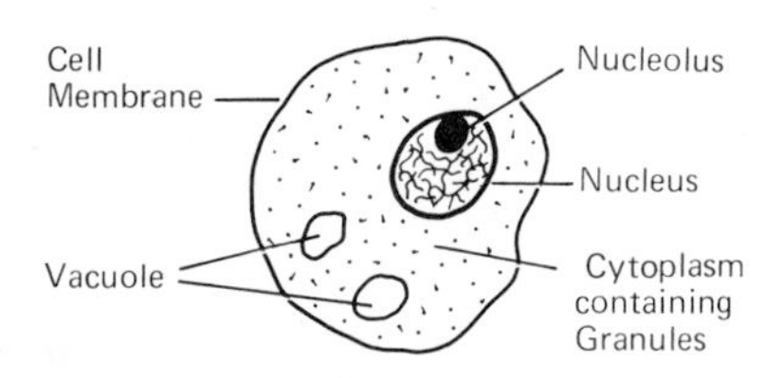

Fig. 3 The general structure of a cell.

found in the body, but masses of similar cells may be collected together to form special organs and tissues and have special functions.

Again, going back a stage further, it may be said that the bricks of which cells are made are the molecules of the various chemical substances present in them.

The general characteristics of the cell are similar, whether we consider a single-celled organism like amoeba or a cell which is but an infinitesimal part of a highly developed animal such as man.

A cell may be defined as a small mass of protoplasm containing a nucleus. It has the following characteristics which indicate its signs of life:

1. Assimilation of nourishment.

2. Growth and repair.
3. Reproduction.
4. Excretion of waste products.

Some cells have the power of movement.

Although we do not understand the source of life or why protoplasm lives, we know that **protoplasm** consists of the following substances:

1. The very complicated organic compounds known as proteins (white of egg is a familiar instance of a protein).
2. Fatty substances.
3. Carbohydrates (sugars and starches).
4. Inorganic salts, especially the phosphates and chlorides of calcium, sodium and potassium.
5. Water (a compound of hydrogen and oxygen—H_2O).

In the centre of the cell protoplasm is the nucleus which is essential for the life, growth and reproduction of the cell and, in fact, appears to control its activity. The nucleus differs from ordinary protoplasm in its chemical reactions and stains more deeply with chemical dyes than the rest of the cell. In it is found a special type of protein called nucleoprotein, which contains a larger proportion of phosphorus than other varieties.

Chromosomes

The nucleoprotein of the nucleus actually consists of a number of minute threads which are called chromosomes. Except when the cell is dividing, the chromosomes look like a mass of darkly staining material, which is called chromatin. However, when the cell divides the chromatin arranges itself into a number of rod-shaped bodies, each of which has a characteristic shape and size. By using special techniques these can be arranged in groups of two, three or more and, in the human, have been given letters and numbers to distinguish the groups.

The number of chromosomes are constant for all the cells in the body of each species of animal. Thus, human cells contain 46 chromosomes, whereas those of a mouse only have 24.

When a cell divides, the chromosomes of the nucleus split longitudinally so that half of each chromosome goes into each new cell. This process is called mitosis.

Each tiny chromosome actually consists of a number of smaller portions rather like a string of beads. Each of these small units is called a gene. It is thought that the genes are composed of molecules of desoxyribonucleic acid (DNA). It is the genes which pass on the characters of the parent cell. In fact the various human characteristics are all transmitted by the genes of the sex cells of the parents.

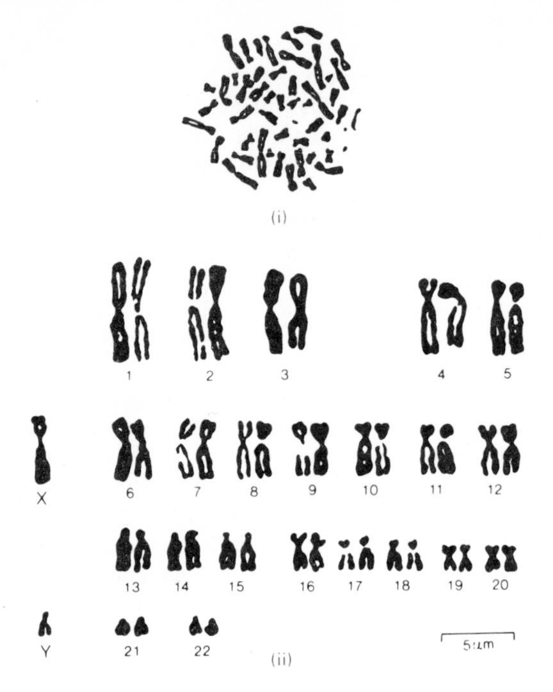

Fig. 4 Typical chromosome culture (i) as it appears in a jumbled form under the microscope and (ii) after the chromosomes have been individually photographed and arranged in pairs in accordance with their size and configuration.

Not only are the more obvious features such as the colour of hair and eyes so transmitted but also the factors which influence the blood groups and, in some instances, certain congenital defects and hereditary diseases.

Genetic information in a cell flows from chromosomal DNA to messenger RNA (m RNA) and thence to the cell's proteins.

Viruses can enter cells and substitute their own type of DNA for that of the cells. Further, the structure of DNA can also be altered by X-rays and atomic radiation (see page 5).

The **sex** chromosomes, called X and Y, can be distinguished from the others which are known as **somatic** (body) chromosomes. A female has two X chromosomes (XX) whereas a male has an X and a Y chromosome (XY constitution). The sex of a child clearly depends on whether it inherits an X or Y chromosome from its father.

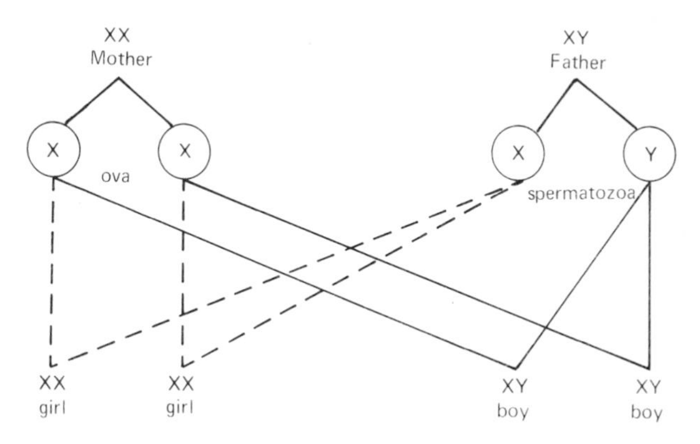

Fig. 4a From this diagram it is seen that the chances of having a boy or a girl child are 50:50.

We know also that, from one point of view, life involves a process of combustion or the burning up of fuel supplied in the form of food. One substance, oxygen, is necessary for combustion of any kind. Its utilization in this way is known as the process of oxidation.

A candle consists principally of carbon and hydrogen. When this burns it consumes oxygen and forms carbonic acid (carbon dioxide CO_2) and water. At the same time it liberates energy in the form of light and heat. Living protoplasm is more complex in composition than a candle but, so far as its carbon and hydrogen are concerned, the final products of combustion are the same and the process of oxidation results in the production of energy (heat), with the formation of the same waste products, namely, carbon dioxide and water.

A living cell, therefore, takes in or assimilates food in the form of carbohydrate and fat which it burns or oxidizes with the production of energy. As this process goes on, waste products are formed which the cell excretes from its body.

The cell is also subject to wear and tear and, in order to repair this and to have the power of growth, protein is necessary. This is also assimilated from outside sources. In utilizing protein for these purposes other waste products, including urea which contains nitrogen, are formed and excreted.

The power of reproduction is centred in the nucleus. This first of all divides into two portions which wander to opposite ends of the cell. The protoplasm then divides also and two complete cells

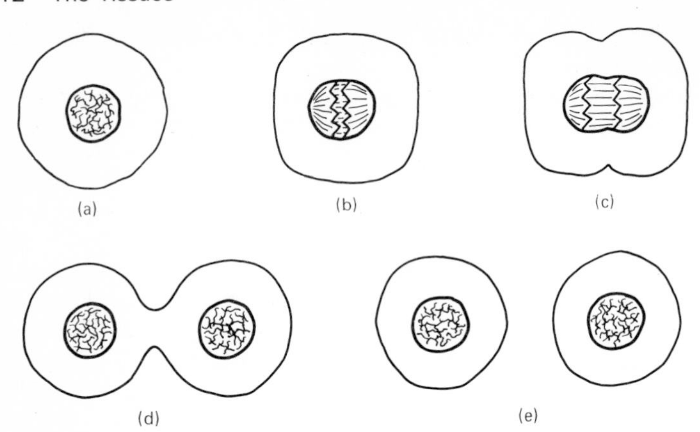

Fig. 5 The process of cell reproduction by mitosis. (*a*) Cell with nucleus and chromosomes. (*b, c*) Rearrangement and splitting of chromosomes. (*d*) Division of protoplasm. (*e*) Two daughter cells.

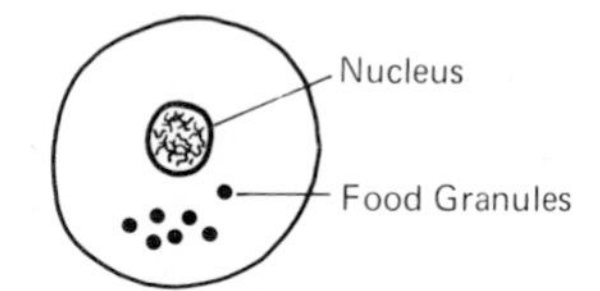

The amoeba at rest

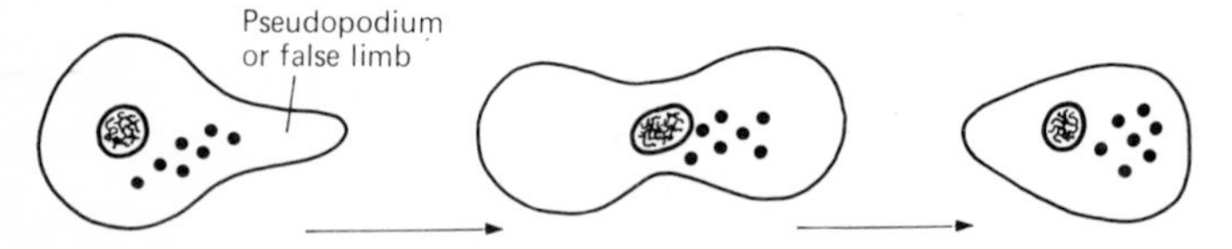

Three stages in the movement of the amoeba

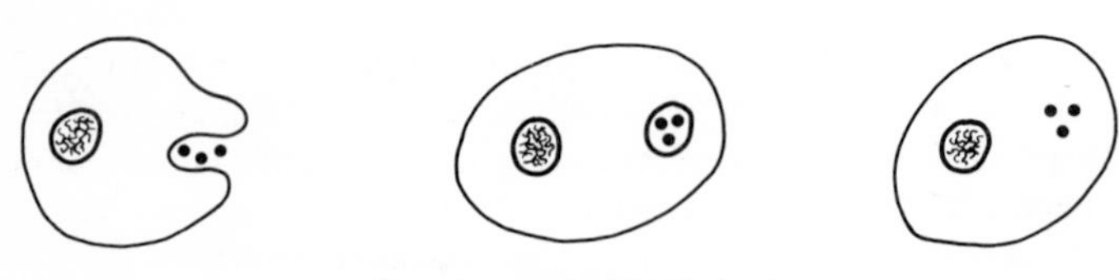

How the amoeba takes in food

Fig. 6 The amoeba.

result, each of which has the same properties and powers of reproduction as the parent cell (Fig. 5).

Single-celled animals like amoeba have the power of movement, which they achieve by pushing out a process of protoplasm (pseudopodium) at one end and withdrawing at the opposite end of the cell. This process of amoeboid movement is also exhibited by the white cells or leucocytes of the human blood when they ingest bacteria or dead tissue (page 169).

We have seen that cells have the power of absorbing substances necessary for their nourishment from their surroundings and, in using them, they act as small chemical laboratories. In addition to absorbing beneficial substances, it is possible for harmful poisons to be taken in, which will either kill the protoplasm or so depress its activity that the cell can no longer function normally. If the cell is not killed, the poison may be excreted as a waste product and the cell will then be able to resume its usual activity. This has a practical bearing on medicine. The activities of the cells of the body may be modified beneficially or otherwise by the action of certain drugs upon them. Also the poisons (toxins) of bacteria may kill the cells or they may be destroyed by corrosive substances and disinfectants. Viruses, which are organisms too small to be seen by the ordinary microscope, may enter cells and cause damage and disease. When invaded by a virus, a cell produces **interferon** as a first line of defence. Interferon is a protein with antiviral properties and is capable of protecting other cells of the organism against the virus.

THE TISSUES

The cells of the human body differ in appearance according to the particular type of tissue to which they belong and the functions which they have to perform.

The last statement illustrates an important law in Anatomy and Physiology, namely that **The structure of an organ or part is adapted to the function which it is called upon to perform.**

The following are the principal kinds of tissue found in the body. Each consists of individual types of cells which can be seen and recognized under the microscope. The cells are held together by some form of cement substance.

1. *Epithelium*	squamous (simple, stratified and transitional)
	columnar or glandular
	ciliated (columnar)
2. *Connective tissue*	fibrous tissue
	areolar tissue
	adipose tissue
	lymphoid tissue
	cartilage
	bone
3. *Muscular tissue*	voluntary or striated
	involuntary or plain
	cardiac or heart muscle
4. *Nervous tissue*	nerve cells or neurons and neuroglia
5. *Blood cells*	red cells or erythrocytes
	white cells (page 168)

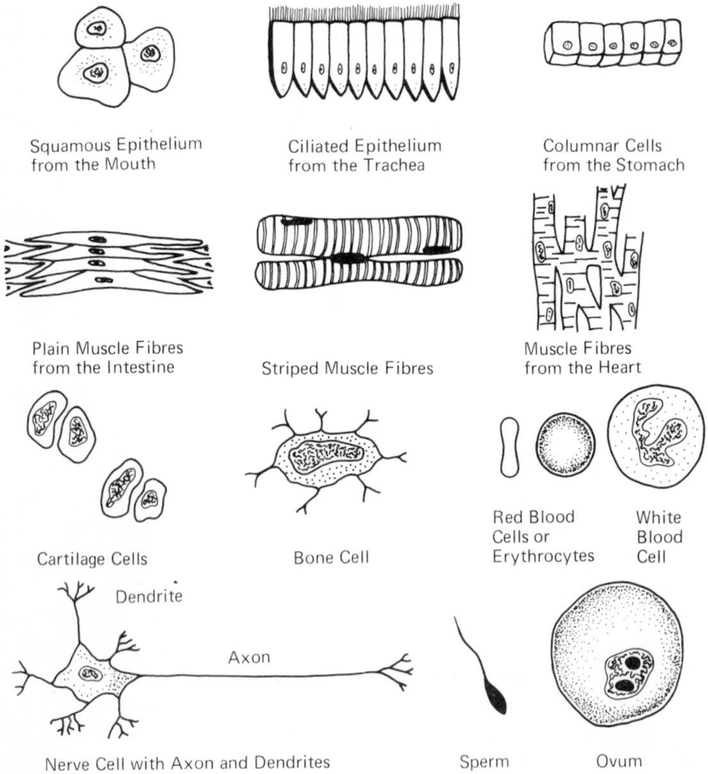

Squamous Epithelium from the Mouth

Ciliated Epithelium from the Trachea

Columnar Cells from the Stomach

Plain Muscle Fibres from the Intestine

Striped Muscle Fibres

Muscle Fibres from the Heart

Cartilage Cells

Bone Cell

Red Blood Cells or Erythrocytes

White Blood Cell

Dendrite

Axon

Nerve Cell with Axon and Dendrites

Sperm

Ovum

Fig. 7 Different kinds of cell found in the human body.

Epithelium

Epithelium is a tissue composed of cells, generally arranged to form a membrane or lining, covering either an internal or external surface. The following types are found:

Squamous epithelium

(*a*) Simple, (*b*) stratified.

Simple squamous epithelium (sometimes known as scaly or pavement epithelium) consists of a single layer of flattened cells (Fig. 8). It is found lining the air cells of the lungs (alveoli), the interior

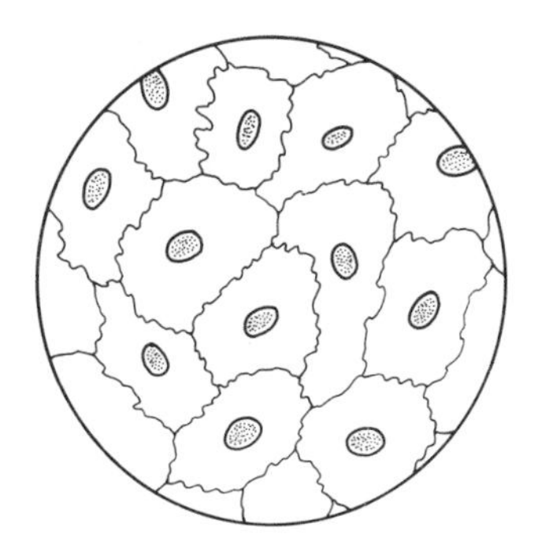

Fig. 8 Simple squamous epithelium.

of the heart (endocardium), the interior of the blood vessels and lymphatics, the pleura and the peritoneum. It forms a smooth flat membrane and, when lining an internal surface, is sometimes called endothelium.

Stratified squamous epithelium consists of similar cells arranged in layers or strata, one on top of the other, so that the membrane is at least twelve layers thick. The surface cells are flattened like the simple variety, but the deeper ones are rounded (Fig. 9).

The skin or epidermis consists of stratified epithelium, the outer layers of which become hard and horny because they contain a substance known as keratin. Stratified epithelium is also found in the mouth, pharynx, oesophagus, vagina and part of the urethra. Its structure is specially adapted to the wear and tear of the surface of the body. The superficial layers are constantly being shed and replaced by the growth of the deeper layers.

The detailed structure of the skin is described on page 254 in

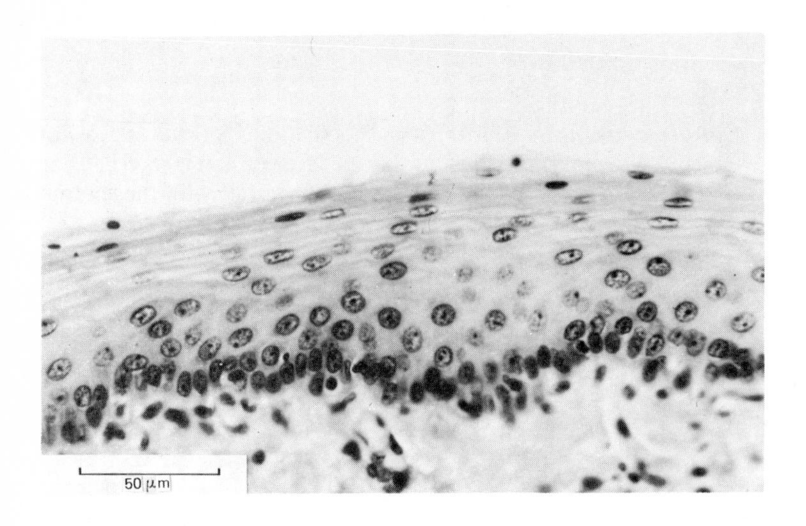

Fig. 9 Stratified squamous epithelium from the pharynx.

connection with one of its important functions, viz. the secretion of sweat and the regulation of body temperature.

Transitional epithelium is a type of stratified epithelium about four cells in thickness and is found in the bladder and ureters. It is adapted to withstand the action of urine with which it is constantly in contact.

Columnar or glandular epithelium.

This consists of cylindrical-shaped cells, only one layer thick, and is found in the secreting glands of the body, such as the salivary glands and the breast. The stomach and intestines area also lined with columnar epithelium through which the absorption of fluids and foodstuffs takes place. Some of these cells are also specially adapted to secrete digestive juices (Fig. 10).

Ciliated columnar epithelium. This is a special form of columnar epithelium. The free surface of the cell is surmounted by a

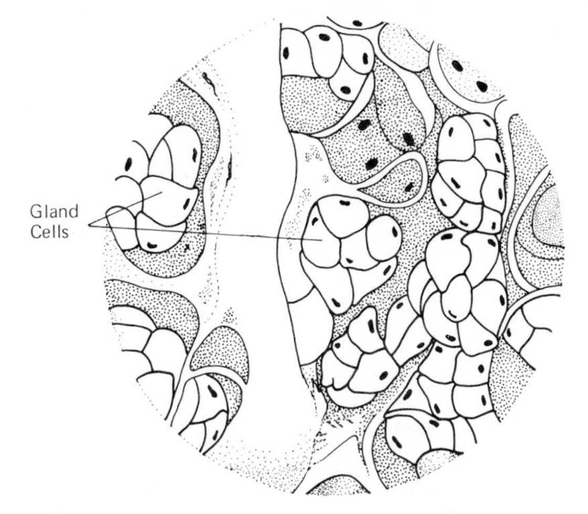

Fig. 10 Glandular epithelium from a salivary gland (submandibular gland).

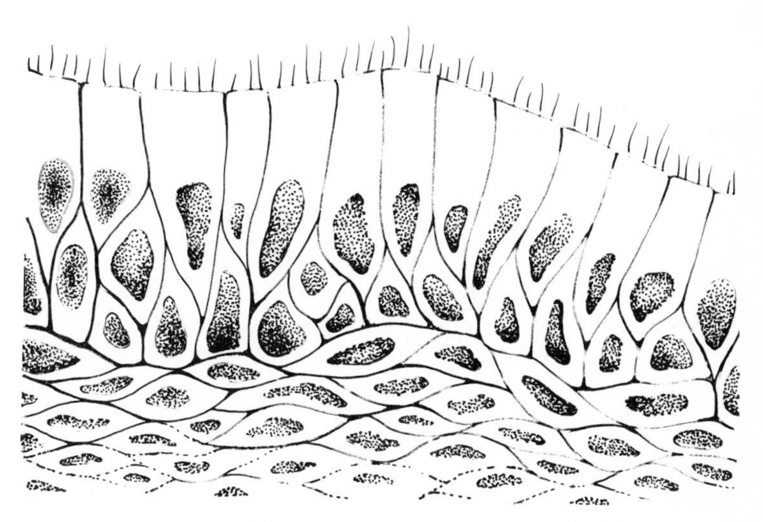

Fig. 11 Ciliated columnar epithelium from the trachea.

bunch of fine hair-like processes or cilia which, during life, are in active movement. They bend rapidly to one side and then straighten again. This whipping movement takes place about ten times a second and has the effect of sweeping onwards in one direction any substance or fluid in contact with the surface of the cell.

Ciliated epithelium is found especially in the respiratory system and lines the nasal cavities, the trachea and bronchi (Fig. 11). By its movement it conveys mucus, dust, etc., from the deeper parts towards the exterior. It is also found in the uterus and Fallopian tubes where it assists the ovum in its progress towards the uterine cavity.

SUMMARY OF FUNCTIONS OF EPITHELIUM

1 Protection (simple and stratified)
2 Secretion (columnar)
3 Absorption
4 Movement of mucus, etc. (ciliated)

Connective tissues

All the connective tissues have certain features in common. They contain a number of cells, the types varying according to the tissue. Fibres and fibrous tissue cells may be present, and there is a greater or less amount of supporting substance (ground substance) between

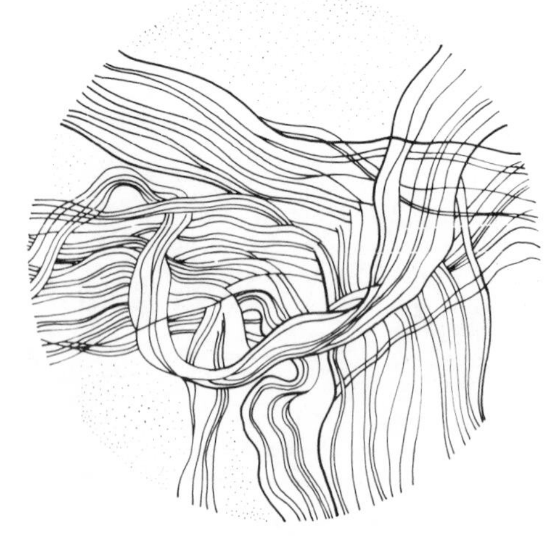

Fig. 12 Fibrous tissue, showing bundles of white fibres.

the cells. The latter may be jelly-like, as in the umbilical cord at birth and in the vitreous humour of the eye, firm as in hyaline cartilage, or hard from the presence of lime salts as in bone.

Fibrous tissue

There are two types: ordinary firm **white fibrous tissue**, which does not stretch, consisting of bundles of white-coloured fibres containing a few cells; tendons and ligaments consist of this material, also the dura mater, the outer layer of the pericardium, fascia and the fibrous covering of organs. Gelatine can be extracted from white fibrous tissue, which also contains a substance known as collagen.

The second variety is **yellow elastic tissue** and consists of yellow fibres which stretch. This is found in the walls of the arteries, in the bronchi and alveoli of the lungs and in a few special ligaments in the spine where, on account of their elasticity, they help to maintain the erect posture (ligamenta flava).

Areolar tissue

This may be described as the general packing and supporting tissue of the body. It is found under the skin and mucous membranes and surrounding blood vessels and nerves. It is a loosely woven tissue containing white fibres, various cells and a gelatinous substance between the cells.

Adipose or fatty tissue

This is a variety of areolar tissue the bulk of which is made up of globules of fat contained in thin membranous envelopes. It is found in all parts of the body, especially where fat is deposited, such as under the skin and around the heart and kidneys.

Lymphoid tissue

This consists of masses of special round cells, known as lymphocytes, supported by a small amount of fibrous and areolar tissue. The lymph glands are formed of this tissue, which is also found in the tonsils, the Peyer's patches of the intestine, and in the spleen.

Cartilage or gristle

This is a firm bluish-white tissue found mainly in connection with the skeleton. It is made up of animal matter and contains only a very small amount of mineral matter. The outer surface is covered

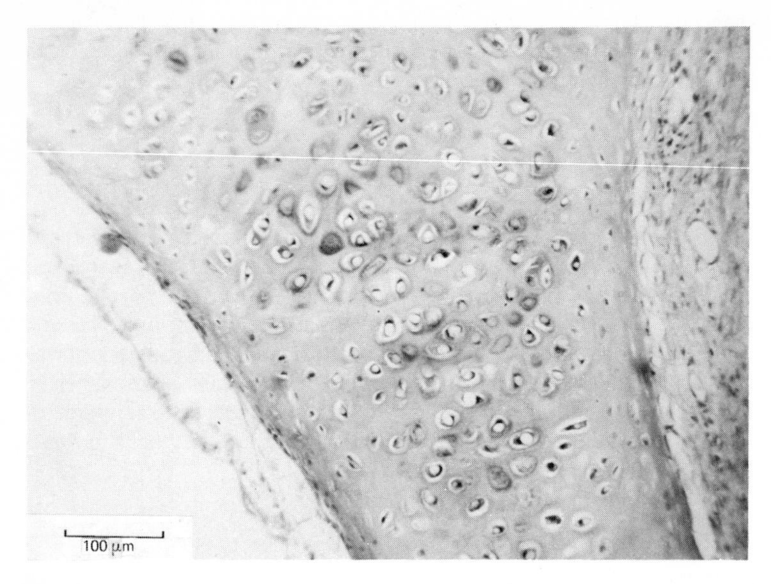

Fig. 13 Section across nasal septum, showing hyaline cartilage between two layers of mucous membrane.

by a fibrous membrane, the perichondrium, which is supplied with blood vessels. No blood vessels, however, enter the cartilage itself, which is nourished by lymph. Cartilage is strong and slightly elastic. There are three types: (*a*) hyaline cartilage, (*b*) fibro-cartilage, (*c*) elastic cartilage. In the first the substance or matrix between the cells is clear and free from fibres. In the second type there is a mixture of hyaline cartilage and white fibrous tissue. **Hyaline** cartilage is found covering the ends of bones where they form joints (articular cartilage), in the costal cartilages, trachea and larynx. **Fibro-cartilage** is found specially in the intervertebral discs and semilunar cartilages of the knee joint, where great strength combined with a certain amount of elasticity is required. **Elastic cartilage,** containing yellow elastic fibres, is found in the epiglottis and pinna of the ear.

Bone

This is the hardest of the connective tissues. The property of hardness is due to the impregnation of its ground substance with mineral salts, chiefly phosphate of lime (calcium phosphate) and carbonate of lime (calcium carbonate).

Composition, The composition of bone is approximately:

50% water

50% solid matter:
- inorganic or mineral matter (calcium carbonate / phosphate) 67% or $\frac{2}{3}$
- animal or organic matter (gelatine, etc.) 33% or $\frac{1}{3}$

General structure. When a longitudinal section of growing long bone from a young person is examined by the naked eye, the following will be observed (Fig. 14):

1. The periosteum
2. Compact tissue } (bone substance proper)
3. Cancellous or spongy tissue } (bone substance proper)
4. The bone marrow

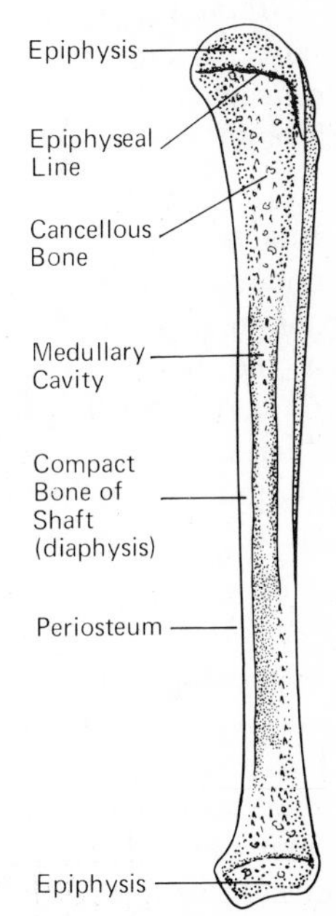

Fig. 14 Section of long bone before growth is complete.

The periosteum. This is a tough membrane of fibrous tissue containing some blood vessels and covers the surface of the bone (except where the latter is covered by hyaline cartilage in the formation of a joint).

Compact bone. This is the surface layer of bone and is found immediately under the periosteum. It is a hard dense substance resembling ivory and forms the shaft of the bone.

Cancellous bone. Although of the same microscopic structure as compact bone, it appears as a spongy, porous tissue containing red bone marrow, which fills in the ends of the bones. The surface of this part is, however, still formed by a thin sheet of denser compact bone.

Bone marrow. The interior of a dried long bone is hollow except for the cancellous tissue at its ends. In adult life, however, this hollow is filled with a soft yellow material containing fatty tissue, the **yellow** bone marrow. The spaces in the sponge-like cancellous part contain **red** marrow tissue which derives its colour from the special marrow cells from which the cells of the blood are developed. The marrow contains a large number of small blood vessels. (In infancy red marrow extends into the shaft.)

Blood supply of bone. Bone receives its blood supply from two sources: the surface of the bone from the periosteum; the interior from an artery which enters the shaft, generally about its middle, through a canal known as the nutrient foramen.

Microscopic structure of bone. Compact bone is built up of units which are called **Haversian systems.** Each of these units contains a minute circular canal—the **Haversian canal**—which runs in a longitudinal direction parallel with the surface of the bone. In the bone substance surrounding the Haversian canal are a number of small spaces, called lacunae, arranged in concentric rings which contain the bone cells.

Minute canals (canaliculi) joint up the lacunae and also communicate with the central Haversian canal.

Small blood vessels and lymphatics run in the Haversian canals and nourish the bone substance.

Development of bone. It has been noted that bone is a specialized type of connective tissue. In very early fetal life no actual bone is apparent, but the bones of the human skeleton are outlined by other forms of connective tissue, either cartilage or fibrous tissue (membrane). It will be seen later that the majority of bones in the skeleton are described as long bones and these are preceded by carti-

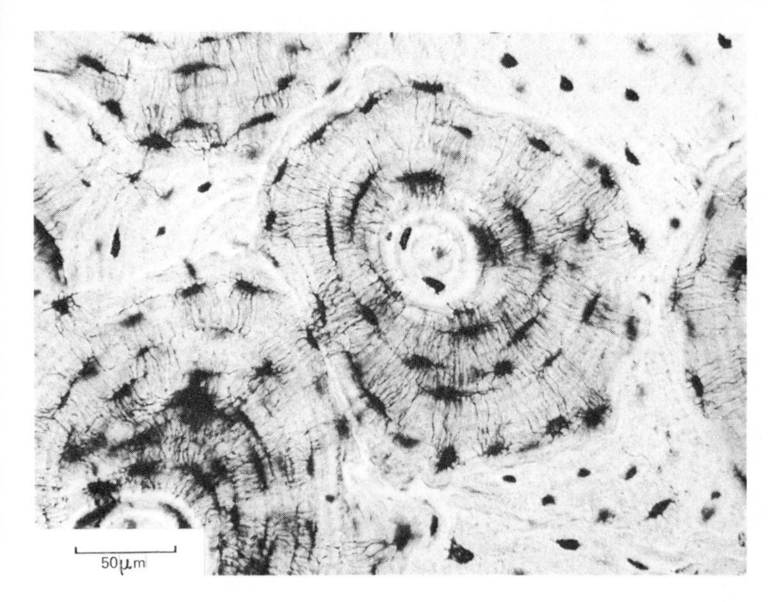

Fig. 15 Transverse section of compact bone (microscopic).

lage in the embryo; while certain flat bones, such as those of the skull, are developed in membrane.

First of all, therefore, the bones are represented by rods or blocks of cartilage, or sheets of membrane, which resemble in shape the mature bone. The next stage in development is the deposit of lime (calcium) salts in the cartilage or membrane (calcification). Later, bone cells or osteoblasts enter the calcified cartilage together with other cells which remove the cartilage. The bone cells proceed to lay down bone in the place of the cartilage which is absorbed. The area in which this process commences is called a **centre of ossification.**

At birth only a proportion of the skeleton is represented by actual bone. The remainder consists of ossifying cartilage, or membrane.

Growth of bone. Most bones go on growing in size for at least twenty years.

Centres of ossification are necessary for growth in length of a bone, i.e. areas in which bone cells are replacing cartilage. In most long bones there are two such centres, one at each end of the bone. An additional centre exists for the shaft of the bone.

The main part or shaft of the bone is called the **diaphysis**. The portion at each end containing a centre of ossification is called the **epiphysis**. Each epiphysis is joined to the diaphysis by a layer of cartilage known as the epiphyseal cartilage. This is gradually replaced by bone, and by adult life has disappeared and the epiphysis and diaphysis are completely fused into a single bony structure, but so long as some epiphyseal cartilage remains, growth in the length of the bone is possible.

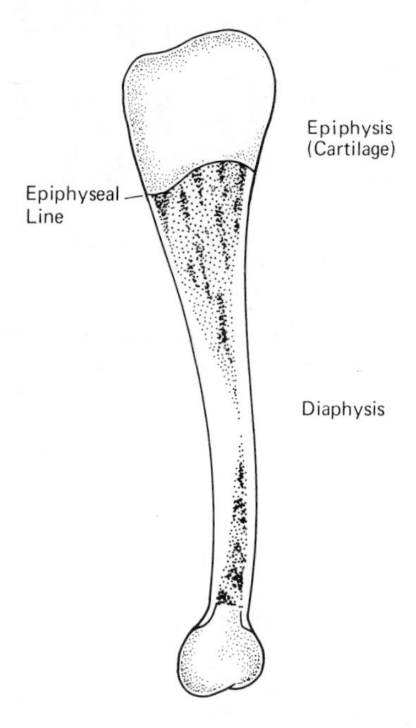

Fig. 16 Section of young bone before ossification of epiphyses has started.

So far reference has only been made to growth in the length of a bone; increase in circumference takes place by new bone being laid down under the periosteum by osteoblasts situated in this position.

New bone may be formed from time to time in order to repair fractures, without any increase in size of the bone as a whole.

The three main factors which influence the normal growth of bone in children are:

(*a*) Sex hormones (oestrogens and androgens).

(*b*) Pituitary growth hormone.
(*c*) Thyroid hormone.

Any defect in these hormones may cause stunting of growth.

PRACTICAL CONSIDERATIONS. In children, the bones do not contain quite as much lime salts as in the adult, so that the bones are more elastic and instead of complete fractures resulting from injury, a partial fracture known as the greenstick variety sometimes occurs.

Injuries to the ends of bones in children often result in damage to the epiphysis, which may become displaced rather than the bony substance being broken.

Deficiency in the amount of lime salts in rickets results in the bones becoming abnormally soft, so that bending with a corresponding deformity of the limbs may take place. Changes in the epiphysis are also apparent.

When a fracture of bone occurs Nature attempts to repair it, and the repair is most efficient when the surgeon can replace the broken fragments in their natural position. The following process then goes on: The space between the broken ends is filled with blood clot, in which fibrous tissue develops. Osteoblasts then migrate from the damaged ends of the bones and from the periosteum and gradually fill up the gap with new bone, known as callus. Eventually the continuity of the bone is restored and the callus becomes hard and firm by the deposit of lime salts in the damaged area. In a healthy individual the repair is usually complete in one to three months.

In cases of lead poisoning much of the lead is deposited in bones. Strontium-90, a product of atomic fission, is also deposited in the bones. This is radioactive, and continues to be so for a long time after it has been absorbed.

Muscular tissue

The muscles are structures which give the power of movement. Examination of a piece of raw meat shows it to be red in colour—due to contained blood—and to consist of a number of bundles of fibres placed side by side, all running in the same direction and bound together by a thin membrane of connective tissue.

Three types of muscular tissue are found in the body:

(*a*) Voluntary (striped), found in the muscles attached to the skeleton.
(*b*) Involuntary (unstriped or plain), present in various internal organs and structures.
(*c*) Cardiac, a special type found only in the heart.

Muscular tissue consists of elongated cells containing a small nucleus. These elongated cells, unlike the fibres of connective tissue, have the power of contraction, in the process of which each one becomes shorter and thicker. The muscle as a whole, therefore, becomes shorter and thicker when it contracts; this can be demonstrated by contracting the biceps muscle of the arm when the fingers are raised to touch the shoulder.

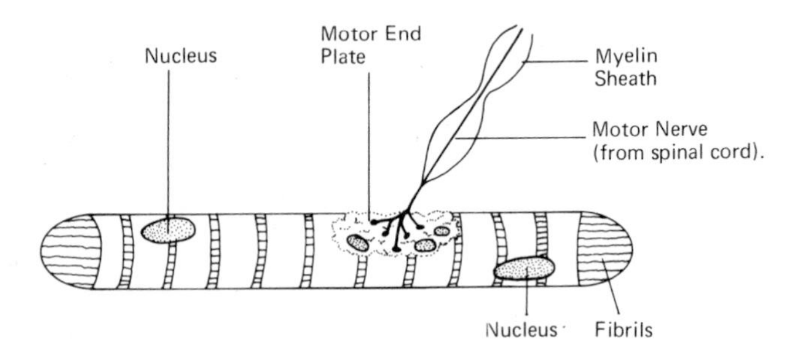

Fig. 17 Diagram of a single striated muscle fibre.

Voluntary muscle

As its name implies, skeletal muscle is under the control of the will and, from its microscopic structure, is sometimes referred to as striped or striated muscle (Fig. 18). All the muscles attached to the skeleton are of this type and their functions are to move the bones at their respective joints and to help in maintaining the posture of the limbs and the body as a whole.

The voluntary muscles are connected to the cells of the motor cortex of the brain by nerve fibres which pass down the spinal cord, from which a second set of fibres from the anterior horn cells run in the peripheral nerves to the muscles (page 332).

These motor nerve fibres terminate in special structures in the muscle fibres known as motor end-plates. It is through these end-plates that the impulse to contract is conveyed to the muscle cells.

The neuro-muscular mechanism. This term implies the structural arrangement just described, together with the complicated chemical and electrical changes which take place at the motor end-plates when the nerve impulse passes through them to the muscle fibres to cause contraction.

Without going into the complicated details it may be stated that acetyl choline liberated from the nerve endings causes certain electrical changes during which there are movements of the electrolyte ions of

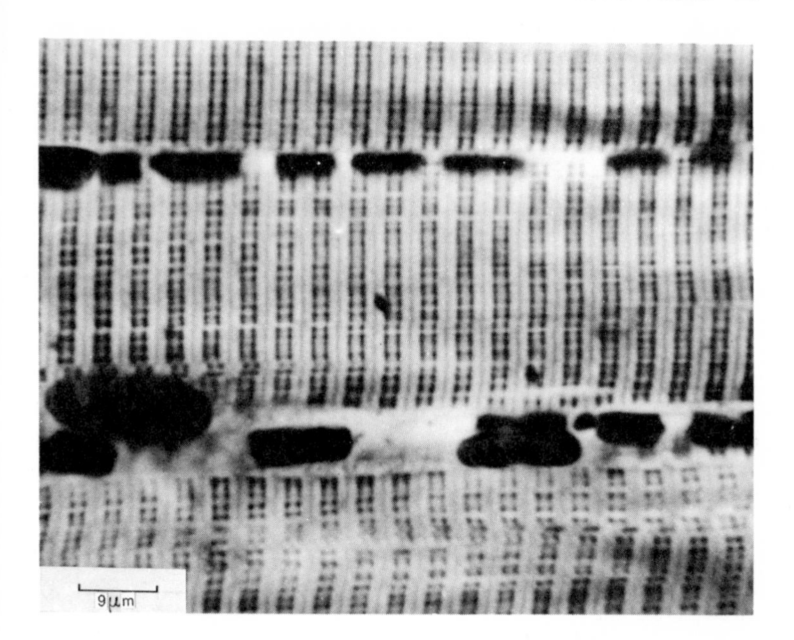

Fig. 18 Section of striped muscle. (See also Fig. 7.)

potassium and sodium and also, to a lesser extent, those of calcium and magnesium.

These facts help to explain certain rare cases of muscular weakness associated with disorders of potassium metabolism and also the muscular spasms of tetany in calcium deficiency.

The muscular relaxation produced by the injection of drugs such as curare, Flaxedil and Scoline used in anaesthesia is due to their action on the motor end-plates interfering with the acetyl choline mechanism.

Involuntary or plain muscle (unstriped)

The cells found in involuntary muscle are somewhat different in appearance from those of voluntary muscle. In particular, they do not show the characteristic striation or striping of the latter.

Involuntary muscle is found in the various internal organs and structures of the body such as the stomach, intestines, bladder, uterus, bronchi and blood vessels and, therefore, is sometimes called visceral muscle. It is usually arranged in a series of layers, i.e. circular and longitudinal, thus constituting part of the wall of the structure concerned.

It is not consciously controlled and its nerve supply comes from the involuntary or autonomic nervous system (page 336).

Cardiac muscle

The muscle of the heart is a special form of involuntary muscle. It is not under the control of the will, but has a form of striation resembling that seen in striped muscle (see Fig. 7).

It has the special property, not observed in the other varieties, of automatic rhythmic contraction which can occur independently of its nerve supply.

Composition of muscle. Muscle consists of 75 per cent water and 25 per cent solids, of which the most important (20 percent) is a protein known as myosin. The process of muscular contraction is associated with certain electrical and very important chemical changes; an acid known as lactic acid is formed and finally converted in the presence of oxygen into the end-products carbon dioxide and water which are excreted from the body. This chemical process results in the formation of heat and it will be seen that muscular activity is an important factor in the maintenance of body temperature (page 257).

Properties of muscle. *(1) The power of contraction.* Voluntary muscles contract as a result of messages (stimuli) reaching them from the nervous system and many nerves have their endings in muscles. Messages passing down a nerve from the brain cause the muscle to contract when they reach it. Other stimuli such as electricity, applied direct to the muscle or its nerves, will also cause the muscle to contract.

(2) *Elasticity.* Muscle tissue is elastic and can be stretched by a weight. When this is removed the muscle returns to its normal length.

(3) *Fatigue.* When a muscle contracts it uses energy. This energy is derived mainly from sugar (glucose) stored in the muscle as glycogen and brought to it by the blood. The blood also conveys oxygen which the muscle uses to burn up the glucose with the formation of lactic acid, which in its turn is ultimately broken down into carbon dioxide and water. After a number of contractions the supply of glucose immediately available is used up and a certain amount of lactic acid accumulates. The muscle then becomes tired and is unable to contract with the same degree of efficiency. It requires rest in order to replenish its supply of glucose and to remove the lactic acid.

(4) *Muscular tone.* Even when a muscle appears to be at rest it is

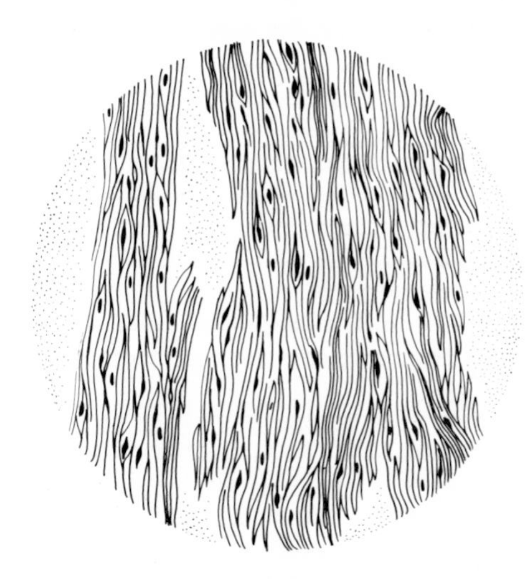

Fig. 19 Longitudinal section of unstriped muscle. (See also Fig. 7.)

always partially contracted and therefore ready for immediate action. This state of partial contraction is called muscle tone. If a muscle lost its tone, it would be compelled to take up a slack portion on the receipt of a stimulus, which would result in delayed and inefficient movement.

Smooth (involuntary) muscle, however, retains its tone when all connections with the central nervous system have been severed.

PRACTICAL CONSIDERATIONS. When death occurs the protein of muscle gradually coagulates, and the muscle loses its elasticity and becomes rigid. This condition, which appears in all the muscles of the body about four hours after death, is known as *rigor mortis*, and accounts for the fact that after a time a corpse becomes stiff and rigid. It is important, therefore, to lay out a body in an appropriate and reverent attitude before the onset of *rigor mortis* renders it impossible.

The fact that a muscle contracts when stimulated by an electric current is used in medicine to give the muscles work and exercise when they have become weak through lack of use (disuse atrophy) or paralysed by a defect in their nerve supply.

Nervous tissue

It has been seen that the muscles are the structures which give the body the power of movement. Almost every movement, however, is governed by some portion of the nervous system which acts as a medium between the mind and muscles.

Three types of tissue are found in the nervous system:

1. The nerve cell.
2. Nerve fibres which convey either motor or sensory impulses.
3. Neuroglia, or supporting fibres resembling connective tissue fibres, which act as a framework or support for the nervous tissue, but take no other part in the activities of the nervous system.

Nerve cells vary very much in size, some being the largest cells met with in the body. The nerve cells are collected together in the nerve

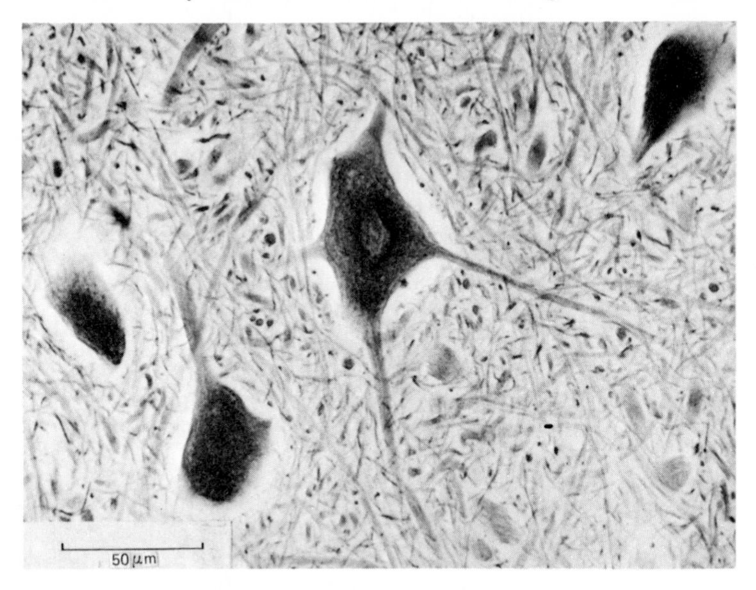

Fig. 20 A nerve cell in its supporting tissue.

centres of the body, the largest of which is the brain. The nervous tissue composed of cells is sometimes referred to as **grey matter**.

Each nerve cell contains a nucleus, and consists of highly specialized protoplasm. From the body of the cell extend various types of fibres.

Nerve fibres

(*a*) *Axon.* From the body of each nerve cell passes one main fibre known as an axon. It is along this fibre that the impulses from the nerve cell pass, in one direction only. The axon may be of considerable length; for example, the axons of certain cells in the lower part of the spinal cord extend in the nerves as far as the foot.

(*b*) *Dendrites.* These are small fibres extending from the body of the nerve cell to which they carry impulses from other nerve cells. The name implies 'branching like a tree'. When impulses pass from one neuron to another they go from the axon of the first to the dendrites of the second. The point at which dendrites meet is called a **synapse.**

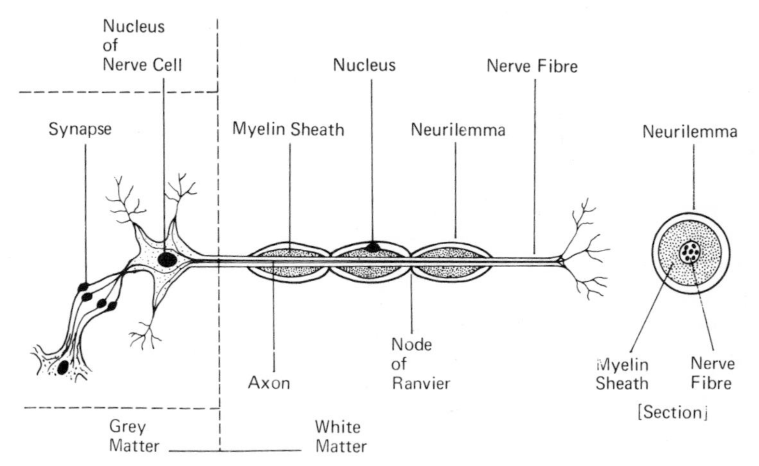

Fig. 20a Diagram of a neurone.

The tissue composed of nerve fibres is referred to as **white matter.** There are two types of nerve fibres: (*a*) those which are surrounded by a fatty sheath, called medullated fibres, and (*b*) those which are without a sheath, or non-medullated fibres.

The medullated sheath contains myelin. (This may degenerate in the absence of vitamin B_{12} as in the degeneration sometimes occurring in the spinal cord in pernicious anaemia.)

The unit with which the nervous system is constructed is called a **neurone** and consists of a nerve cell, its axon and dendrites. The neurone may be compared to an electric battery. The nerve cell generates the impulse and the axon conveys it in the same way as wires lead the current from the battery.

Membranes

The term membrane is applied to any thin expansion of tissue which usually forms an enveloping or lining layer for various structures or organs. The most important types of membrane are (i) mucous, (ii) serous, (iii) synovial.

Mucous membranes

These are found lining the alimentary, respiratory and genito-urinary tracts. They are covered by epithelium of various types, e.g. stratified in the mouth, columnar in the stomach and intestines, ciliated in the respiratory tract.

A mucous membrane is usually supported by a submucous layer of connective tissue. It has the following functions: (*a*) protection, (*b*) secretion of mucus which moistens its surface, (*c*) providing a surface for the absorption of foodstuffs in the alimentary canal.

Serous membranes

These are lined by simple squamous epithelium and line the important cavities of the body and the organs which lie within them. The most important are: the **pericardium** covering the heart; the **pleura** lining the pleural cavity and covering the lungs; the **peritoneum** lining the abdominal cavity and covering the abdominal organs.

Their surfaces are smooth and moistened by serous fluid which enables the walls of the cavity or the organs contained within it to glide smoothly over each other without friction.

Synovial membranes

These are a specialized form of serous membrane which line the capsules of joints, the sheaths of tendons and the interior of bursae. They are moistened by synovial fluid and their function is to prevent friction. Serous fluid is thin and watery. Mucus is thicker and rather sticky. Synovial fluid is thicker than either of the other two. Each type of fluid is therefore adapted to the functions it has to perform.

3 The Body as a whole

All the various cells and tissues which have been described form the basis of the separate systems of the body. A system may be defined as a group of structures or organs which carry out an essential fundamental function of the individual. Although, to some extent, each system works and can be considered on its own, the functions of the various systems are very closely connected and are dependent on each other. For example, the bones, joints, ligaments and muscles are all concerned with the function of movement which in turn is controlled by the activity of the nervous system. The vitality of the nervous system is dependent on an adequate circulation of blood and a supply of oxygen which reaches the blood via the respiratory system, and so on.

It is essential to be familiar with certain **terms used in anatomical description.** The body is considered in the upright or erect position with the palms of the hands facing to the front and the toes pointing forwards. The following terms are then applied:

anterior (ventral)	towards the front of the body or limbs
posterior (dorsal)	towards the back of the body or limbs
medial	the side nearest the mid-line of the body
lateral	the side farthest away from the midline
superior	above any point referred to
inferior	below any point referred to
plantar	belonging to the sole of the foot
palmar	belonging to the palm of the hand

NOTE: The terms *internal* and *external* should only be used to describe the relationship to the inner and outer surfaces of the body and not as alternatives to *medial* and *lateral.*

In describing the limbs, the upper part nearest the trunk is called the **proximal** part; the lower portion farthest away from the trunk is called the **distal** part.

The body as a whole is built up round the bony framework of the skeleton and consists of three main parts:

1. The head and neck
2. The trunk: (*a*) the chest or thorax
   (*b*) the abdomen and pelvis
3. The limbs: (*a*) the upper or arms
   (*b*) the lower or legs

The head is separated into two parts, the cranium or brain case

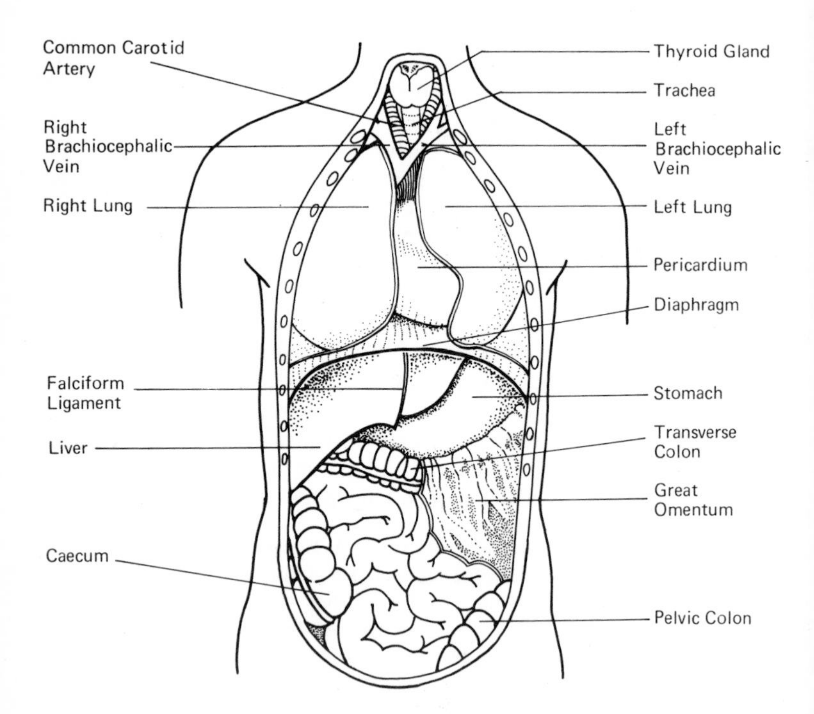

Fig. 21 The viscera of the body from the front.

and the face. The trunk is divided into the thorax or chest and the abdomen or belly. There are two pairs of limbs, the upper and lower, each divided into parts which roughly correspond:

Upper limb:	upper arm,	elbow,	forearm,	wrist,	hand,	fingers
Lower limb:	thigh,	knee,	leg,	ankle,	foot,	toes

The fingers and toes are referred to as digits, the bony parts of which, separated by joints, are called phalanges.

Covering the bones and giving rise to the general contour of the body are numerous muscles, and blood vessels and nerves traverse all parts of the human frame.

The chief organs of the body are contained in special cavities:

1. The brain in the cavity of the skull (cranium).
2. The lungs and heart in the cavity of the thorax.
3. The digestive organs in the abdominal cavity.

It is of importance to know the boundaries of these cavities and the structures which they contain. All those mentioned will be referred to later, but are included at the present for purposes of classification.

The skull

This consists of the cranium which protects the brain and the eyes, and the mandible or lower jaw which is hinged to it. The movements of the mandible are essential for the chewing of food and for the production of speech.

The thorax

This important cavity is situated in the upper part of the trunk and its walls consist of a bony framework supporting various muscles.

Boundaries of the thorax

Anterior: the sternum, costal cartilages and front end of the ribs.

Posterior: the thoracic or dorsal part of the vertebral column or backbone, consisting of twelve individual vertebrae and the intervertebral discs of cartilage between them.

Lateral: the twelve ribs and the intercostal muscles.

Superior: the root of the neck with its muscles and blood vessels.

Interior: the diaphragm—a large dome-shaped muscular structure separating the cavity of the thorax from that of the abdomen—through which pass the oesophagus, aorta and inferior vena cava.

Contents of the thorax

(*a*) *The lungs* occupy the greater part of the thoracic cavity except for a central portion behind the sternum which extends backwards to the vertebral column and is known as the **mediastinum**.

(*b*) *The heart.* This is situated in the mediastinum in the central part of the thoracic cavity and is enclosed in a fibrous bag or sac known as the pericardium.

(*c*) *The trachea* or windpipe enters the thorax through its upper opening from the neck and passes down in the posterior part of the mediastinum until it divides into the two main bronchi at the level of the fourth thoracic (dorsal) vertebra. A bronchus passes to each lung.

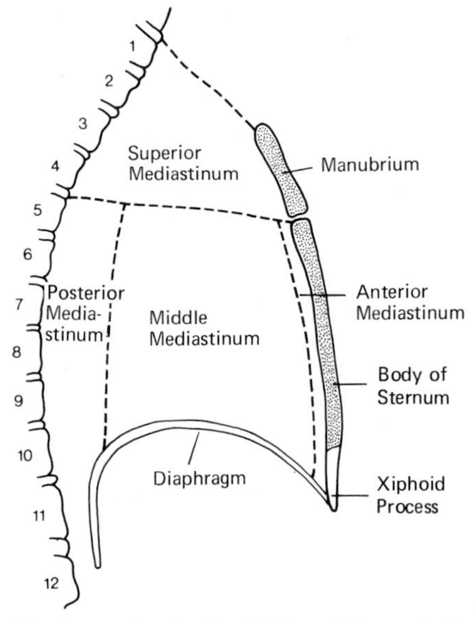

Fig. 22 Section of the thorax, showing the divisions of the mediastinum.

(*d*) *The oesophagus* or gullet. This also enters the superior opening of the thorax from the neck where it commences as a continuation of the pharynx. The oesophagus lies just in front of and to the left of the vertebral column and behind the trachea. It leaves the thorax by passing through an opening in the diaphragm and enters the abdominal cavity where it shortly joins the stomach.

Other structures in the thorax include the thymus gland, the phrenic and vagus nerves, the aorta, some important veins (the superior vena cava and inferior vena cava), the thoracic duct and lymphatic glands.

The abdomen

The abdominal cavity is the largest cavity in the body. It is described in two parts, the abdomen proper and the pelvic cavity.

The latter is directly continuous with the rest of the abdomen but is situated in the space bounded by the sacrum behind, the ischium on each side and the pubic bones in front. (see Fig. 141, page 203.)

Boundaries of the abdominal cavity

Anterior: the muscles of the abdominal wall (the rectus, internal and external oblique and transversus on each side).
Posterior: The lumbar part of the vertebral column in the mid-line and the psoas, quadratus lumborum and iliacus muscles on either side.
Superior: the diaphragm.
Inferior: the abdominal cavity is continuous with the superior opening of the pelvic cavity.

Contents of the abdominal cavity

(*a*) The stomach and intestines.
(*b*) The liver, gall-bladder and spleen.
(*c*) The pancreas, kidneys and adrenal glands. The abdominal aorta and inferior vena cava. All of these structures lie outside the peritoneum and are situated on the posterior wall of the cavity.

Boundaries of the pelvic cavity

Anterior: the pubic bones and symphysis pubis.
Posterior: the sacrum.
Lateral: the ischium on each side.
Superior: the pelvis is continuous with the abdominal cavity.
Inferior: the muscles of the pelvic floor (levator ani, etc.). Through the pelvic floor pass the lower part of the bowel (rectum) in the hollow formed by the sacrum behind; the urethra and, in the female, the vagina towards the front.

Contents of the pelvic cavity

(*a*) The lower part of the large intestine (the pelvic colon and rectum).
(*b*) The bladder.
(*c*) The female organs of reproduction (uterus, ovaries and Fallopian tubes).
(*d*) Some loops of small intestine may also be present in the pelvis.

SYSTEMS OF THE BODY

As a matter of convenience the various systems of the body are generally described separately but it must be remembered that functionally they are all closely related and interwoven. The major systems are:

The locomotor system (bones, muscles and joints)
The nervous system (central, peripheral and autonomic)
The skin and organs of special sense (sight, hearing, smell and taste)
The cardiovascular system (heart, blood vessels and lymphatics)
The blood (the haemopoietic system)
The respiratory system
The digestive system
The endocrine system (the ductless glands)
The urinary system
The reproductive system

Anatomically there is a close relationship between the urinary and reproductive systems. Hence they are frequently described together under the heading 'The Urogenital System'.

The nervous system receives and utilises information from the sense organs and controls voluntary and involuntary (reflex) movements of muscles. The higher centres regulate thought, memory and social behaviour. The cardiovascular or circulatory system enables nutrients to be conveyed to and waste products removed from all parts of the body by means of the blood. The respiratory system provides for the intake of oxygen and discharge of carbon dioxide. The digestive system provides for the intake and breakdown of food into relatively simple chemical compounds suitable for absorption. The ductless glands influence metabolism growth and reproduction. The urinary system removes waste products circulating in the blood and helps to keep its chemical and physical properties constant.

It should be noted that there is a variety of glands in various parts of the body. Broadly speaking there are exocrine glands which pour their secretions into the alimentary canal and the endocrine glands (ductless glands) whose secretions enter the blood stream. Lymph nodes, sometimes called lymph glands, are not really glands since they do not secrete hormones or enzymes but they produce lymphocytes and help to protect the body from the spread of infection. The ducts or lymphatic channels which connect these structures and ultimately enter the blood stream convey tissue fluids which surround the various tissue cells and act as a 'middle man' conveying nourishment to and removing waste products from them.

QUESTIONS (Chapters 1–3)

1. Describe a cell, a tissue and an organ. Give examples of each.
2. Describe the different types of muscle tissue. State the special work of every type and say where it is found in the body.
3. Describe the structure of bone. What are the functions of the skeleton?
4. Describe the boundaries which enclose the thoracic cavity. Enumerate the contents of this cavity.
5. Describe the boundaries of the abdominal cavity and enumerate its contents.
6. Describe the contents either of the male or female pelvis.

4 The Skeleton

The skeleton as a whole

The composition and general structure of bone have already been discussed (page 21).

The skeleton is the framework of the body consisting of bones and, strictly speaking, the cartilages and ligaments which bind them together. It serves to support the soft structures which are grouped around it and to protect the organs of the body. It is so jointed that various parts move on each other and many bones act as levers for the muscles which are attached to them.

The main **functions** may be summarized as:

1. to act as a framework and to support the soft tissues.
2. To enable free movement by the action of muscles (i.e. to combine stability with mobility).
3. To protect delicate organs and structures.
4. The formation of blood cells in the bone marrow (page 165).

Types of bone

The bones of the skeleton are classified according to their shape into long, short, flat and irregular bones.

1. **Long bones** are found in the limbs and consist of an elongated shaft with two extremities. The bones of the arm, forearm, thigh and leg are typical examples. The shaft consists of a cylinder of compact bone containing yellow bone marrow. The extremities are formed by a thin outer shell of compact tissue with an interior network of spongy or cancellous bone containing red marrow (see Fig. 14).

2. **Short bones** have no shaft but consist of smaller masses of spongy bone surrounded by a shell of compact bone. They are roughly box-like in shape. Examples are found in the small bones of the wrist (carpus) and ankle (tarsus).

3. **A flat bone** consists of two layers of compact bone between which is a layer of cancellous bone. Examples are the scapula, innominate bone and bones of the skull.

4. **Irregular bones** cannot be placed strictly in any of the previous categories and include the vertebrae and most of the bones of the face.

Sesamoid bones are small bones which are developed in the tendons around certain joints. The patella, in the front of the knee joint, is the largest and most important.

Descriptive terms. The following terms are used especially in the description of bones:

Articulation, a joint between two bones, e.g. the humerus articulates with the scapula and the femur with the tibia.

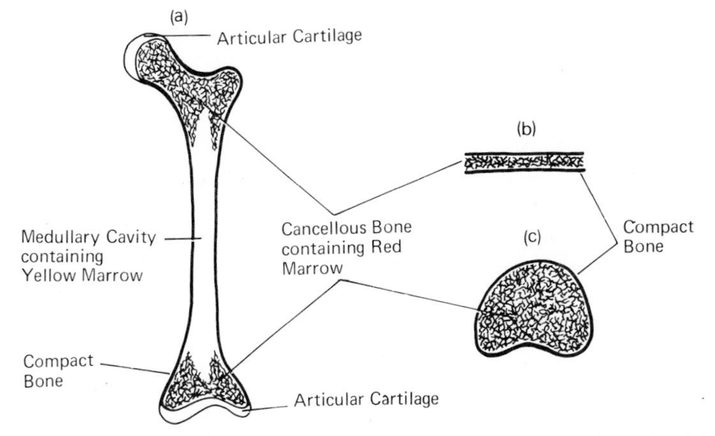

Fig. 23 Diagram illustrating the structure of (*a*) a long bone (femur), (*b*) a flat bone (skull) and (*c*) an irregular bone (body of vertebra).

Border, the edge separating two surfaces of a bone, e.g. the vertebral and axillary borders of the scapula separating the anterior from the posterior surfaces.

Condyle, a rounded enlargement at the end of a bone, usually covered by cartilage, e.g. the condyles of the femur.

Crest, an elevated ridge on a bone, e.g. the crest of the ilium and the crest of the tibia.

Facet, a small articulating surface.

Foramen, an opening or hole perforating a bone, e.g. the obturator foramen in the innominate bone and the nutrient foramen present in all bones.

Fossa, a hollowed-out area or depression in the surface of a bone, e.g. the olecranon fossa and coronoid fossa of the humerus.

Process, a projection from a bone, e.g. the spinous process of a vertebra.

Trochanter, tuberosity, tubercle, all terms used to describe various types of process. A tubercle is a small rounded prominence, e.g. the tubercle of the tibia. A tuberosity is a larger protuberance of bone, e.g. the tuberosities of the humerus. A trochanter is a large, rough eminence, e.g. the trochanter of the femur.

The human skeleton

The human skeleton, which contains approximately 200 individual bones, is made of of the following parts:

The skull, viz. the bones of the cranium, face and lower jaw.

The bones of the trunk, viz. the spinal column, ribs and sternum.

The bones of the limbs with the shoulder and pelvic girdles.

The spinal column consists of 33 vertebrae and is divided for purposes of description into cervical (7 vertebrae), thoracic or dorsal (12 vertebrae), lumbar (5 vertebrae), sacral (5 vertebrae) and coccygeal (4 vertebrae) from above downwards. The five sacral vertebrae are fused together to form a single bone, the sacrum; the coccygeal are similarly joined to form the coccyx.

The ribs are twelve in number; they articulate behind with the thoracic vertebrae and, in front, the upper seven articulate with the sternum.

The bones of the upper limb include the shoulder girdle (clavicle and scapula) and those of the arm (humerus), forearm (radius and ulna), the wrist (carpus), the hand (metacarpals), and the digits or fingers (phalanges).

Included in the lower limb are the innominate bones which with the sacrum form the pelvic girdle, the thigh bone (femur), the leg bones (tibia and fibula), the ankle bones (tarsus) and those of the foot (metatarsals) and toes (phalanges).

The order in which the bones are studied is immaterial and a convenient one has been selected. There are, however, certain points of great importance in the study of Osteology. The first, on which it is impossible to lay too much emphasis, is that bones cannot be learnt from books alone. Students must read the description of a bone with that bone in front of them and they must handle the bone so that they are thoroughly familiar with it. Secondly, bones must be described in a systematic manner.

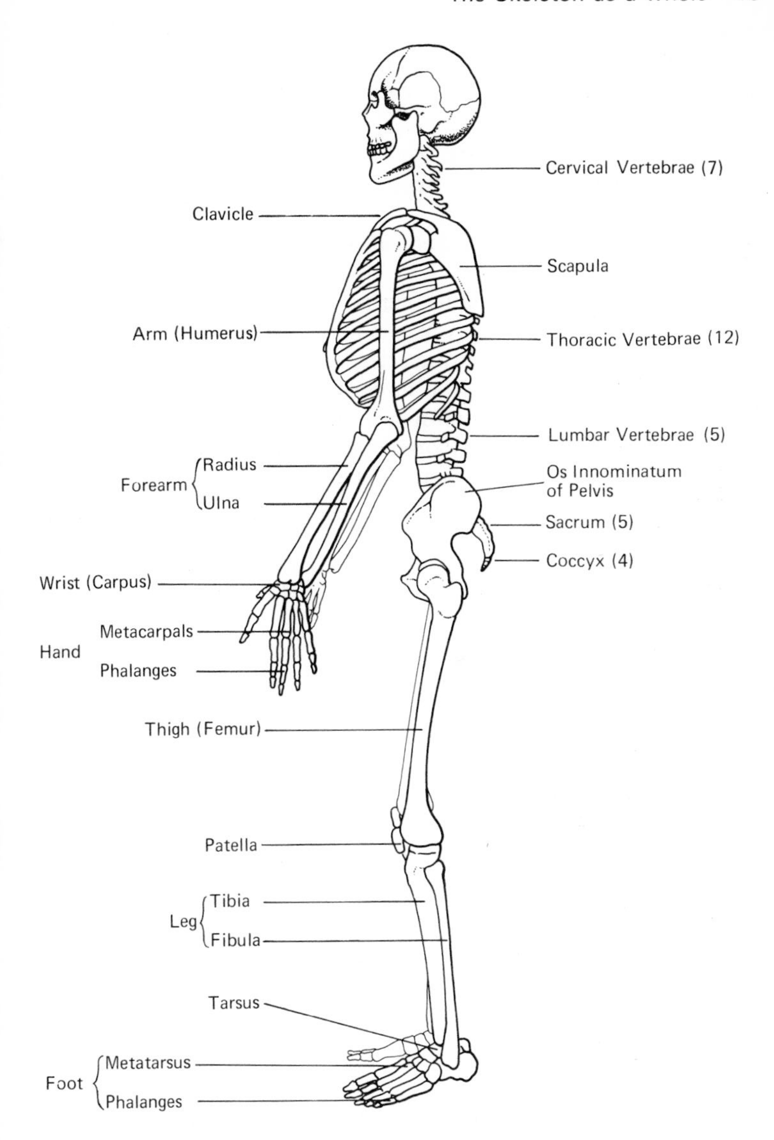

Fig. 24 The human skeleton.

1. State the name and then the type of bone, i.e. long, short, flat, or irregular.
2. State the part of the body from which it comes, e.g. arm, fore-arm, leg or shoulder girdle.
3. Indicate its main parts, e.g. upper part, shaft and lower part, or if it is a flat bone its surfaces and borders, and describe these in detail.
4. Point out any processes, tuberosities, grooves, fossae and foramina.
5. Mention the articulating surfaces and the other bones with which it articulates. Name the joints thus formed.
6. Mention any important muscles, blood vessels, nerves or other structures which may be in close relationship with the bone, and any special function it may have.

If a list of this sort is thoroughly memorized it will be found possible to reproduce a good account of any bone the student may be asked to describe.

BONES OF THE SHOULDER GIRDLE AND UPPER LIMB

The scapula

The scapula or shoulder-blade is a flat bone, triangular in shape, which forms part of the shoulder girdle. It is situated on the posterior aspect of the thorax and is superficial to the ribs.

Being flat and triangular, it has two surfaces—anterior and posterior—and three borders—the upper or superior, the lateral or axillary, and medial or vertebral. The three borders meet to form three angles. The superior border and vertebral border meet at the superior angle. The vertebral border and axillary border meet at the inferior angle.

The point at which the superior and axillary borders should meet to form a lateral angle is replaced by a special formation in the bone known as the glenoid cavity.

Surfaces

The anterior surface is slightly hollowed or concave and is known as the **subscapular fossa.** It is the surface which lies nearest to the ribs and gives attachment to the subscapularis muscle.

The posterior surface is slightly convex and is divided into two unequal parts by a large ridge of bone known as the **spine** of the scapula. The upper and smaller of these parts is called the **supraspinous fossa** and gives origin to the supraspinatus muscle. The larger **infraspinous fossa** below affords an attachment for the infraspinous muscle.

If the spine of the scapula is traced from the vertebral border it will be seen to have a surface which, as it passes outwards, becomes expanded into a flattened process, directed forward at its extremity. This expanded, broad, free end is called the **acromion process** and the extreme anterior end will be seen to have a smooth surface or facet for articulation with the clavicle. The upper part of the spine gives attachment to the trapezius muscle of the neck and the lower portion to part of the deltoid muscle of the arm.

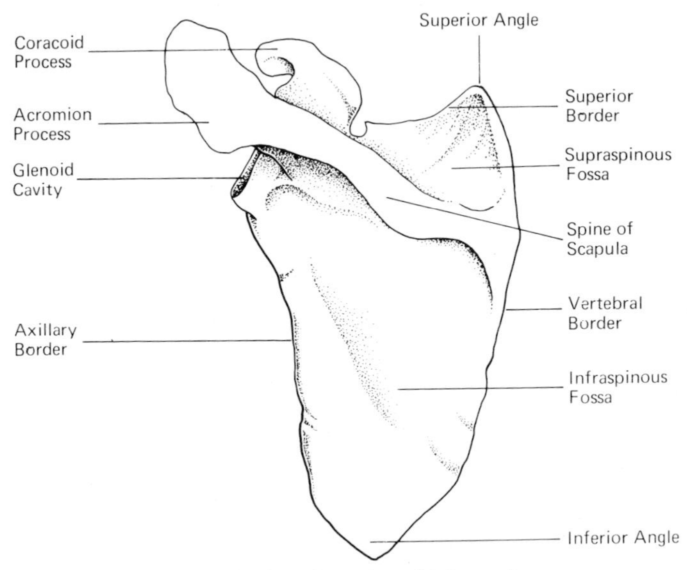

Fig. 25 Posterior surface of left scapula.

Borders

On tracing the thin superior border outwards from the superior angle a small notch will be observed. This is the suprascapular notch for the passage of an artery and nerve. At the outer end of the superior border another big and irregular process of bone, not so large as the spine on the posterior surface, will be seen projecting forwards. This is the **coracoid process** (for attachment of short head of biceps and pectoralis minor muscles).

The **vertebral border** is long and thin, but is interrupted by a flat surface where the spine arises. Various muscles are attached to it.

The **axillary border** is thick and strong. At its upper part, where the lateral angle of the bone would be, is a smooth, shallow, pear-shaped cavity of considerable size, the face of which is directed outwards. This is the **glenoid cavity** which articulates with the head of the humerus to form the shoulder joint. The edge of the cavity is roughened for attachment of the ligament which forms the capsule of the shoulder joint. At the upper part is a small tubercle for the

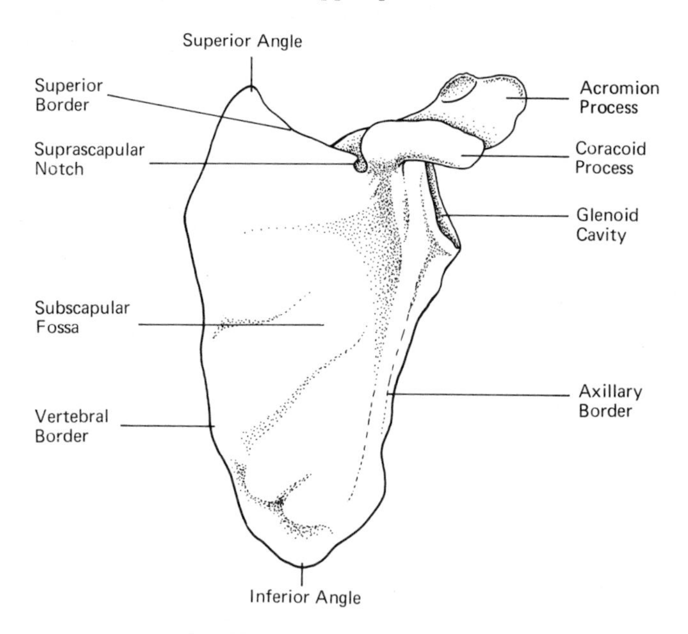

Fig. 26 Anterior surface of left scapula.

attachment of the long head of the biceps muscle. From the lower part the triceps muscle of the back of the arm arises.

The portion between the glenoid cavity and the rest of the bone is sometimes called the neck.

The scapula is so attached by muscles that it can glide across the surface of the thorax. If the shoulders are shrugged backwards and forwards a degree of transverse movement will be felt, and if the arms are raised above the head the inferior angle will be observed to move outwards towards the lower part of the axilla.

The scapula, being covered by muscles and lying flat against the chest wall, is not often exposed to injury and only a direct blow of considerable severity is likely to produce fracture.

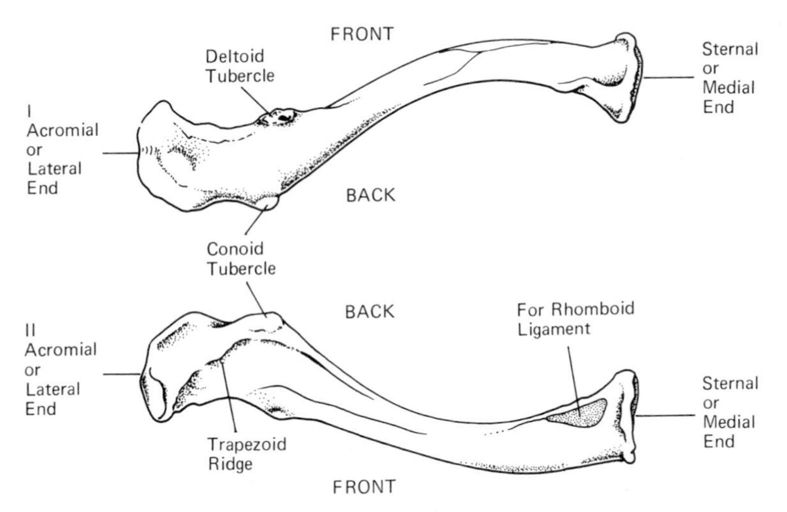

Fig. 27 Left clavicle from above (*top*) and from below.

The clavicle

The clavicle or collar bone is a long bone situated at the lower part of the neck and forming the front portion of the shoulder girdle. It lies above and in front of the first rib and extends outwards from the sternum to the acromion process of the scapula. It has a curved **shaft** with two enlarged extremities. The bone may be described as having anterior and posterior borders and superior and inferior surfaces. The anterior border and superior surface are important because they are subcutaneous and can be easily felt under the skin. The position in which the bone should be held to represent its position in the body is indicated by the comparative smoothness of the upper surface and the roughness of the lower surface.

The **medial** or **sternal extremity** is an expanded portion of the shaft having at its extreme end a smooth surface which is covered with cartilage and articulates with a corresponding facet on the upper part of the sternum.

The **lateral** or **acromial extremity** is enlarged and flattened. This also has an articular facet, directed backwards, to form a joint with the acromion process of the scapula.

The under surfaces of both extremities have rough elevations of bone for the attachment of ligaments which hold the clavicle firmly in position. It will be seen that the clavicle acts as a brace between the sternum and the scapula and helps to hold the shoulders back.

The clavicle is a bone which is commonly fractured. In spite of its superficial and exposed situation it rarely receives a direct blow, but in a fall on the shoulder or even on the outstretched arm the force is transmitted to the clavicle, which, being a slender bone, often snaps as a result.

The humerus

The humerus is a long bone of the arm, extending from the shoulder to the elbow. It consists of an elongated cylindrical **shaft** having upper and lower extremities.

The enlarged upper extremity consists of the head and two tuberosities. The **head** is a smooth hemispherical structure placed at an angle on the medial side of the bone and articulates with the glenoid cavity of the scapula to form the shoulder joint. It is covered with hyaline articular cartilage and is surrounded by a slight groove where it joins the shaft which is called the **anatomical neck** of the humerus. It is in this area that the capsule of the shoulder joint is attached. Immediately to the lateral (outer) side of the head is a prominence of bone called the **greater tuberosity**. In front of this is a smaller prominence known as the **lesser tuberosity**. Between the two tuberosities on the anterior aspect of the bone facing almost directly forwards is a deep groove about 8 cm (3 in) long known as the **bicipital** (intertubercular) **groove** in which lies the tendon of biceps muscle.

Below the head and tuberosities the shaft begins to narrow and an imaginary line drawn transversely across the shaft at this point is called the **surgical neck** of the humerus because it is a common site for fractures to occur.

Nearly halfway down the outer side of the shaft is a roughened elevated area for attachment of the deltoid muscle known as the **deltoid tuberosity**. On the posterior surface, passing from above downwards and outwards and having its lower end just below the deltoid tubercle, is a shallow groove, the **spiral groove** for the radial (musculospiral) nerve.

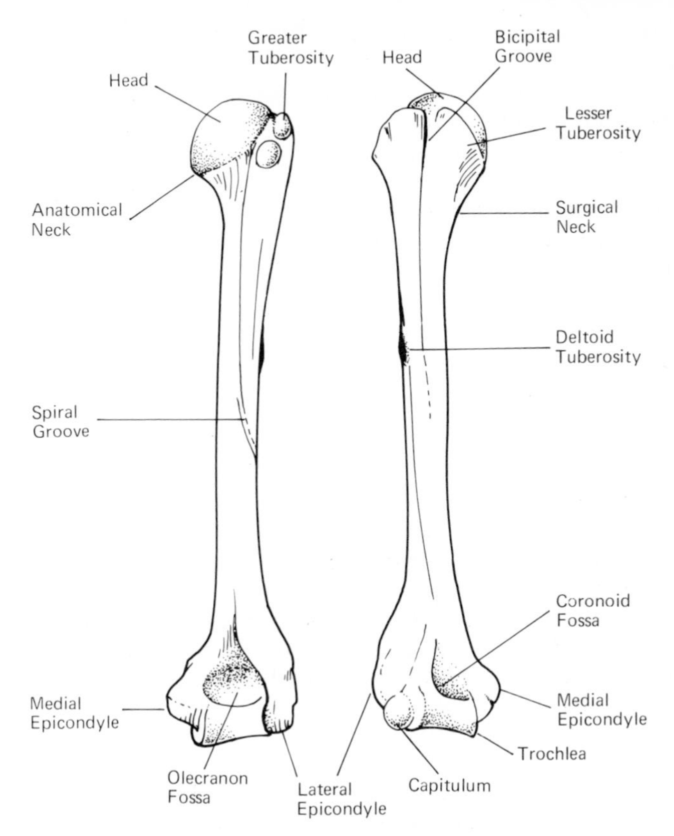

Fig. 28 Posterior (*left*) and anterior views of right humerus.

The lower end of the shaft loses its cylindrical character and becomes flattened, having anterior and posterior surfaces with medial and lateral borders. The medial border, if traced downwards, ends in a projection called the **medial epicondyle**, while on the outer side of the bone the corresponding but smaller protuberance is called the **lateral epicondyle**. The ulnar nerve passes down behind the medial epicondyle. Between the two epicondyles is an irregularly shaped but smooth articular surface.

This articular surface projects downwards a little more on the medial than the lateral side. The medial portion is called the **trochlea** and articulates with the trochlear (great sigmoid) notch of the ulna.

The lateral part or **capitulum** (capitellum) articulates with the head of the radius.

Immediately above the articular surface deep depressions are present both on the anterior and posterior surfaces of the bone. On the anterior surface is the **coronoid fossa** into which the coronoid process of the ulna fits when the elbow is flexed. On the posterior surface is the larger **olecranon fossa** for the olecranon process of the ulna when the elbow is extended.

The humerus is commonly fractured in three sites: the surgical neck, already mentioned, the shaft usually about its middle, or at the lower end when the fracture may extend into the elbow joint or one of the epicondyles may be separated.

Bones of the forearm

The forearm contains two bones, the radius on the lateral (outer) side and the ulna on the medial (inner) side. When the arm is placed in the position for anatomical description, viz. with the palm of the hand facing forwards (page 33), the radius and ulna are parallel, and the forearm and hand are said to be in the position of **supination.** The forearm can, however, be rotated so that the back of the hand is directed forwards. This position is called **pronation,** and the radius then lies partly across the ulna.

The radius

The radius is the lateral bone of the forearm. It is a long bone having a shaft and upper and lower extremities.

The upper extremity is smaller than the lower and consists of a circular **head** on the top of which is a depression for articulation with the capitulum of the humerus. The circumference of the head articulates with the radial notch (lesser sigmoid cavity) of the ulna and is held in position against the ulna by a circular ligament (annular ligament). Immediately below the head is a narrower portion, the **neck.** Below the neck and on the medial side of the bone facing the ulna is a rough elevation called the **bicipital tuberosity** for the insertion of the tendon of biceps muscle.

The **shaft** is narrower at its upper end and expands towards the lower extremity. It has a sharp medial border facing the ulna to which is attached the interosseous membrane which stretches between the two bones. In this way the forearm is divided into anterior and posterior compartments. The former contains the

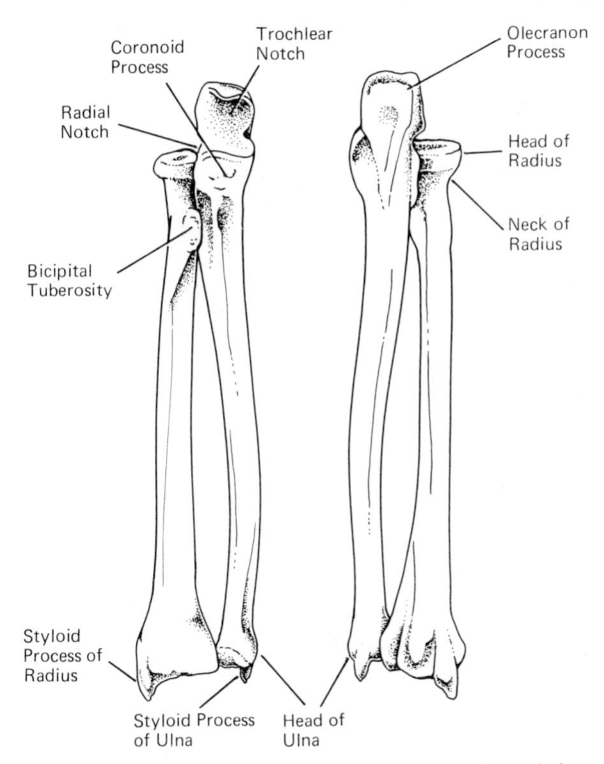

Fig. 29 Anterior (*left*) and posterior views of right radius and ulna.

flexor muscles of the wrist and fingers, the latter the extensor muscles.

The lower extremity presents an articular surface for the bones of the wrist, viz. the scaphoid and lunate (semilunar). At the lateral edge of this surface is a downward projection of the bone called the **styloid process** which can be felt on the lateral aspect of the wrist just above the base of the thumb. On the medial aspect of the extremity the articular surface is continued upwards for a short distance for articulation with the ulna.

The ulna

The ulna is the medial bone of the forearm. It is a long bone with a shaft, upper and lower extremities. It is slightly longer than the radius and, whereas the lower end of the radius is the largest part,

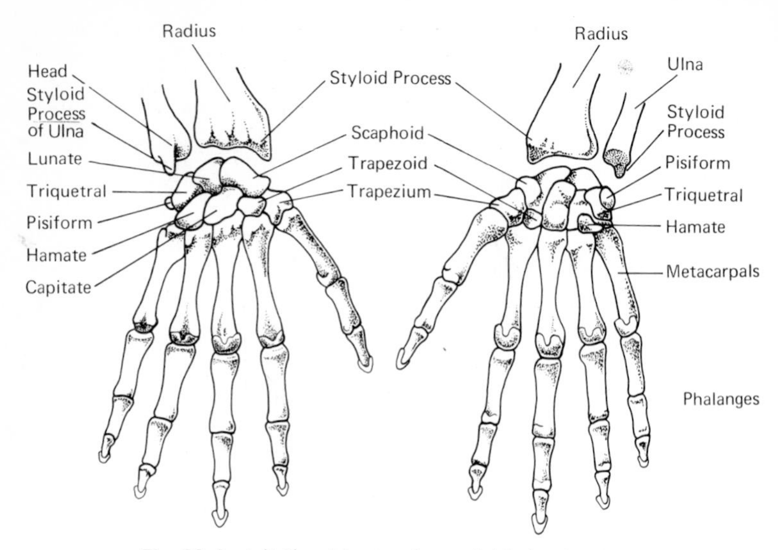

Fig. 30 Back (*left*) and front surfaces of right hand and wrist.

it is the upper end of the ulna which is expanded. The ulna extends from the elbow joint to the wrist.

The upper end presents a large C-shaped cavity, the **trochlear notch**. The large mass of bone forming the posterior wall of the trochlear notch and overhanging its upper part rather like a beak is the **olecranon process** which fits into the olecranon fossa on the posterior surface of the lower end of the humerus.

The floor of the trochlear notch projects forwards in a small process known as the **coronoid process** which fits into the coronoid fossa on the anterior aspect of the humerus. The trochlear notch articulates with the trochlear surface of the humerus. The articular surface is continued downwards on the lateral side of the bone to form the radial notch which articulates with the circumference of the head of the radius, and it is in this cavity that the head of the radius rotates in the movements of pronation and supination.

The **shaft** of the ulna narrows as it passes downwards but expands slightly at its lower end to form the **head** of the ulna. It will be noticed that contrary to the general description of other bones the head of the ulna is at its lower end. There is a small downward process at the extreme lower end on the medial side called the **styloid process**. The portion which can be seen projecting on the medial and posterior aspect of the wrist is the head of the ulna.

The bones of the forearm are a common site of fracture. Either may be involved in fractures occurring in the region of the elbow joint. Sometimes the olecranon process is separated from the rest of the bone. Either or both bones may be broken in the shaft. A very common type of fracture occurs about an inch above the lower end of the radius known as Colles' fracture, and at the same time the styloid process of the ulna is frequently torn off. This fracture usually results from a fall on the outstretched hand.

The bones of the wrist and hand

The **carpus** or wrist consists of eight bones arranged in two rows, proximal and distal, with four bones in each.

First (proximal) row	*Second (distal) row*
scaphoid	trapezium
lunate (semilunar)	trapezoid
triquetral (cuneiform)	capitate (os magnum)
pisiform	hamate (unciform)

They are situated from the lateral to the medial (radial to ulnar) side of the wrist in the above order.

They are irregular bones which articulate with each other and are held in position by ligaments.

The scaphoid and lunate bones articulate with the lower end of the radius (see Figs 30 and 67).

The **metacarpal bones**, or bones of the palm of the hand, are long bones consisting of a base, a shaft and a head. The bases articulate with the distal row of carpal bones, and the heads with the proximal or first row of phalanges.

The **phalanges** (singular = phalanx). These are long bones. The thumb has only two phalanges while the fingers have three, proximal, middle and distal, the proximal being the longest.

The joints between the metacarpals and the phalanges are called the metacarpo-phalangeal joints, those between the phalanges themselves, the interphalangeal joints.

BONES OF THE PELVIC GIRDLE

The innominate bone

The innominate bone on each side, together with the sacrum, which is part of the vertebral column, form the pelvic girdle. It is a large, irregular flat bone which in the child consists of three parts separated by cartilage. In the adult these parts are fused together but their names are retained for purposes of description. The uppermost bone is called the ilium (not to be confused with ileum, a part of the small intestine). The part situated in front is the pubis, and that posteriorly, the ischium. All three bones unite and take part in the formation of the large cup-shaped cavity on the outer surface of the bone, known as the **acetabulum**, into which fits the head of the femur forming the hip joint.

The ilium

This is the upper, expanded and flat part of the bone. It has external and internal surfaces which are separated along the upper margin of the bone by the rough **crest** of the ilium. The external surface is marked by three gluteal ridges, the superior, middle and inferior. The gluteal muscles arise from this surface of the bone (the gluteus maximus above the superior gluteal line, the gluteus medius between the superior and middle lines, and the gluteus minimus below the middle line).

The slightly concave internal surface gives attachment to the iliacus muscle and forms part of the floor of the iliac fossa. The posterior part of the internal surface is roughened for the attachment of ligaments which bind the ilium to the sacrum and immediately below the area for the ligaments is an ear-shaped facet where the sacrum articulates.

If the iliac crest is traced forwards it will be seen to end in a process of bone called the **anterior superior spine** to which the outer end of the inguinal (Poupart's) ligament is attached. The spine can easily be felt under the skin and is an important landmark in examination of patients. The posterior end of the iliac crest ends in the **posterior superior spine**.

The anterior inferior spine and the posterior inferior spine are situated just below the corresponding superior spines. If the bone is examined, immediately below the posterior inferior spine a large notch will be seen. This is the **great sciatic notch** through which passes the sciatic nerve, an important nerve which extends down

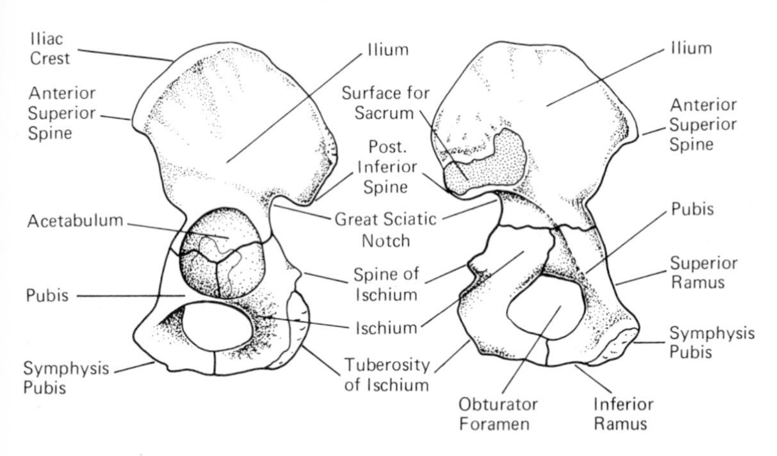

Fig. 31 The left innominate bone. External (*left*) and internal surfaces.

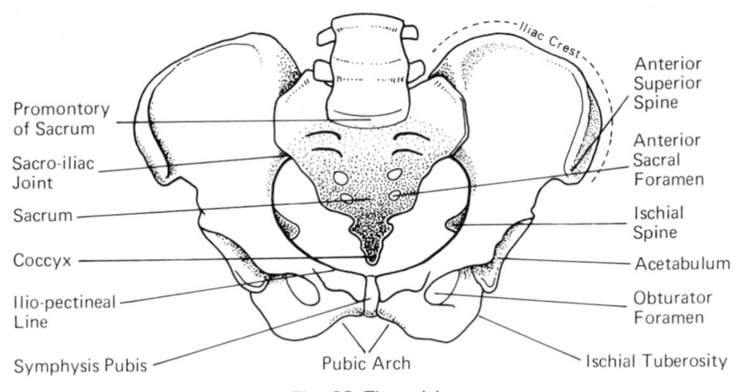

Fig. 32 The pelvis.

the back of the leg. Damage to this nerve causes a painful condition known as sciatica. When seen from the internal side the great sciatic notch is situated immediately below the area for articulation with the sacrum. (See also Fig. 69.)

The pubis

The front portion of the innominate bone is called the pubis. This consists of a more or less square **body** which articulates with its fellow of the opposite side at the **symphysis pubis**. Projecting from the outer part of the body and joining it to the ilium is a bridge of bone called the **superior ramus**. From the lower part of the body and joining it to the ischium is the **inferior ramus**. The upper border of the body of the pubis is called the crest of the pubis and at its junction with the superior ramus is a process named the **tubercle** or spine of the pubis, to which is attached the medial end of the inguinal (Poupart's) ligament.

The urinary bladder lies directly behind the body of the pubic bone and therefore may be injured in fractures of the pelvis involving the pubis.

The ischium

The solid broad portion at the lower and back part of the innominate bone is the ischium. It consists of a body, tuberosity and spine. The **body** takes part in the formation of the acetabulum on the outer side. The ramus passes forwards to join the inferior ramus of the pubis.

The roughly triangular opening bounded in front and above by the pubic bone and behind and below by the ischium is called the **obturator foramen**. In life it is closed by the obturator membrane through which pass blood vessels and nerves.

The rough **tuberosity** is the portion of bone which supports the body weight when sitting. It gives origin to the hamstring muscles.

Inspection of the whole pelvis will show that it is divisible into two parts. The upper part formed by the two iliac bones is called the 'false pelvis'. The smaller cavity below, which is bounded by the rest of the innominate bone (ischium and pubis) on each side and in front and the sacrum behind, is called the 'true pelvis'. The upper opening of the true pelvis is called the pelvic brim and a well-marked line, the **iliopectineal line**, marks the junction of the ilium with the ischium.

There are some differences in shape between the male and female pelvis. The latter is adapted to facilitate the passage of the baby's head during childbirth.

Fractures of the pelvis are liable to be complicated by severe concealed internal haemorrhage and by damage to the bladder and male urethra. Rupture of the male urethra must be suspected if blood issues from the external urinary meatus following pelvic injury.

THE BONES OF THE LOWER EXTREMITY

The femur or thigh bone

This is the longest and strongest bone in the skeleton, and consists of a shaft and two extremities.

The upper extremity has a head, a neck and two trochanters.

The **head** is hemispherical in shape and is covered with hyaline articular cartilage. It fits into the acetabulum of the innominate bone forming the hip joint. Just below the centre of its surface is a small depression for the attachment of the ligamentum teres.

The **neck** extends upwards and medially from the shaft at an angle (of about 125°) and terminates in the head.

The **great trochanter** is a large process projecting upwards from

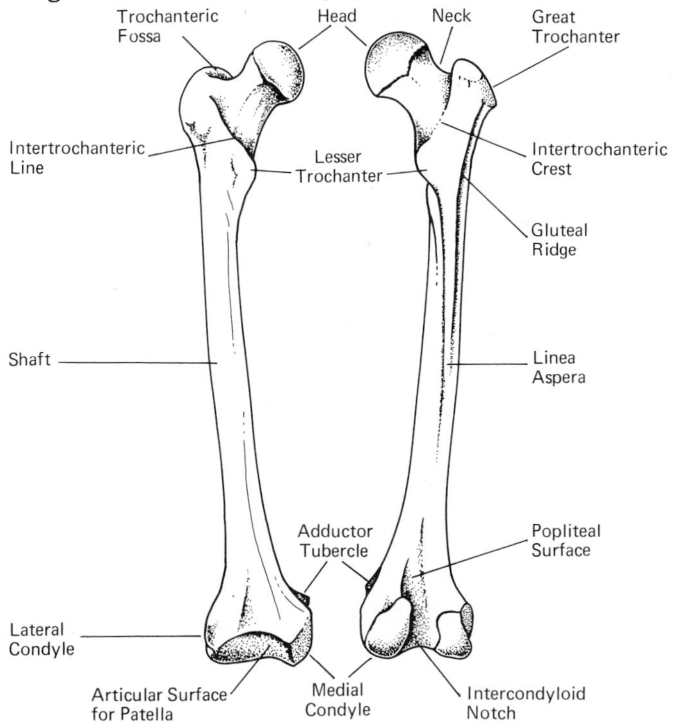

Fig. 33 Anterior (*left*) and posterior views of right femur.

the top of the shaft so that it is situated to the lateral side of the neck. Its upper portion curves inwards and overhangs the neck, producing a small hollow called the **trochanteric** (digital) **fossa**. Its outer surface is for the attachment of muscles.

The **lesser trochanter** is a rounded knob situated at the top of the shaft at its inner and posterior part where it joins the neck. To it is attached the tendon of the psoas muscle. A ridge of bone joining the great and lesser trochanters on the posterior aspect of the femur is called the intertrochanteric crest.

The **shaft** is almost cylindrical in shape except in its lower third, which becomes flattened and expanded. The anterior and lateral parts of the shaft are smooth, but on the posterior aspect is a prominent vertical ridge for the attachment of muscles, called the **linea aspera** (rough line). In the lower third of the bone the linea aspera divides into two smaller ridges, one of which passes to each of the condyles. The triangular area enclosed by these ridges is called the popliteal surface of the femur.

The lower extremity consists of two (medial and lateral) **condyles**, separated behind by the intercondylar fossa, which articulate with the corresponding tuberosities of the tibia to form the knee joint. The articular cartilage covering each condyle is continuous on the anterior aspect of the bone and forms a surface for articulation with the patella or knee-cap. Immediately above the medial condyle is a small tubercle, the **adductor tubercle**, to which part of the adductor magnus muscle is attached.

PRACTICAL CONSIDERATIONS. Fractures of the femur are common, and may occur through the neck, shaft or lower extremity. Those of the neck are common in elderly persons. Fractures of the lower end may extend into the knee joint.

The patella or knee-cap

This is really a sesamoid bone, developed in the tendon of the quadriceps extensor muscle of the thigh. It is flat and roughly triangular in shape, with its point of apex directed downwards. From this point the patellar ligament extends to the tubercle of the tibia.

The anterior surface is rough, the posterior smooth and covered with articular cartilage. The posterior surface articulates with the patellar surface of the femur, taking part in the formation of the knee joint.

The patella is quite commonly fractured, (*a*) being torn across into upper and lower halves, (*b*) shattered into fragments by a direct blow—stellate (like a star) fracture.

The tibia or shin bone

The tibia consists of a shaft and upper and lower extremities. It is the medial of the two bones of the leg.

The upper extremity or **head** is composed of a **medial** and **lateral condyle**, the upper surfaces of which articulate with the corresponding condyles of the femur and are covered by articular cartilage. Attached to each condyle, in life, are the two semilunar (half-moon shaped) cartilages which assist in the formation of the knee joint.

Corresponding to the intercondylar notch of the femur is the intercondylar or **popliteal notch** which separates the condyles of the

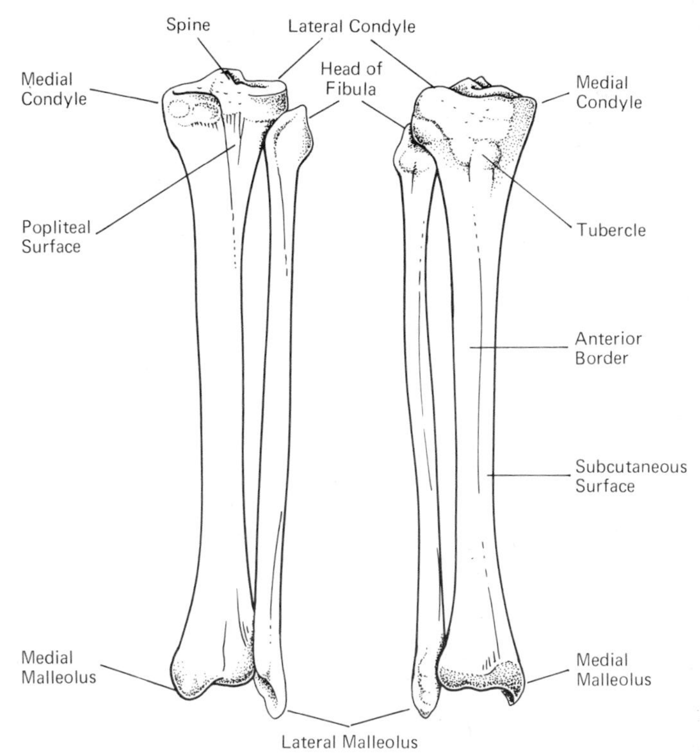

Fig. 34 Posterior (*left*) and anterior views of right tibia and fibula.

tibia behind. On the under surface of the lateral condyle, but taking no part in the formation of the knee joint, is a small articular facet for the head of the fibula.

The **shaft** of the tibia is triangular in shape, except at its lower end, which becomes more rounded. It has, therefore, three borders and three surfaces.

The anterior border or crest commences above in a roughened elevation called the **tubercle** of the tibia to which is attached the tendon of the quadriceps muscle or patellar ligament (ligamentum patellae). It passes downwards as a sharp ridge which can be felt under the skin gradually fading out as it passes towards the medial malleolus at its lower end.

The medial surface lies between the anterior and medial borders. It is smooth and is situated directly beneath the skin, hence it is referred to as the **subcutaneous surface**. When traced downwards it will be seen to end in the **medial malleolus**.

The posterior surface is roughened for the attachment of muscles. The upper part is triangular in shape and, together with the similar area of the posterior aspect of the lower end of the femur, forms the floor of the popliteal space.

The lower extremity is slightly expanded and has an articular surface which, with the talus (astragalus) takes part in the formation of the ankle joint. The medial aspect projects downwards as the medial malleolus.

The fibula

The fibula is the lateral of the two bones of the leg. It is a long slender bone, consisting of a shaft with upper and lower extremities, which lies parallel with the tibia.

The upper extremity or **head** is expanded and has a facet for articulation with the under surface of the lateral condyle of the tibia. (NOTE: This does not take any part in the formation of the knee joint.)

The lower extremity is expanded to form the **lateral malleolus** on the inner aspect of which is an articular surface for the talus.

The lower ends of the tibia and fibula are firmly held together by ligaments and are also joined throughout their shafts by the interosseous membrane which divides the leg into anterior and posterior compartments.

Fracture of the lower ends of the tibia or fibula, or both, is very common and is sometimes called Potts' fracture. Fracture of the shaft of the tibia also occurs. A fracture of a condyle at the upper end may extend into the knee joint.

Bones of the foot

The framework of the foot consists of:

1. The tarsus: (*a*) talus, navicular, three cuneiform bones; (*b*) calcaneum or os calcis, cuboid.

2. The metatarsus, five metatarsal bones.
3. The phalanges.

The **tarsus** may be considered as consisting of a medial and lateral series of bones.

(*a*) The medial series is formed by the talus, the navicular and the three cuneiform bones which, in their turn, articulate with the three medial metatarsals.

(*b*) The lateral series, viz. the calcaneum and cuboid, which articulate with the two lateral metatarsals.

If the general architecture of the foot is examined it will be seen that it is not flat but consists of a **longitudinal arch** which is most marked on its medial aspect. It will be observed also that this arch results from the fact that the talus is placed on top and, therefore, above the level of the calcaneum.

In addition, the foot is also **arched transversely** and this arch is most marked at the level of the bases of the metatarsal bones. These arches are very important in walking and are maintained by strong ligaments aided by muscles. If the arches become weakened and collapse the condition known as 'flat foot' results. It is interesting to note that these arches can withstand the very considerable strain caused by an adult jumping from the height of several feet.

The **talus** or astragalus. The talus is a very irregular bone having articular surfaces: (*a*) above, for the tibia and on either side for the medial and lateral malleoli forming the ankle joint; (*b*) below, for the calcaneum; (*c*) in front, for the navicular.

The **calcaneum** or os calcis. This also is an irregular bone situated below the talus. It projects backwards behind the talus forming the heel and to its posterior margin is attached the important tendon of Achilles (tendo Achillis). In front it articulates with the cuboid.

The **navicular** or scaphoid bone is a disc-shaped bone having a convex anterior surface and a concave posterior surface. It articulates

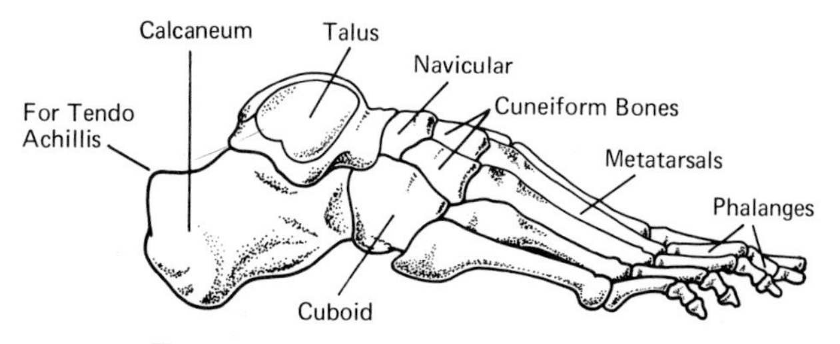

Fig. 35 Bones of the right foot—lateral view.

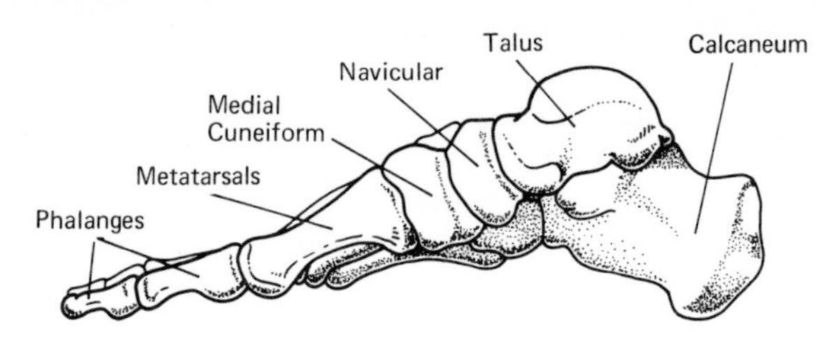

Fig. 36 Bones of the right foot—medial view.

behind with the talus and in front with the three cuneiform bones. Its lateral surface rests against the cuboid.

The **cuboid** bone articulates behind with the calcaneum, in front with the bases of the two lateral (fourth and fifth) metatarsals, while its medial aspect is in contact with the navicular and lateral cuneiform bones.

The three (medial, middle and lateral) **cuneiform**, or wedge-shaped bones, are placed between the navicular behind and the three medial matatarsals in front.

The **metatarsal bones** are five in number and correspond to the five toes. They are long bones and each consists of a base, shaft and head. The bases of the first three articulate with the cuneiform bones, those of the lateral two with the cuboid. It will be observed that the first metatarsal (of the great toe) is stouter than the others, while that of the fifth (little toe) has a marked projection on the lateral side of its base.

The **phalanges** are also long bones. There are two for the first or great toe (as in the case of the thumb), and three for each of the others. They are arranged in three rows, viz. the first or proximal phalanges which articulate with the heads of the corresponding metatarsals, the middle phalanges and the third or distal phalanges.

BONES OF THE SKULL

The bones of the skull are divided into two groups:

(*a*) Those of the **cranium** or brain-box, eight in number:

1 frontal	1 ethmoid
2 parietal	1 sphenoid
2 temporal	1 occipital

(*b*) Those of the **face** (including jaws, nose and orbit), fourteen in number:

2 maxillae (upper jaw)
2 zygomatic or malar (cheek) bones
2 nasal
2 lacrimal
2 palate
2 inferior turbinate
1 vomer
1 mandible (lower jaw)

All the bones of the skull, except the lower jaw, are joined together by sutures or immovable joints. The joint of the lower jaw or temporo-mandibular joint is a condyloid joint (see page 92) allowing both hinge-like and side-to-side movements.

Before discussing the individual bones, the skull as a whole must be examined. As a rule, the top or vault can be removed so that the skull can be seen from without and within. It must be definitely understood that the following description is meant to be read with the skull at hand and interpreted by the aid of diagrams.

Anterior aspect. When looked at from the front, the facial bones and lower jaw can be seen surmounted by the frontal bone which is placed above the orbits or sockets for the eyes.

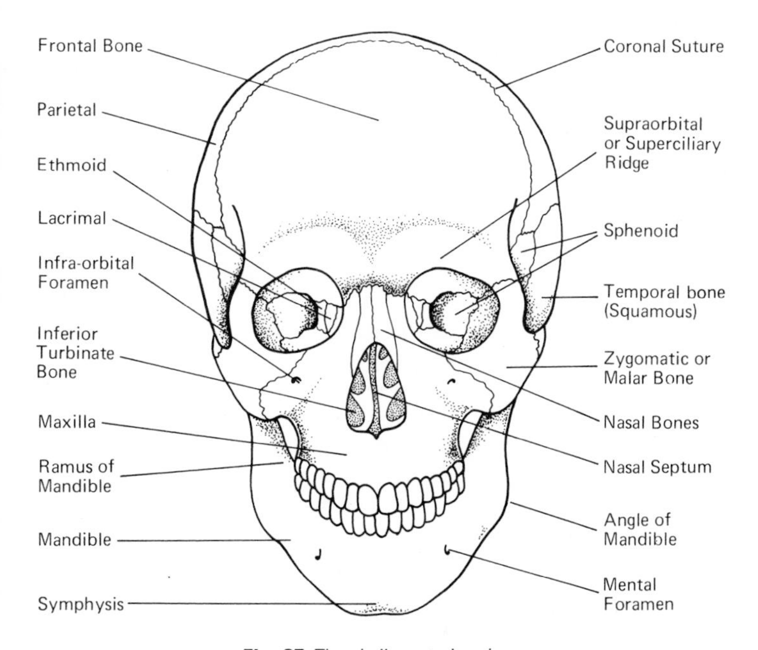

Fig. 37 The skull—anterior view.

The upper margins of the orbits and the ridges which lie beneath the eyebrows are formed by the frontal bone; their lateral wall by the zygomatic or malar and part of the sphenoid bone; while several bones, including the maxilla, form the medial wall (see also page 75). The portion of the maxilla which forms the medial margin of the orbit projects upwards to meet the frontal bone. Between these projections of the maxillae are the two nasal bones. Passing vertically downwards from these bones in the mid-line, the bony nasal septum can be seen dividing the nasal cavity into right and left

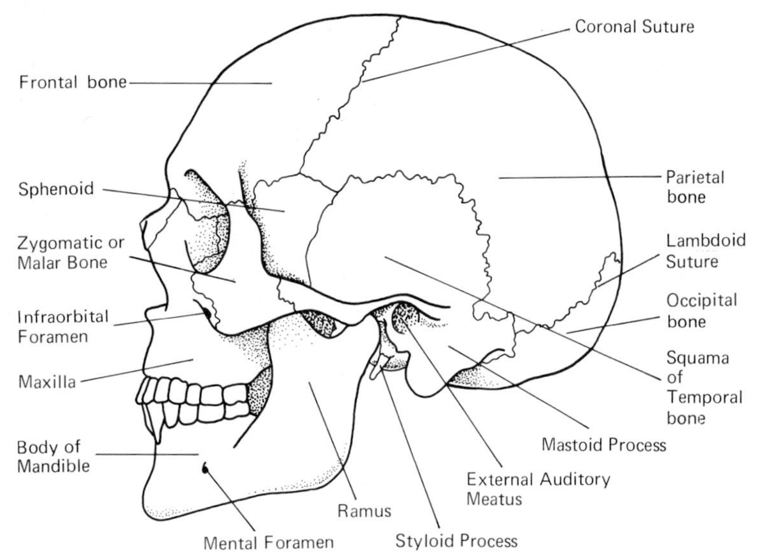

Fig. 38 The skull—lateral view.

halves. On the side walls of the nasal cavity the delicate turbinate bones (conchae) can be seen.

Surrounding the opening of the nasal cavity and meeting in the mid-line below it are the right and left superior maxillae or upper jaws which carry the upper set of teeth.

Lateral aspect. When viewed from the side the greater mass of the cranial bones can be seen behind the facial bones.

The central part of the lateral aspect is occupied by the temporal bone. A process from this bone passes forward to join the zygomatic (malar) bone, thereby enclosing a fossa (the temporal fossa) in which can be seen a portion of the sphenoid bone.

Above the temporal bone, from before backwards, are placed

portions of the frontal and parietal bones. At the posterior part of the cranium, the occipital bone will be seen.

Marked features of the lateral aspect of the skull are the presence of the external auditory meatus, behind which is the prominent mastoid process of the temporal bone.

Superior aspect. On looking at the skull from above, the frontal bone will be observed to occupy the front of the vault of the cranium and to be separated from the two parietal bones by a suture, which runs transversely, the **coronal suture**. The two parietal bones together form the greater part of the vertex and are separated from each other in the mid-line by a suture placed at right angles to the coronal suture. This is called the **sagittal suture**.

Posterior aspect. This view shows the posterior portions of the two parietal bones separated in the mid-line by the sagittal suture and below them a suture dividing them from the occipital bone. This suture consists of two limbs which meet at an angle at the posterior end of the sagittal suture and is called the **lambdoid suture**. (It is sometimes complicated by a series of small irregular bones called Wormian bones.)

Inferior aspect (base of the skull). The lower jaw should be removed. In the mid-line in front will be seen the under surface of the maxilla carrying the upper teeth and forming the hard palate, the posterior portion of which is formed by part of the palate bones. Immediately behind the hard palate, the posterior openings of the nasal cavities (posterior nares) will be seen.

On either side of the maxilla and palate can be seen the fossa or hollow bounded laterally by the zygomatic (malar) bone and process of the temporal bone (described in the lateral aspect as the temporal fossa). Part of the sphenoid bone can be seen here, immediately to the lateral side and slightly behind which is the area for the articulation of the mandible or lower jaw.

In the centre of the inferior aspect is a large oval opening, the **foramen magnum**, through which passes the spinal cord. This is situated in the occipital bone, part of which projects forwards to meet the sphenoid close to the level of the posterior nares. On either side of the foramen magnum are two smooth condyles for articulation with the atlas vertebra. Behind the foramen magnum is the main portion of the occipital bone.

On either side of the occipital bone are portions of the temporal bone including the mastoid process and an irregular portion projecting forwards and medially in the direction of the mid-line—the petrous portion.

Interior of the skull. The interior of the vault or vertex which has been removed requires no special description and will be seen to consists of the frontal, parietals and of part of the occipital bones. The base is clearly divided into three compartments: the anterior, middle and posterior cranial fossae.

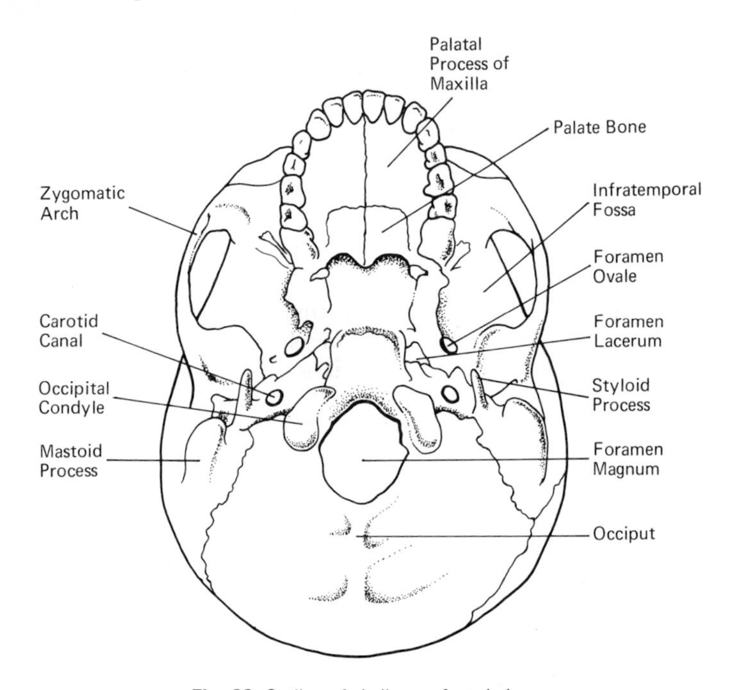

Fig. 39 Outline of skull seen from below.

Anterior cranial fossa. In the mid-line, in front, is a small vertical projection (the crista galli), on either side of which there is a narrow plate of bone perforated by numerous minute holes and forming part of the roof of the nasal cavity. This is the cribriform plate of the ethmoid and the holes are for the passage of the olfactory nerves. On either side of these plates, part of the frontal bone forms the roof of the orbit. The posterior margin of the anterior fossa is formed by a sharp edge of bone which is part of the small wing of the sphenoid.

Middle cranial fossa. In the mid-line is a small irregular space—the sella turcica (like a Turkish saddle)—which, in life, contains the pituitary gland and is part of the body of the sphenoid bone. On either side the fossa widens out and is formed partly by the sphenoid and partly by the temporal bones. The middle fossa is bounded posteriorly on either side by a broad ridge of bone, the petrous portion of the temporal bone which contains the internal ear.

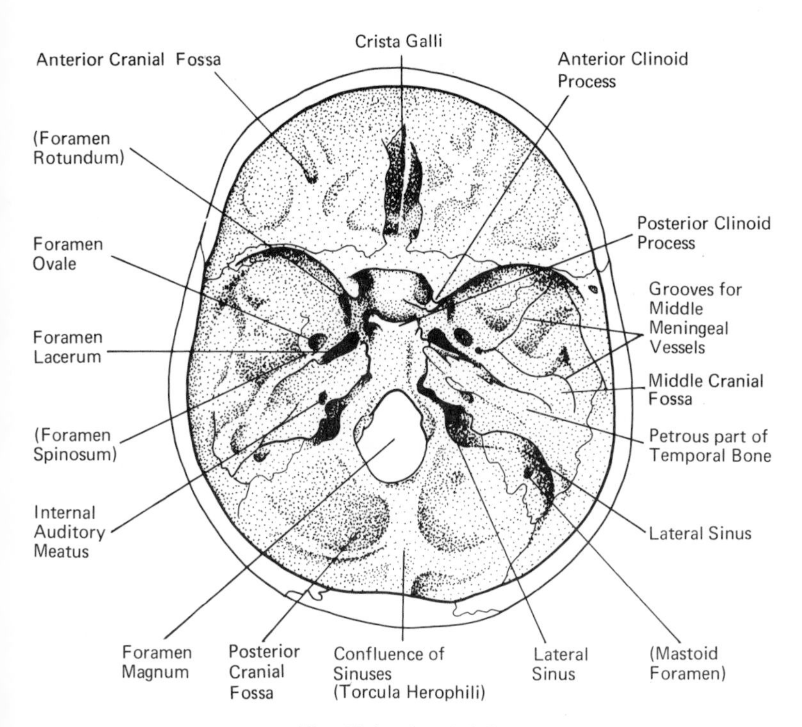

Fig. 40 Interior of skull.

Posterior cranial fossa. This is bounded in front by the petrous portions of the temporal bones, with part of the occipital bone in the mid-line. Its most conspicuous feature is the foramen magnum. Most of this fossa is formed by the occipital bone and it contains the cerebellum or lesser brain. On either side of the foramen magnum, and immediately below the petrous portion of the temporal bone, deep S-shaped grooves will be observed. These are for the transverse (lateral) sinuses. (NOTE: Observe their proximity to the mastoid processes on the outer aspect of the skull.)

A number of holes or foramina will be observed in the base of the skull for the passage of various nerves and blood vessels.

Having observed the general arrangement of the bones of the skull, some of the more important ones will be described in detail.

The frontal bone

This is an irregular flat bone which forms the front of the vault of the skull, most of the floor of the anterior cranial fossa and the roofs of the orbits. Posteriorly it is separated from the right and left parietal bones by the coronal suture. In the mid-line in front it articulates with the nasal bones and, in this area, is hollowed out to contain the frontal air sinuses.

The parietal bone

The right and left parietal bones are separated from each other in the mid-line by the sagittal suture. Each is a four-sided or quadrilateral

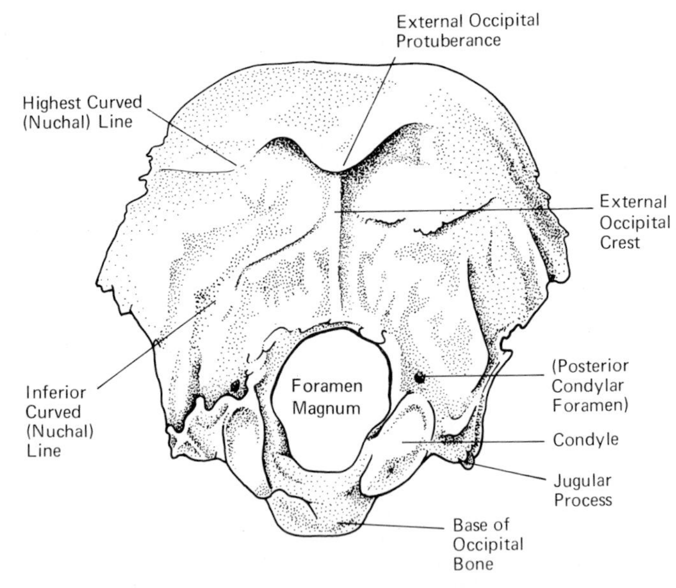

Fig. 41 Outer surface of occipital bone.

bone. The anterior margin articulates with the frontal bone at the coronal suture; the posterior with the occipital bone at the lambdoid suture; the lower and outer with the temporal and the upper or inner margin with the parietal bone of the opposite side at the sagittal suture. The parietal bones form the greater part of the roof and side walls of the vault of the skull.

The occipital bone

The occipital is an irregular flat bone situated at the back and base of the skull. It forms the greater part of the floor of the posterior cranial fossa.

It is separated in front by the lambdoid suture from the two parietal bones and on either side it articulates with the temporal bone. Its most marked feature is the large **foramen magnum** for the passage of the spinal cord. The **condyles** for articulation with the atlas vertebra on either side of the foramen magnum have already been noted. The portion in front of the foramen magnum is called the base of the occipital bone and unites with the sphenoid.

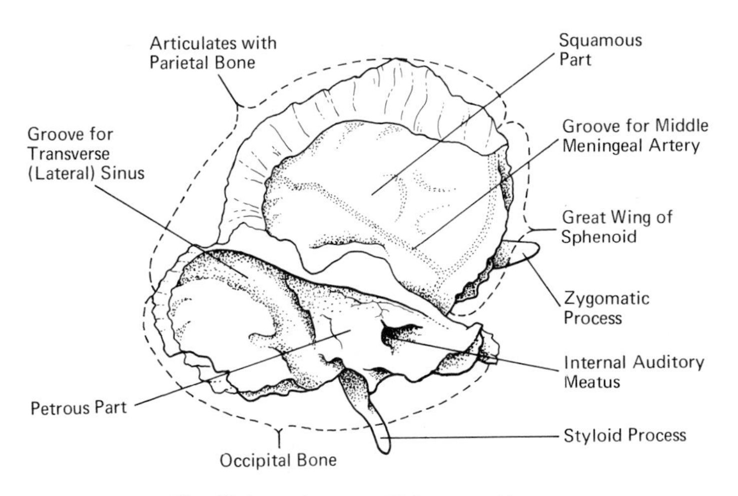

Fig. 42 Internal aspect of left temporal bone.

The temporal bone

This important bone consists of three main parts:

1. The squamous portion.
2. The mastoid portion.
3. The petrous portion.

Squamous portion. This occupies part of the side wall of the skull. It articulates above mainly with the parietal bone, in front with part of the sphenoid and behind and below is continuous with the mastoid portion. In its lower part is a circular opening—the **external auditory meatus**—which leads by a canal to the cavity of the middle ear. Immediately above the opening of the external auditory canal is a process of bone which projects forwards to join the zygomatic (malar) bone (zygomatic process).

Mastoid portion. Immediately behind the external auditory meatus is a dense downward projection of bone called the **mastoid process.** This contains the mastoid air cells. The close relationship of the mastoid process to the groove for the sigmoid portion of the transverse (lateral) sinus in the posterior cranial fossa has been observed. The sternomastoid muscle arises from its lateral surface.

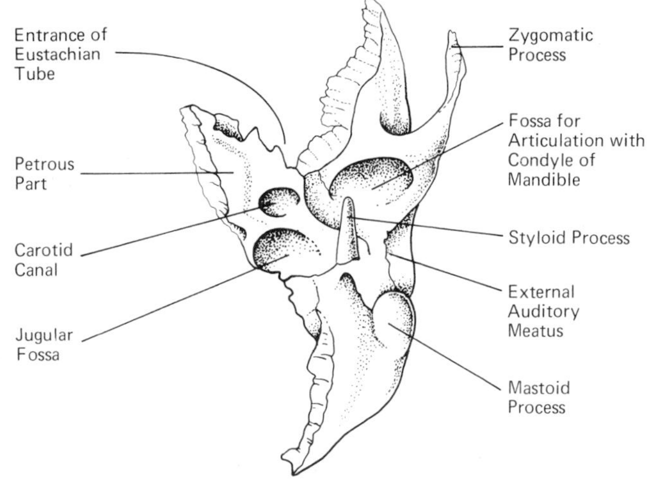

Fig. 43 Under-surface of left temporal bone.

Petrous portion. This projects inwards and forwards at an angle and can be seen on the base of the skull from below and from the interior. As the ends of the petrous portion approach each other they are separated in the mid-line by the base of the occipital bone. Close to the anterior extremity is a canal (not always easy to make out on the whole skull) by which the Eustachian tube from the nasopharynx passes to the middle ear.

On the under surface, close to the junction with the mastoid process, the long slender **styloid process**, for the attachment of muscles, may be seen. When viewed from the interior of the skull, the groove for the transverse (lateral) sinus has already been noted on the posterior aspect. Close to the anterior extremity of the petrous portion in the posterior fossa is a well-marked foramen—the **internal auditory meatus**—for the passage of the auditory or VIIIth cranial nerve. The petrous portion of the temporal bone contains the internal ear and semicircular canals.

The sphenoid bone

This is an irregular bone which takes part in the formation of the middle cranial fossa. It has a body situated in the mid-line from which project outwards (*a*) the great wings and (*b*) the small wings.

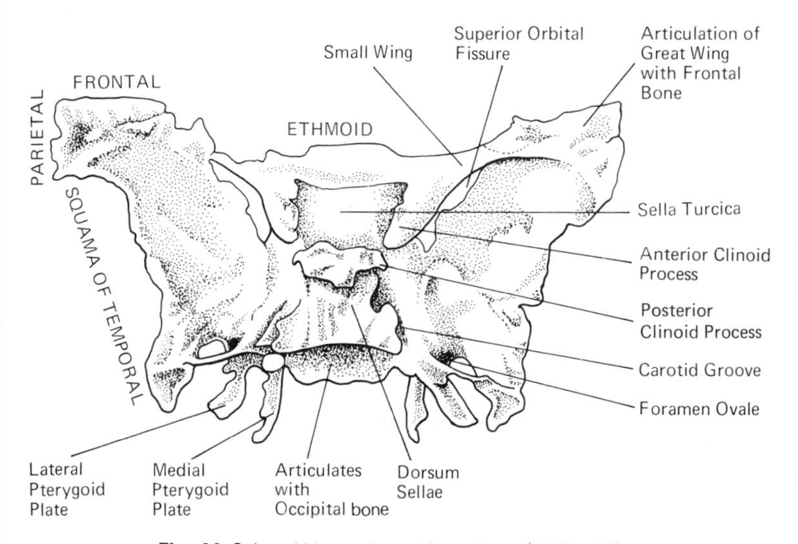

Fig. 44 Sphenoid bone viewed from above (cf. Fig. 40).

The **body** is hollow and contains the sphenoidal air cells. Its upper surface is shaped like a Turkish saddle—the **sella turcica**, which contains the pituitary gland. From the anterior part of the body the **small wings** spread out on either side to form the posterior margin of the anterior cranial fossa. From the sides of the body the great wings extend outwards between the frontal and temporal bones, and close to their origin is a well-marked foramen for the mandibular branch of the trigeminal or Vth cranial nerve (foramen ovale).

The ethmoid bone

This is a box-shaped bone of delicate structure, hollowed out to contain the ethmoidal air cells. Its lateral walls help to form the medial wall of each orbit. Its roof is formed by the **cribriform plate** which can be seen between the two portions of the frontal bone in the anterior cranial fossa and is perforated by the olfactory nerves. The inferior surface enters into the formation of the roof of the nose.

From this surface projects a **perpendicular plate** which forms the upper part of the nasal septum, dividing the cavity of the nose into right and left halves. Also from its under surface project the upper and middle **turbinate bones** (conchae) which are situated in the side wall of the nasal cavity.

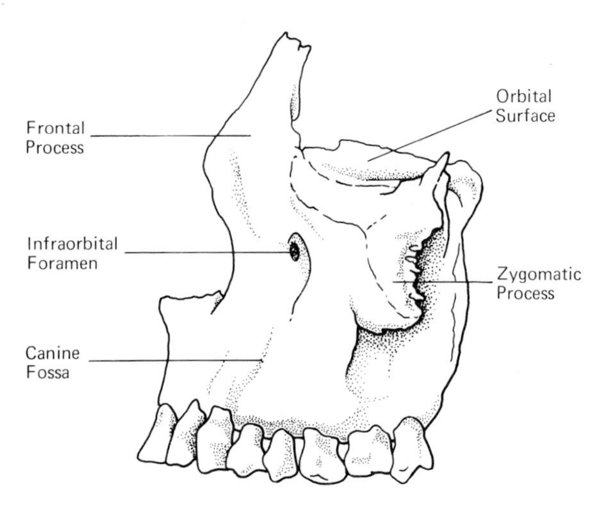

Fig. 45 Left maxilla—external or lateral view.

The maxilla (upper jaw)

This is an irregular bone which, with its fellow, carries the teeth of the upper jaw. Its upper part forms the greater part of the floor of the orbit; its lower part, the major portion of the hard palate. Laterally it articulates with the zygomatic (malar) bone. Medially it helps to make up part of the side wall of the nasal cavity and to it is attached the **inferior turbinate bone** (concha).

The body of the maxilla is hollow and contains the important **maxillary antrum** (of Highmore). It will be noticed that the roots of the first and second molar teeth are closely related to the floor of the antrum.

Inflammation within the nose, e.g. a common cold, may spread to the maxillary antrum, so that the cavity becomes filled with mucus or mucopus causing discomfort in the area. Unless this drains spontaneously it may be necessary to wash out the antrum by perforating its medial wall. Similar inflammation may affect the frontal sinus and, much less commonly, the ethmoid and sphenoid air cells.

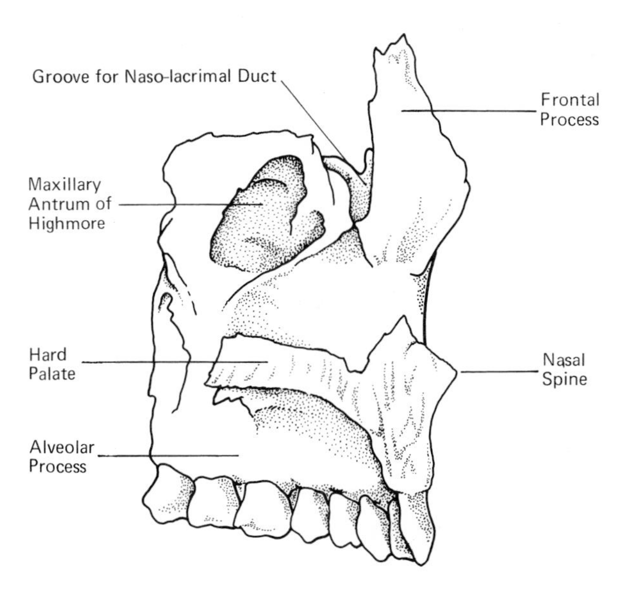

Fig. 46 Left maxilla—medial aspect.

The zygomatic (malar) bone

The zygomatic bone (cheek bone) forms the prominence of the cheek and part of the floor and lateral part of the wall of the orbit. It articulates medially with the maxilla, above with the frontal bone and posteriorly with a process from the temporal bone.

The lacrimal bones

These are two small bones which take part in the formation of the medial wall of the orbit (Fig. 49).

The palate bones

These are irregular in shape and fit in between the maxilla and the sphenoid bones, taking part in the formation of the hard palate, the walls of the nasal cavity and a small area of the floor of the orbit. The **vomer** is situated in the mid-line and rests on the hard

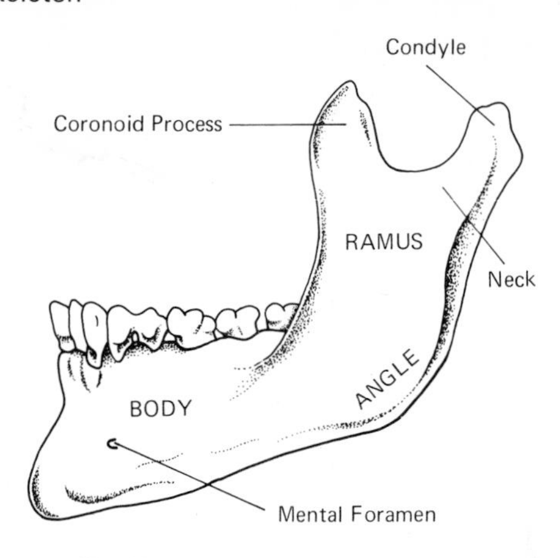

Fig. 47 Mandible—outer view of left half.

palate. It forms the lower and posterior part of the bony nasal septum. The rest of the nasal septum is formed above by the perpendicular plate of the ethmoid bone and anteriorly by cartilage.

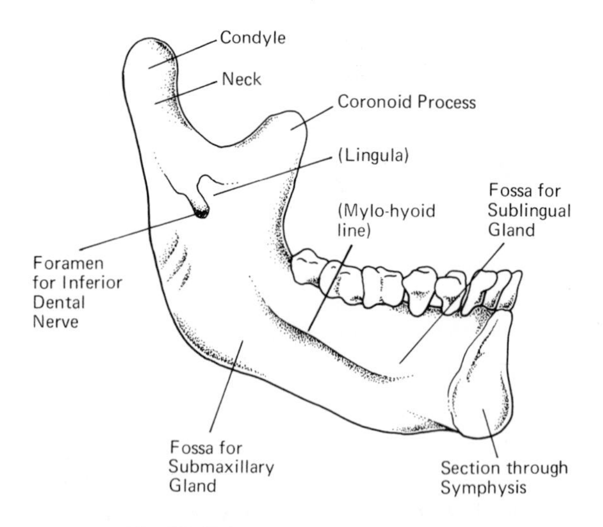

Fig. 48 Mandible—inner view of left half.

The mandible or lower jaw

The mandible is the only movable bone in the skull. It consists of a curved **body** which carries the teeth of the lower jaw and has outer and inner surfaces. The outer surface forms the chin in front

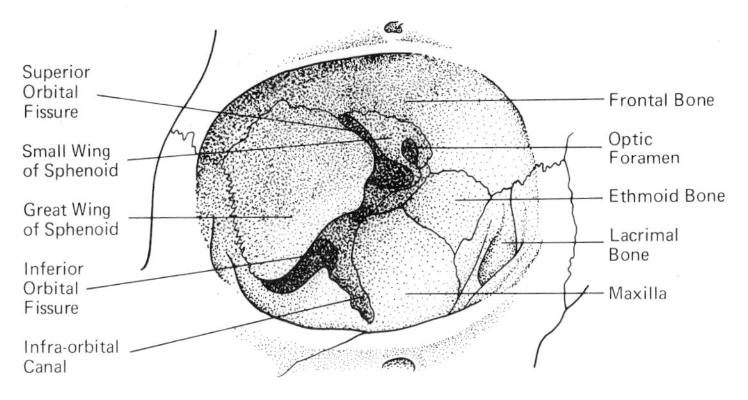

Fig. 49 Boundaries of right orbit viewed from in front.

and its central point is called the **symphysis**. About an inch to the side of the symphysis is the mental foramen for the exit of the inferior dental nerve.

From the posterior ends of the body the **rami** pass almost vertically upwards on each side. The lower posterior corner of the ramus is called the **angle** of the jaw. On the inner surface of the ramus is a foramen for the entry of the inferior dental nerve. The ramus terminates above in two processes: the sharp **coronoid process** in front, and the strong **condyle** of the jaw behind, which articulates with the temporal bone to form the temporomandibular joint.

The orbit

The orbit is the bony cavity which contains and gives protection to the eye. It is cone-shaped, having its wide open base in front and its narrow apex behind. At its apex or deepest part is a circular foramen for the passage of the optic nerve. A deep groove will be seen extending from the lateral side of this foramen and extending into the floor of the orbit.

The roof of the orbit is formed by the frontal bone; the floor mainly by the maxilla; the lateral wall by the zygomatic and great wing of the sphenoid bone. Its medial wall contains the lacrimal bone and parts of the maxilla, ethmoid and sphenoid bones.

The fontanelles

It will be recalled that the sagittal suture separating the two parietal bones in the mid-line joins the coronal suture in front and behind meets the lambdoid suture separating them from the occipital bone. At birth, however, the actual bones are not in contact. There is a diamond-shaped space covered only by membrane between the frontal and two parietal bones in the mid-line: this is called the **anterior fontanelle.** It lies immediately above the superior longitudinal sinus.

The **posterior fontanelle** is situated at the area of junction between the sagittal and lambdoid sutures. It is smaller than the anterior fontanelle and closes a little earlier.

After birth the size of the fontanelles gradually lessens and the anterior fontanelle is usually closed by the age of eighteen months.

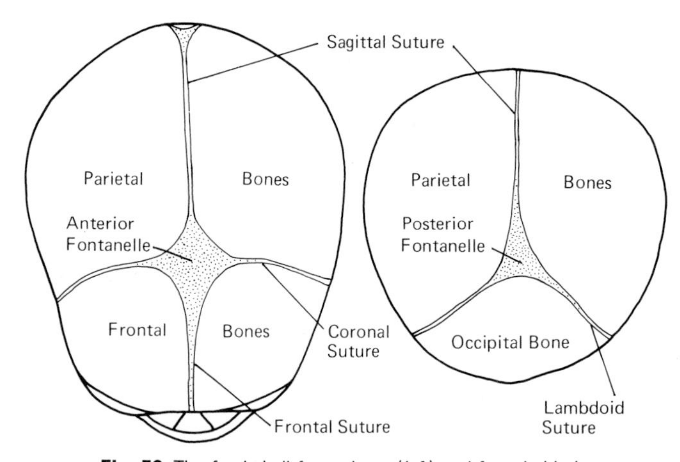

Fig. 50 The fetal skull from above (*left*) and from behind.

The skull is commonly fractured, following a blow on the head. Some such fractures consist of a linear crack either affecting the vault or base of the skull. If such a fracture involves the squamous portion of the temporal bone, the middle meningeal artery may be torn, resulting in haemorrhage within the skull which will cause the intra-cranial pressure to rise. Unless this is treated surgically the condition may prove fatal.

In another type of fracture a portion of bone may be depressed and damage or press on some part of the brain.

THE VERTEBRAL COLUMN

The vertebral or spinal column is the central part of the skeleton which supports the head and encloses the spinal cord. It is constructed to combine great strength with a moderate degree of

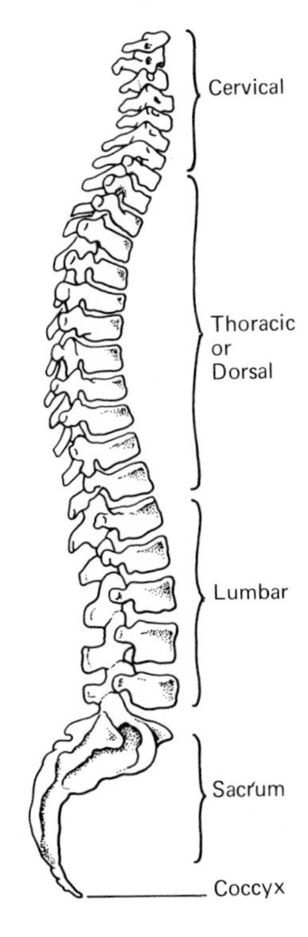

Fig. 51 The vertebral column.

mobility. It attains these objects by consisting of a number of separate bones united by ligaments and by tough discs of fibrocartilage (the intervertebral discs) which, in addition, act as 'shock absorbers' when vigorous movements of the body are carried out. The bones also give origin to a number of muscles. The vertebral column consists of thirty-three bones which are divided into five groups (from above downwards):

7 cervical 12 thoracic or dorsal 5 lumbar	true or movable vertebrae
5 sacral vertebrae forming the sacrum 4 coccygeal vertebrae forming the coccyx	false or fixed vertebrae

In the last two groups, viz. the sacral and coccygeal, the vertebrae are fused together to form two bones—the sacrum and coccyx which enter into the formation of the bony pelvis.

The spine as a whole

It will be noticed that the bones become increasingly larger as the column descends, reaching their maximum width at the upper part of the sacrum, only to become greatly reduced in size as it tapers off into the coccyx.

When looked at from the side it will be seen to have several curves:

1. It is convex forwards in the cervical or neck region.
2. The thoracic region is concave. Excessive curvature is known as **kyphosis** (hump-back).
3. The lumbar region is markedly convex forwards. When it is excessively convex the term **lordosis** is used.
4. The sacrum and coccyx form a marked concavity.

The three main curves in the vetebral column are developed from a single curve in the spine of an infant with the assumption of walking and the erect position.

It will be noticed also that the last lumbar vetebra joins the sacrum at a pronounced angle (the lumbosacral angle.)

Lateral (sideways) curvature of the spine is abnormal and is known as **scoliosis**.

The vertebrae

The individual vertebrae are all built on the same plan, although there are certain variations in different parts of the spinal column and special vertebrae (atlas and axis), the sacrum and coccyx, which require separate description.

A **typical vertebra** consists of:

1. a body.
2. the neural arch: (*a*) 2 pedicles
   (*b*) 2 laminae

(*c*) 2 transverse processes
(*d*) 1 spine or spinous process
(*e*) 4 articular processes

The **body** is the solid box-shaped structure situated anteriorly which has slightly concave upper and lower surfaces. It is separated from the bodies of the vertebrae immediately above and below by the tough intervertebral discs of fibrocartilage.

From either side of the posterior aspect of the body two short, stout bars of bone, the **pedicles**, project backwards.

From the posterior ends of the pedicles, the two **laminae** are directed backwards and towards each other and meet in the mid-

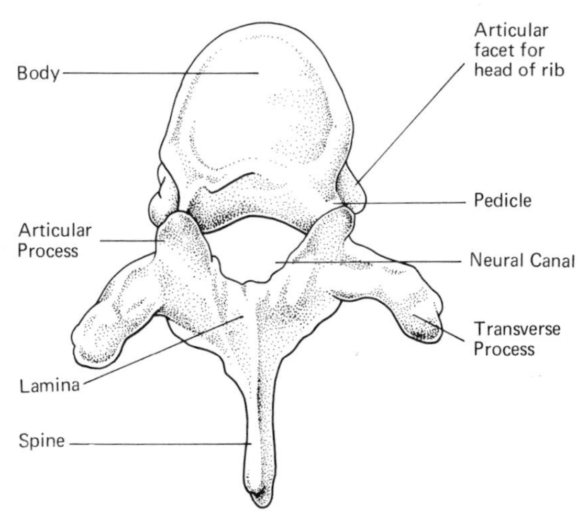

Fig. 52 A thoracic or dorsal vertebra seen from above.

line behind. From the junction of the pedicles with the laminae, the **transverse processes** project outwards on each side of the bone.

Where the laminae unite in the mid-line behind, the **spine** or **spinous process** is formed and projects backwards and, in some parts of the column, downwards.

Two **articular processes** are situated on the upper and lower surface of each vertebra at the junction of the pedicles with the laminae close to the origin of the transverse processes.

The roughly circular opening enclosed by the body in front, the pedicles on either side and the laminae behind, through which passes the spinal cord, is called the vertebral or spinal foramen, or **neural canal**. This bony canal helps to protect the spinal cord from injury.

The cervical vertebrae

The lower five cervical vertebrae have the same general form, but the

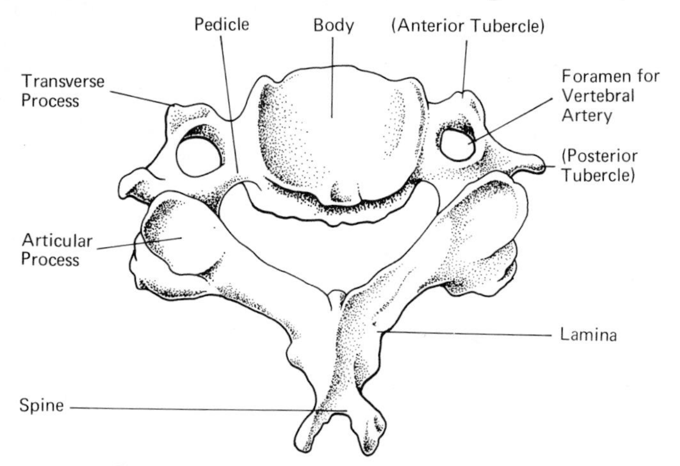

Fig. 53 A typical cervical vertebra viewed from above.

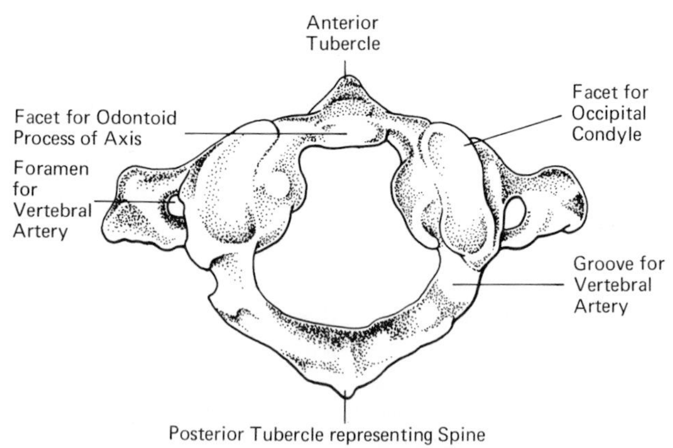

Fig. 54 The atlas viewed from above.

upper two, the atlas and axis, require separate description. A typical cervical vertebra differs from those of the rest of the spinal column:

1. The body is smaller than those of other groups.
2. It is oblong in shape, being broadest from side to side.

3. The neural canal is roughly triangular in shape and larger than in other parts of the spine.
4. The spinous process is bifid, i.e. it has a double end.
5. There is a well-marked foramen in the transverse process for the passage of the vertebral artery.

The atlas. The atlas or first cervical vertebra together with the axis are specially adapted to carry the weight of the head and to facilitate its movements. The atlas is readily recognized by the fact that it has no body. Instead, it consists of a ring of bone enclosing a very large neural canal, on the anterior aspect of which is a small facet for articulation with the odontoid process of the axis. Its characteristics are:

1. Absence of the body.
2. No spinous process.
3. Large concave facets on its upper surface for articulation with the condyles of the occipital bone. (This is the atlanto-occipital joint at which nodding movements take place.)

The axis. The axis or second cervical vertebra is characterized by a peg-like process—the **odontoid process** (dens)—which projects

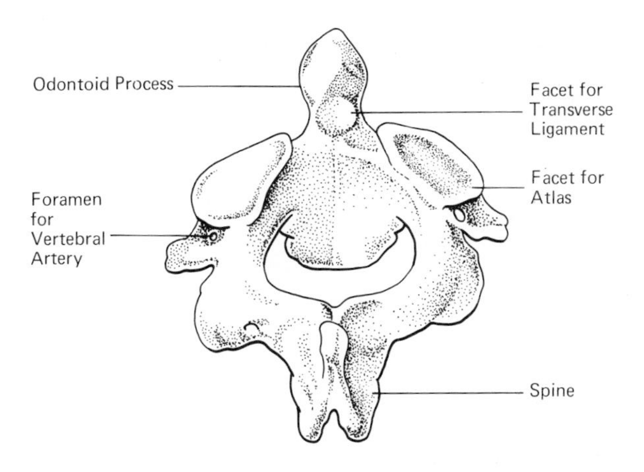

Fig. 55 The axis viewed from above and behind.

upwards from its body. This process actually represents and occupies the position of the missing body of the atlas. On its anterior surface is a small facet which articulates with the anterior arch of the atlas. This arrangement permits the rotation of the head.

The thoracic or dorsal vertebrae

These vertebrae carry the ribs. They are twelve in number and are characterized by:

1. Heart-shaped bodies, intermediate in size between those of the cervical and lumbar types (Fig. 52).
2. Marked downward projection of their pointed spinous processes.
3. Articular facets for the ribs (see Figs 59 and 60).
   (*a*) Two, one above and one below, on each side of the bodies.
   (*b*) A small facet at the tips of the transverse processes articulating with the tubercles of the ribs.

The lumbar vertebrae

1. These are the largest vertebrae and are five in number.
2. Their bodies are kidney-shaped.
3. There are no articular facets for ribs.
4. Their spinous processes are stout, broad and flat, and are directed backwards.

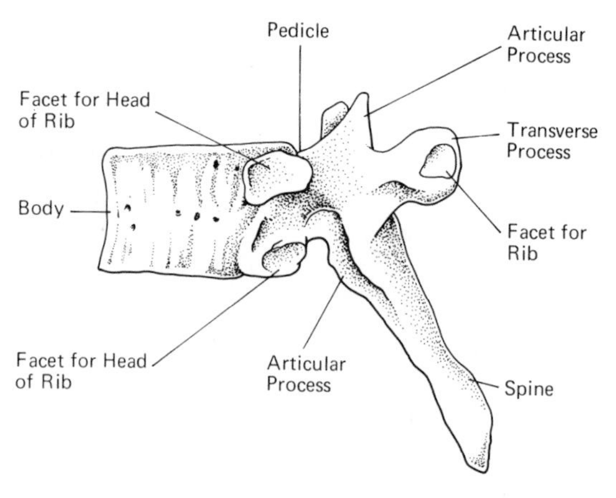

Fig. 56 A thoracic or dorsal vertebra seen from the side.

The sacrum The sacrum is a triangular bone articulating above with the fifth lumbar vertebra forming the lumbo-sacral angle and below with the coccyx (see Fig. 32). It is composed of five fused vertebrae, the individual parts of which can be discerned. It has anterior and posterior surfaces and lateral margins.

The sacrum consists of a central **body** from which project the transverse processes joined together to form the **lateral masses** on either side. The upper part of the body projects forwards and is called the **promontory** of the sacrum. Between the body and the lateral masses are the four anterior sacral foramina for the passage of nerves. The anterior surface of the sacrum is markedly concave.

The upper part of the lateral mass has an ear-shaped surface on its lateral margin for articulation with the iliac part of the innominate bone with which it forms the sacro-iliac joint. Behind this surface is a very rough area for the attachment of strong ligaments (see Figs 31 and 69).

On the convex posterior surface, three or four rudimentary **spines** can be seen in the mid-line, on either side of which are the posterior sacral foramina for nerves.

The neural canal is continued into the sacrum and, at its lower end, opens on to the surface of the bone. The inferior part of the bone articulates with the coccyx.

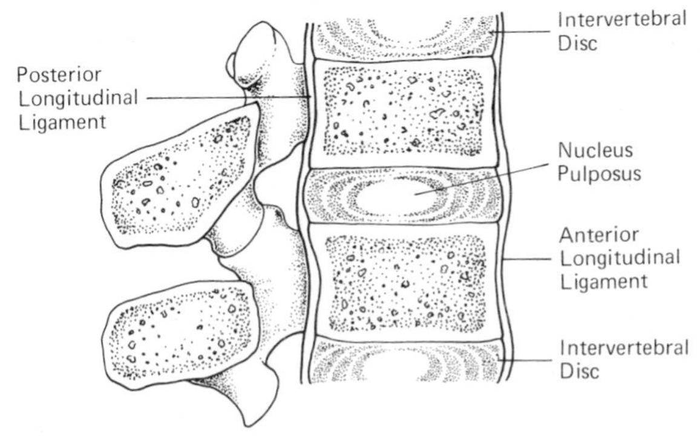

Fig. 57 Section of vertebral column showing ligaments and intervertebral discs.

The coccyx (or tail). This consists of four fused rudimentary vertebrae, the individual characteristics of which are not apparent. It tapers to a point and articulates above with the sacrum.

Ligaments of the vertebral column

The individual vertebrae are held together by strong ligaments which include:

1. Those connecting the bodies; (*a*) the **anterior** (Fig. 60) and (*b*) the **posterior longitudinal ligaments** which run the whole length of the spine, connecting the anterior and posterior aspects of the bodies respectively.

2. Those connecting (*a*) the laminae and (*b*) the spinous processes.
   (*a*) Those connecting the laminae are called the **ligamenta flava** and by their elasticity help to maintain the erect posture.
   (*b*) The **supraspinous ligaments** between the spines (which are penetrated by the lumbar puncture needle).
3. The **intervertebral discs** of fibro-cartilage.

Injury to these ligaments by various types of back strain often causes both acute or chronic back pain ('lumbago').

This may be associated with or followed by damage to an intervertebral disc. This may eventually protrude through the damaged posterior ligament in such a way that the protrusion may press on a nerve root in the vicinity causing pain in its distribution (e.g. 'sciatica').

Movements of the spinal column

Apart from the movements of nodding and rotation of the head which, it has been seen, take place at the atlas and axis respectively, the movements of the spine as a whole are considerable although the actual range of movement between the individual vertebrae is small. They are:

1. (*a*) Flexion or bending forwards and (*b*) extension or bending backwards, in which the maximum movement takes place in the cervical and lumbar regions.
2. Lateral movement or bending from side to side, well marked in the neck and also possible in the dorsal region.
3. Rotation or twisting the spinal column as a whole round its long axis.

NOTES: The following structures are on a level with various vertebrae:

structure	*on a level with*
Oral pharynx	axis vertebra
Opening of larynx	4th cervical vertebra
Bifurcation of carotid artery	3rd cervical vertebra
Upper margin of sternum	2nd thoracic vertebra
Bifurcation of trachea	4th thoracic vertebra
Heart	6th, 7th and 8th thoracic vertebrae
Kidneys	12th thoracic, 1st and 2nd lumbar vertebrae
Pancreas	1st and 2nd lumbar vertebrae
Bifurcation of aorta	4th lumbar vertebra

A line joining the highest points of the iliac crests passes between the spines of the fourth and fifth lumbar vertebrae and is a landmark for lumbar puncture.

Fractures of the spine may be accompanied by damage to the spinal cord. Patients with spinal cord injuries suffer from retention of urine, which may be temporary or permanent. If infection is introduced at catheterization, ascending pyelonephritis may result and may be fatal.

BONES OF THE THORAX

The skeletal framework of the thorax is formed (*a*) behind, by the thoracic or dorsal vertebrae, (*b*) anteriorly, by the sternum and costal cartilages, (*c*) the remainder of the circumference, by the ribs.

The sternum

The sternum or breast bone is a flat bone having anterior and posterior surfaces. To it are attached costal cartilages of the ribs. It is divided into three parts: (*a*) the manubrium, (*b*) the body (gladiolus) and (*c*) the xiphoid (ensiform) process.

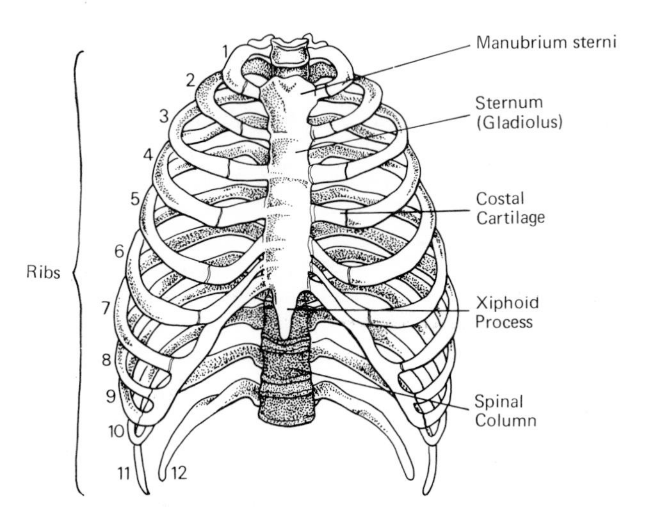

Fig. 58 Skeleton of the thorax.

The **manubrium sterni** is the upper part of the bone which on either side articulates with the clavicle at the sternoclavicular joint. The articular facets are placed at the upper and lateral corners of the manubrium and, between them, is the suprasternal or interclavicular notch. Immediately below the facet for the clavicle, the bone is joined by the cartilage of the first rib. Its lower border is joined to the body.

The **body** is oblong in shape. The cartilage of the second rib joins it at the level of its union with the manubrium. The cartilages of the 3rd, 4th, 5th and 6th ribs are attached at intervals; while the 7th meets it at its junction with the xiphoid process.

The **xiphoid** forms the lower extremity of the sternum. It sometimes remains cartilaginous during life, but in the adult is usually ossified. It is a small triangular plate to which the fibres of the linea alba and the rectus abdominis muscle are attached. A part of the diaphragm is attached to its posterior surface.

The ribs

The ribs, arranged in twelve pairs, are flat, curved bars of bone directed forwards and downwards, which are classified into the following groups:

1. True ribs—the upper seven pairs, extending from the dorsal vertebrae behind to the sternum in front, to which they are joined by the costal cartilages.
2. False ribs—the next five pairs, viz 8th to 12th, that do not join the sternum directly. The costal cartilages of the 8th, 9th and 10th fuse with the cartilage immediately above. The 11th and 12th (the floating ribs) have only a small costal cartilage and only partly surround the circumference of the thorax, so that they are unattached in front (Fig. 58).

The **typical ribs** are long bones each having (*a*) a head, (*b*) a neck, (*c*) a shaft.

The **heads** are the slightly enlarged posterior ends of the bones which articulate with facets on the bodies of the corresponding thoracic vertebrae (see also page 82).

The **neck** is the portion extending outwards from the head which, where it joins the shaft, has a well-marked **tubercle** and a facet for articulation with the transverse process of the corresponding vertebrae.

The **shaft** is flattened, curved and slightly twisted on itself. It has, therefore, internal and external surfaces with superior and inferior borders. The maximum point of curvature is in the posterior part

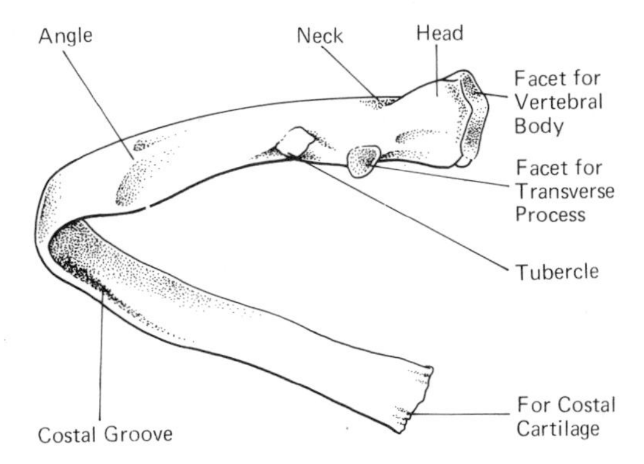

Fig. 59 A typical rib.

a little distance beyond the tubercle. This is called the **angle** of the rib. The anterior end is hollowed out for the attachment of the costal cartilage.

The borders of the ribs give attachment to the intercostal muscles which pass to the ribs immediately above and below. There is a groove close to the inferior border occupied by the intercostal artery and nerve (Fig. 59).

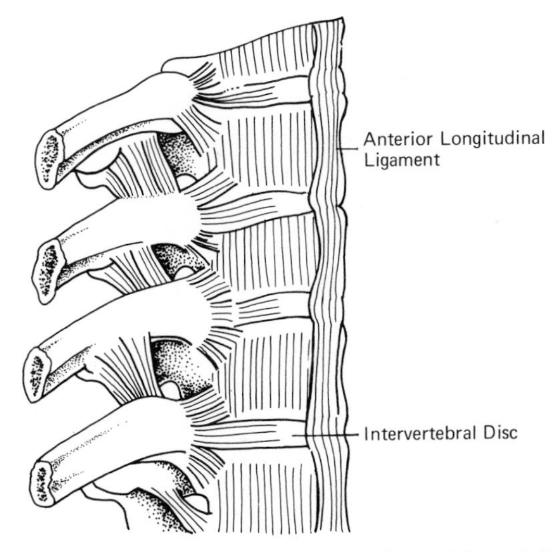

Fig. 60 Thoracic vertebrae with ligaments and articulations of ribs.

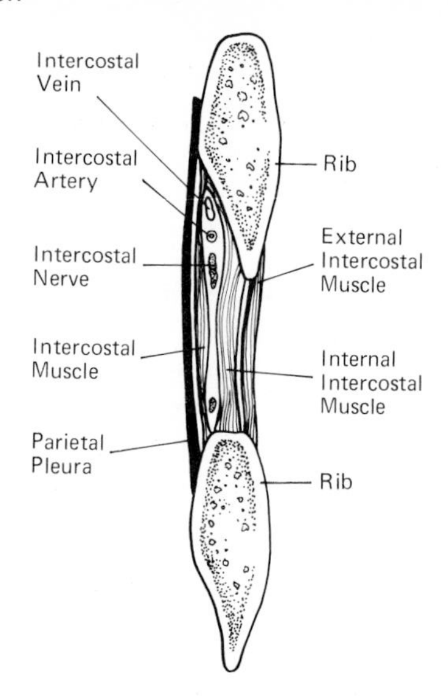

Fig. 61 Section through ribs and an intercostal space.

The inner surface of the rib is lined by parietal pleura, which is firmly adherent to the periosteum of the bone.

The **first rib** is the shortest and stoutest, and differs somewhat from the others. It is flat and not twisted, and has upper and lower surfaces. Its upper surface is crossed by the subclavian artery and brachial plexus. The **eleventh** and **twelfth**, being floating ribs, are short and thin.

The costal cartilages. It has been seen that the costal cartilages unite the terminations of the 1st to 7th ribs to the sternum. The 8th costal cartilage joins the 7th. The 9th and 10th are small and join the cartilage immediately above. The angle formed at the xiphoid process by the costal cartilages diverging on either side is called the **epigastric angle.**

Rib fractures are common. Sometimes the broken end of the bone is driven into the lung and may cause severe bleeding into the pleural cavity (haemothorax).

QUESTIONS

1. Describe briefly the skull and its contents.
2. Describe the skeleton of the thorax.

3. Describe the humerus. What factors influence the growth and formation of bone?
4. Describe the bony framework of the pelvis. Enumerate its chief contents in the male and in the female.
5. Describe the shoulder girdle. What are its movements and what are the chief muscles concerned with them?
6. Describe the femur. Give the names and positions of the chief blood vessels and nerves in the thigh.
7. Describe the mandible (lower jaw). What are its functions?
8. What are the different varieties of bones? Give one example of each type and describe its structure.

5 The Joints or Articulations

A joint or articulation implies the apposition of two or more bones. Joints may be divided, according to their mobility, into three main classes:

1. Fixed or immovable joints (fibrous joints)
2. Slightly movable joints (cartilaginous joints).
3. Freely movable joints (synovial joints).
   (*a*) Gliding joints.
   (*b*) Hinge joints.
   (*c*) Ball-and-socket joints.
   (*d*) Also condyloid, pivot and saddle joints.

Fixed joints. These are exemplified by the sutures between the skull bones. These bones either overlap each other or are fitted together in a jagged line so that their edges interlock.

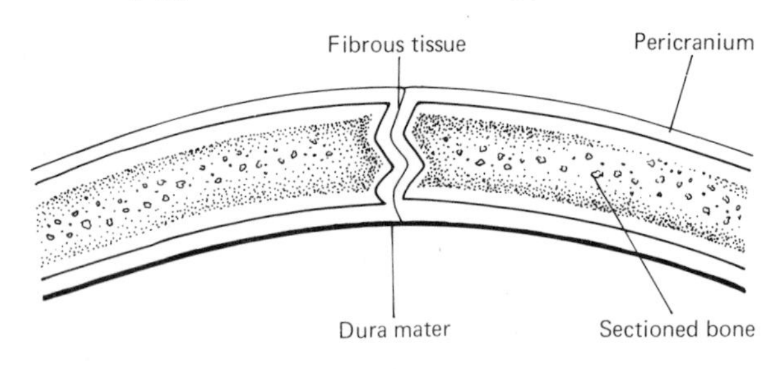

Fig. 62 A fixed joint or suture (skull).

Slightly movable joints. These are represented by the symphysis pubis, the manubrio-sternal joint, and the joints between the vertebral bodies.

Freely movable joints. These are characterized by:

(*a*) The ends of the articulating bones are covered by hyaline cartilage.

(*b*) The joint is enclosed in a fibrous capsule, supported by ligaments.
(*c*) The capsule of the joint is lined by synovial membrane.
(*d*) The cavity of the joint contains synovial fluid for its lubrication.

The names given to the various forms of freely movable joints are self-explanatory, but may be illustrated by examples:

Gliding joints. These permit merely a gliding between two plane surfaces, e.g. between the articular processes of the vertebrae.

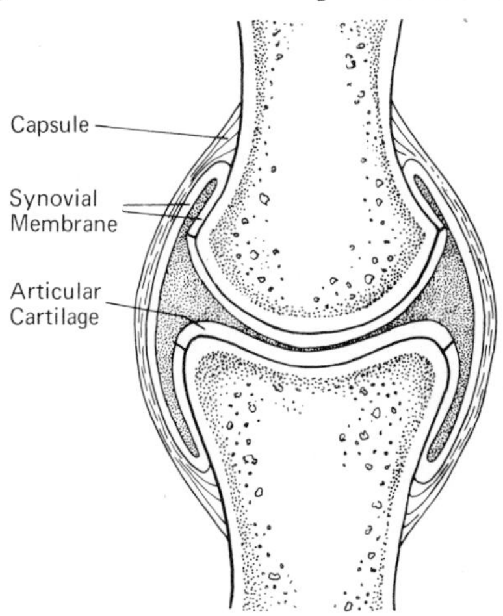

Fig. 63 Typical synovial joint.

Hinge joints. These permit movement on a single axis which is in the line of the bones concerned, e.g. the elbow, knee and ankle joints. The movement of these joints is called flexion and extension.

Ball-and-socket joints. Present in the limbs, e.g. shoulder and hip, these joints are very freely movable in all directions, viz.:

1. Extension (straightening) and flexion (bending).
2. Abduction, or movement of the part away from the mid-line of the body.
3. Adduction, or movement towards the mid-line of the body.
4. Rotation, or movement around the axis of the limb itself.
5. Circumduction, which is seen particularly in the shoulder joint and also in the hip joint, and is illustrated by the swinging of the arm in the act of bowling a cricket ball (overarm). It involves a combination of all the other movements in some degree.

Condyloid, pivot and saddle joints. These are found in the smaller varieties. A condyloid joint is really a hinge joint which also allows some lateral movement, e.g. the articulation of the radius with the carpal bones, the metacarpo-phalangeal joints and the temporo-mandibular joint. A pivot joint permits only rotation, e.g. the rotation of the atlas around the odontoid process of the axis. A saddle-joint has one convex and one concave surface allowing free hinge-like movement, e.g. between the metacarpal bone of the thumb and the trapezium.

JOINTS OF THE SHOULDER GIRDLE AND UPPER LIMB

The scapula has two articulations:

1. The acromio-clavicular joint.
2. The shoulder joint (humero-scapular).

The acromio-clavicular joint

This is surrounded by a capsule and there is frequently a pad of cartilage between the ends of the bones. It allows a limited amount of movement in all directions. The clavicle is also firmly attached to the coracoid process of the scapula by a strong ligament.

The sterno-clavicular joint

The joint formed between the medial end of the clavicle and the upper and lateral angle of the manubrium sterni has a moderate degree of movement and the ends of the bones are separated by a pad of fibro-cartilage.

The shoulder joint

This is an important ball-and-socket joint formed by the articulation of the head of the humerus with the glenoid cavity of the scapula. The outer margin of the glenoid cavity is deepened by a ring of fibro-cartilage (the glenoid ligament). The capsule of the joint is loose on account of the very free movement which takes place and the head of the humerus is mainly kept in position in the glenoid cavity by the action of surrounding muscles.

The tendon of the long head of the biceps muscle passes through the synovial cavity of the joint. It crosses the top of the head of the humerus and helps to maintain the stability of the joint.

Movements. There is a full range of movement, viz.:

1. Flexion, or moving the arm forwards, and extension, or moving it backwards.
2. Abduction away from the trunk, and adduction towards it.
3. Circumduction and rotation.

The elbow joint

This is a hinge joint in which three bones take part, viz. the humerus above, and the ulna and radius below. The trochlear notch of the ulna articulates with the trochlear surface of the humerus and the head of the radius with the capitulum. The tendon of the biceps muscle at its lower end passes in front of the joint and that of the triceps behind.

Movements. (1) Flexion, or moving the forearm towards the shoulder, and (2) extension, or straightening the limb. During flexion the coronoid process of the ulna fits into the coronoid fossa of the humerus. In extension, the olecranon process occupies the olecranon fossa of the humerus.

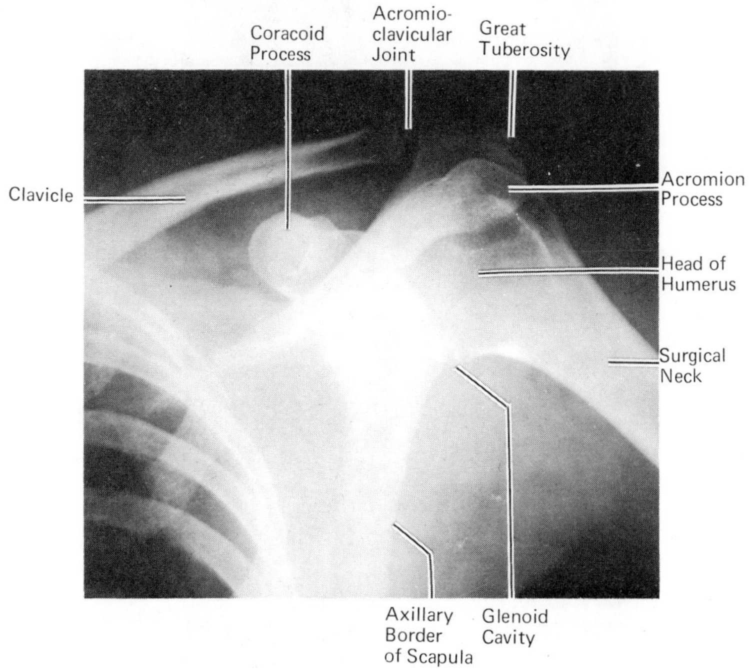

Fig. 64 X-ray of the shoulder (front view).

The radio-ulnar joints

The radius and ulna articulate at their upper and lower extremities, and throughout their shafts are joined by the interosseous membrane which separates the forearm into anterior and posterior compartments.

Movement. The movement between the radius and ulna is one of rotation. The radius rotates about the ulna, carrying the hand with it. When the forearm is in the position of anatomical description, viz. with the palm of the hand facing forwards, the bones are parallel and the forearm is said to be supinated. The movement resulting in this position is called **supination**. Rotation of the radius so that its lower part crosses the ulna, as in placing the palm of the hand on a flat surface, is called **pronation** (see page 50).

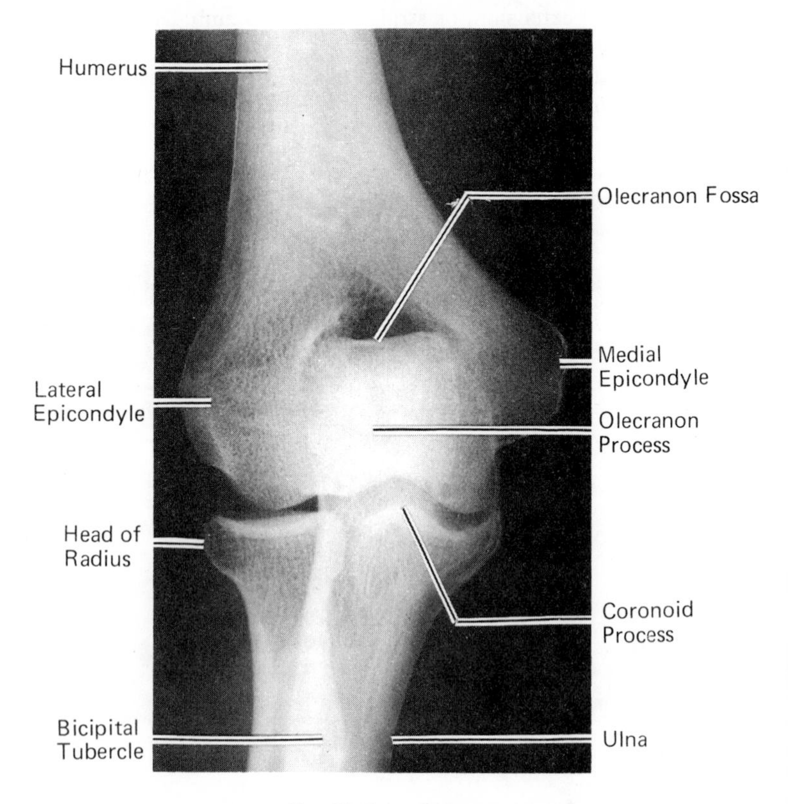

Fig. 65 X-ray of elbow.

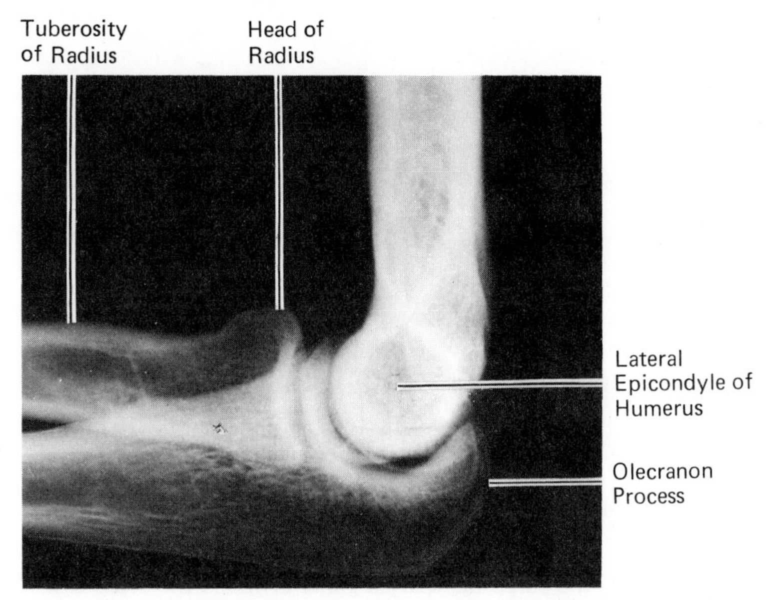

Fig. 66 X-ray of elbow joint.

In the **superior radio-ulnar joint**, the circumference of the head of the radius articulates with the radial notch on the lateral side of the upper extremity of the ulna. It is held in position by the circular, annular ligament which is attached to the anterior and posterior margins of the radial notch.

The **inferior radio-ulnar joint** is a small joint between the head of the ulna and the lower end of the radius. The bones are held in position by ligaments and also by the **triangular cartilage**, a pad which separates the lower end of the ulna from the wrist joint.

The wrist joint

This is situated between the lower end of the radius, and the triangular cartilage, and the upper or proximal row of carpal bones (viz. the scaphoid and lunate articulate with the radius, and the triquetral with the triangular cartilage).

The carpal joints

These include (*a*) the articulation between the individual carpal

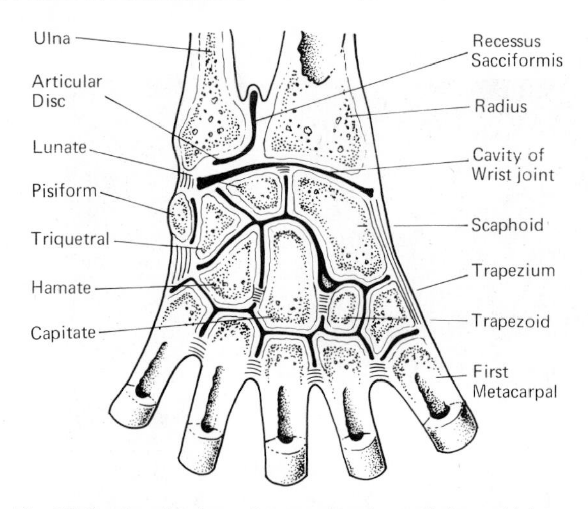

Fig. 67 Section showing wrist, carpal and capophalangeal joints.

bones and (*b*) the carpo-metacarpal joints in which the distal row of carpal bones articulates with the bases of the metacarpal bones.

Movements of the wrist and fingers. Movements take place in both the wrist and carpal joints at the same time. They include:

1. Flexion, or bending the wrist forwards, and extension, or bending it backwards.
2. Adduction, i.e. movement to the ulnar side, and abduction, to the radial side.

The **metacarpo-phalangeal** are condyloid and the **interphalangeal** are hinge joints having movements of flexion and extension. In addition, spreading movements of the fingers take place at the metacarpo-phalangeal joints and consist of abduction and adduction. The movement of applying the tip of the thumb to the tip of the little finger across the palm of the hand is called **opposition**.

JOINTS OF THE PELVIC GIRDLE AND LOWER LIMB

Joints of the innominate bone

Each innominate bone is concerned in the formation of three joints: (*a*) the sacro-iliac joints, (*b*) the symphysis pubis, (*c*) the hip joint.

The sacro-iliac joint. This is an important joint transmitting

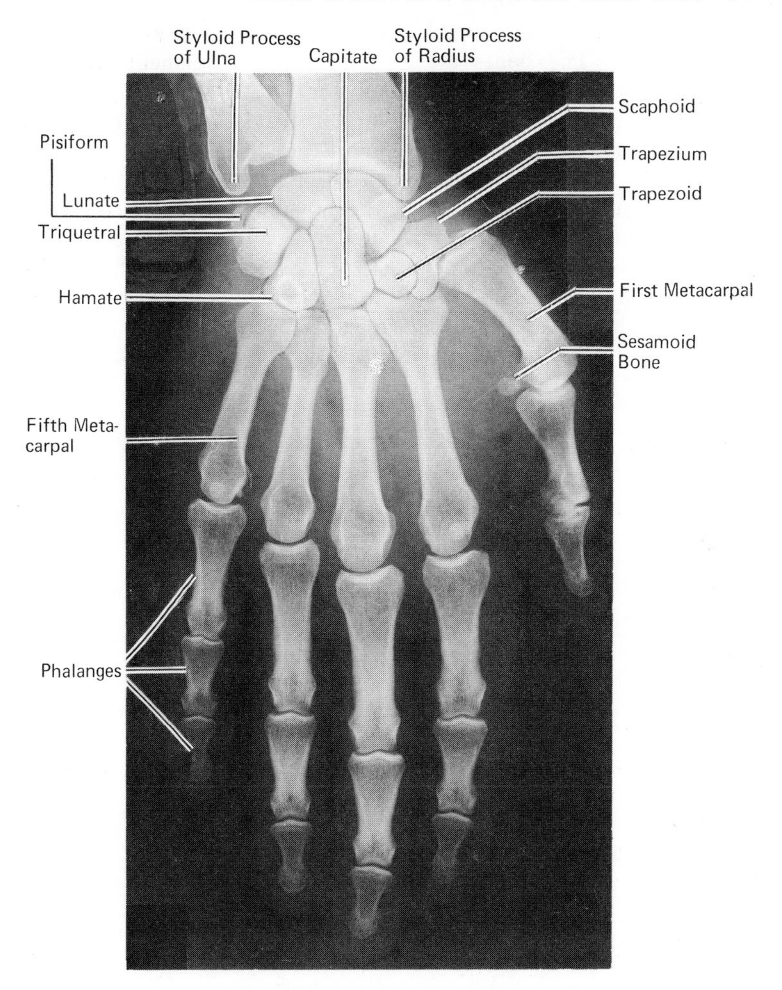

Fig. 68 X-ray of the wrist and hand.

the weight of the body, through the vertebral column, via the pelvis, to the lower limbs.

Very little movement takes place at this joint, which is formed by articulation of the ear-shaped area on the lateral mass of the sacrum with a similar area on the posterior part of the internal surface of the ilium. Strong ligaments hold the bones in apposition.

The symphysis pubis. The symphysis pubis is the articulation between the two pubic portions of the innominate bone in the mid-line in front. The bones are held together by ligaments and are separated by a pad of fibrocartilage. Very little movement normally takes place, but it is interesting to note that this joint and the sacro-iliac joints have increased mobility during pregnancy.

The hip joint. It has been seen that the essential feature of the shoulder joint is mobility. That of the hip joint is **stability** combined with a reasonable degree of movement.

The hip joint is a ball-and-socket joint formed between the head of the femur and the acetabulum of the innominate bone. The socket of the acetabulum, like that of the glenoid cavity of the scapula, is

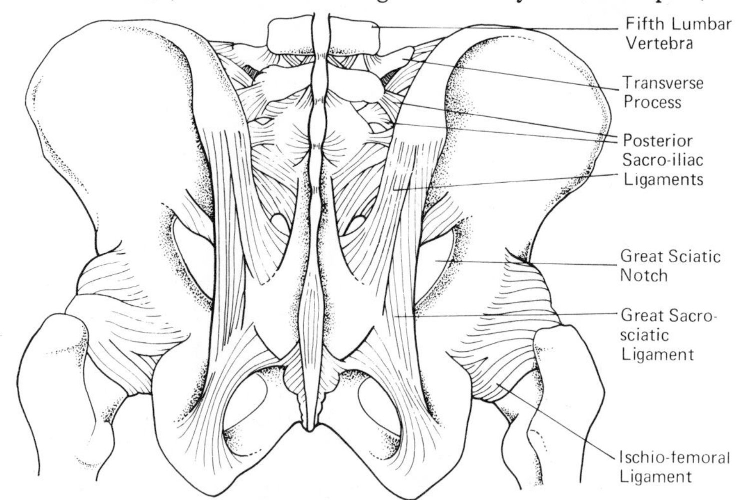

Fig. 69 Ligaments of the pelvis viewed from behind.

deepened by a surrounding ring of fibrocartilage called the **acetabular labrum**. The capsule of the joint is strengthened by three important ligaments:

(*a*) The **ilio-femoral** ligament, in front.
(*b*) The **pubo-femoral** ligament, below.
(*c*) The **ischio-femoral** ligament, behind.

The **ligamentum teres**, a fibrous band extending from a depression in the head of the femur to the notch in the rim of the acetabulum. is of no functional importance.

Movements. These include:

1. Flexion, or bending the thigh upwards to meet the trunk; and extension, or movement of the thigh backwards.

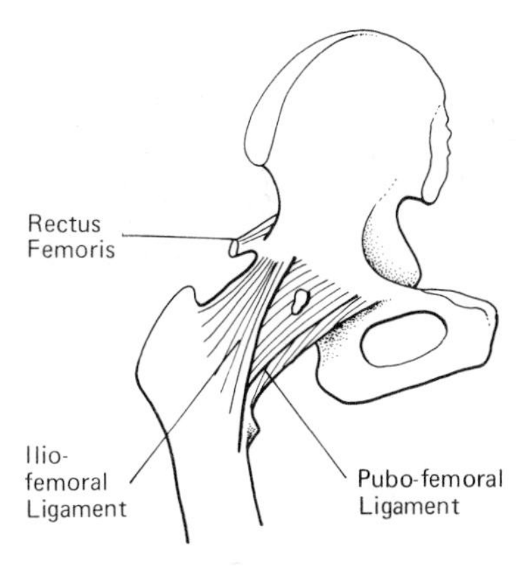

Fig. 70 The anterior ligaments of the hip joint.

2. Abduction and adduction, or movement of the thigh away from and towards the mid-line.
3. Circumduction and rotation.

Sometimes in the fetus there is defective formation of the acetabulum so that the head of the femur is not retained in its cavity. This condition is known as congenital dislocation of the hip and may not always be discovered until the child starts to stand. It requires special orthopaedic treatment.

In late middle and old age, degenerative changes are common in the hip joint. The condition, known as osteoarthritis, may require surgical treatment.

The knee joint

This is the largest joint in the body and is of great importance because of (*a*) movement in walking, etc., (*b*) stability, including maintenance of the erect posture and the transmission of the body weight to the feet.

The knee joint is formed by the articulation of the two condyles of the femur with the condyles of the tibia. The patella also takes part in its formation.

The capsule is strengthened by the important **medial** and **lateral ligaments.**

The following structures are present within the joint:

1. The **cruciate ligaments.** These are attached above to the intercondylar notch of the femur and below to the upper surface of the tibia, in such a way that they cross each other.

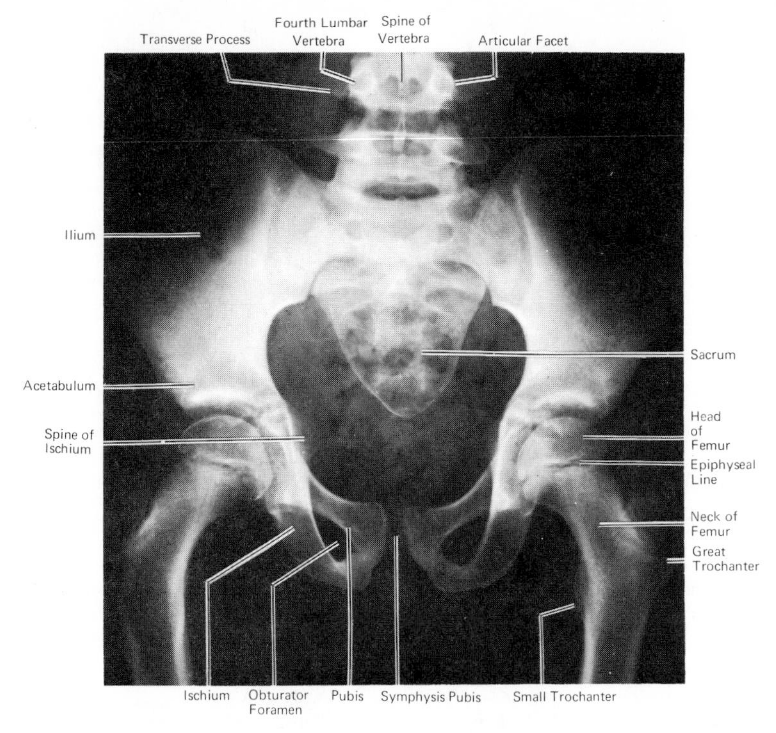

Fig. 71 X-ray of the pelvic girdle.

2. The **medial** and **lateral semilunar cartilages** or menisci. These are half-moon or C-shaped pads of cartilage attached to the upper surface of the tibia and help to deepen the articular surface. They are liable to injury, especially in sports such as football, and may need surgical removal.

The patella articulates with the anterior aspect of the femur between the two condyles.

The knee joint is a typical hinge joint at which the movements of flexion or bending and extension or straightening take place.

The tibio-fibular joints

The **superior tibio-fibular** joint is formed by the junction of the head of the fibula with a small articular facet on the under surface of the lateral condyle of the tibia. In contrast with the mobility of the radius on the ulna, very little movement takes place between the tibia

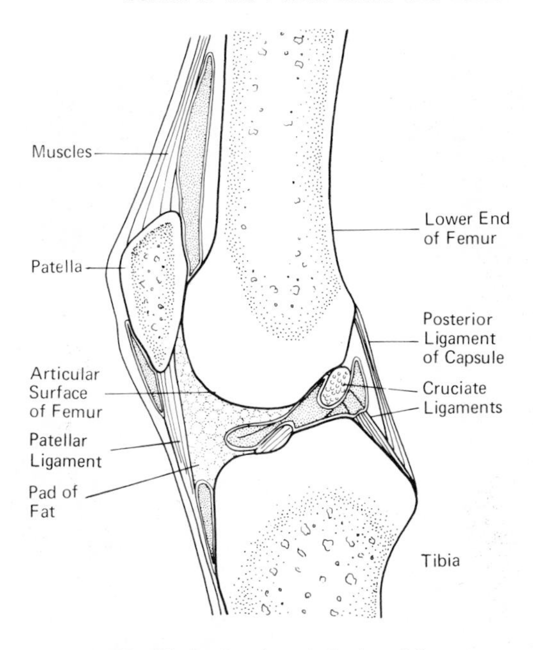

Fig. 72 Section through the knee joint.

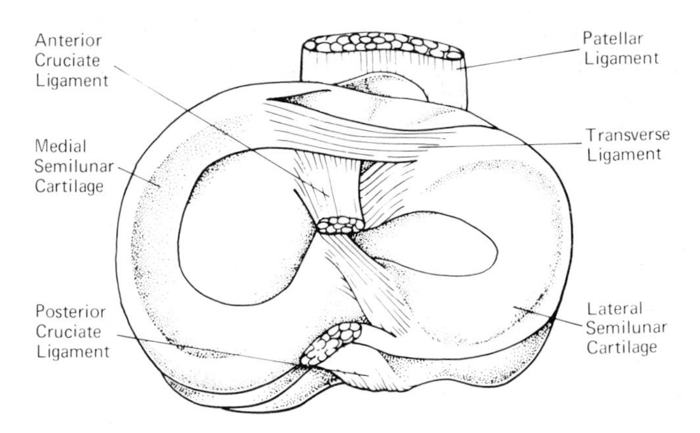

Fig. 73 The semilunar cartilages of the right knee joint and structures attached to the upper surface of the tibia.

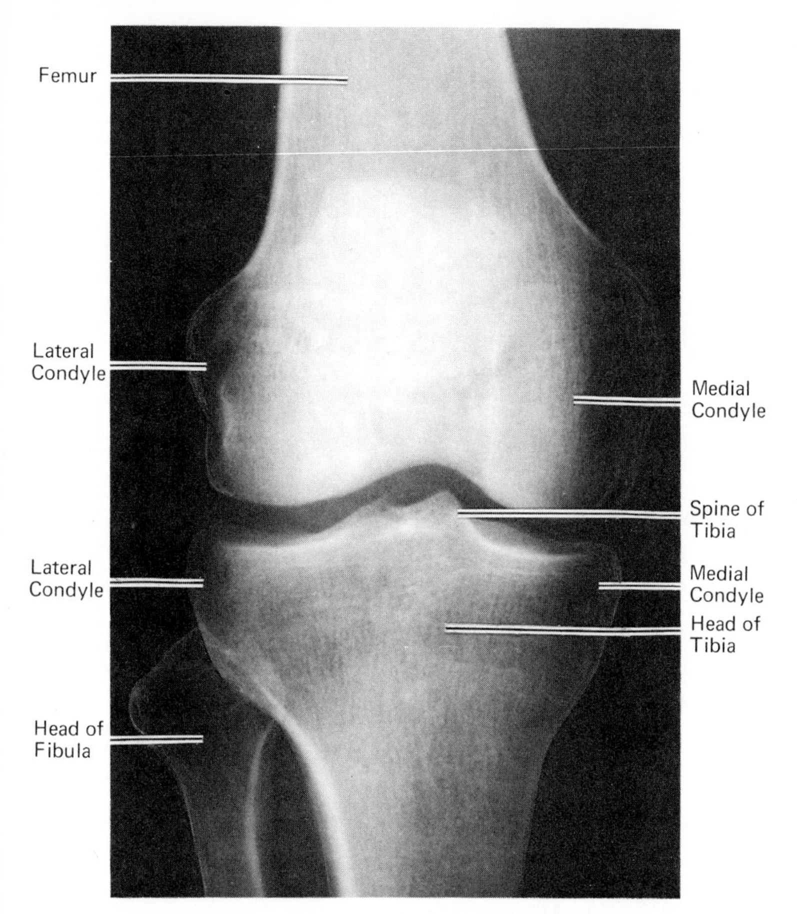

Fig. 74 X-ray of the knee joint—anterior view.

and fibula because the main function of this part of the leg is weight-bearing rather than movement. The **inferior tibio-fibular joint** consists of the firm binding together of the lower ends of the bones by ligaments. Firm union is necessary in order to give stability to the ankle joint below.

Stretching between the shafts of the bones is an **interosseous membrane** which divides the leg into anterior and posterior compartments like the forearm.

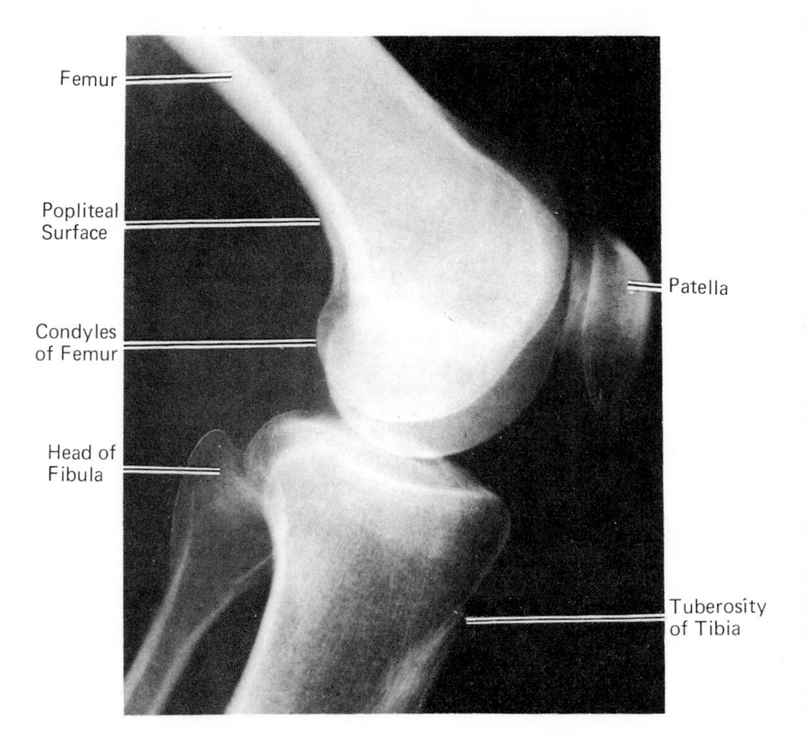

Fig. 75 X-ray of the knee joint—lateral view.

The ankle joint

The ankle joint is a hinge joint formed between the tibia above and on the medial side, by the fibula on the lateral side, and the talus below. The upper surface of the talus articulates with the lower end of the tibia; its medial surface with the articular portion of the medial malleolus, while the lateral part of the joint is formed by the articular surface of the lateral malleolus of the fibula. The capsule of the joint is strengthened by the medial and lateral ligaments (Fig. 76).

Movements of the foot. Flexion of the foot, or bending it upwards towards the leg (dorsiflexion); and extension, or pointing the foot downwards (sometimes called plantar flexion). The foot can also be moved so that the sole faces medially (**inversion**) or laterally (**eversion**).

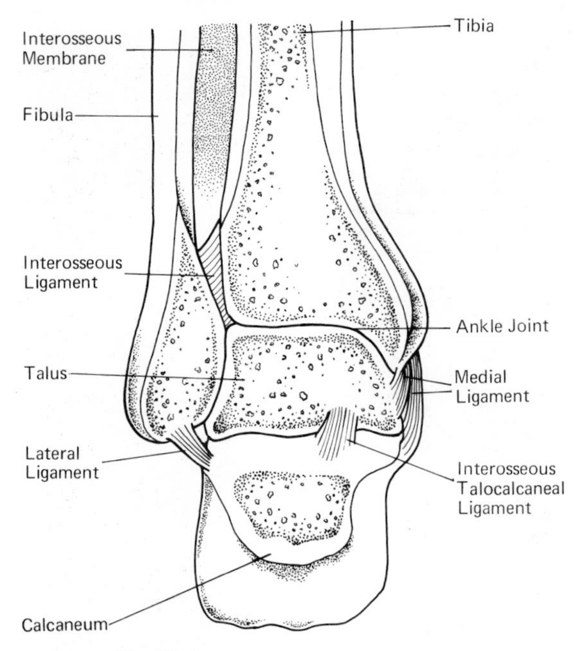

Fig. 76 Section through the ankle joint.

Other tarsal joints

These include the joints between the talus and the calcaneum below, those between the other tarsal bones, and the tarso-metatarsal joints. The metatarso-phalangeal and interphalangeal joints are similar to those of the hand and fingers.

Movements of the toes. Movement of the toes upwards towards the dorsum of the foot is called extension; movement downwards towards the sole is called flexion (i.e. if the ankle and toes are both moved towards the leg, the ankle is flexed and the toes are extended. In the opposite movement, the ankle is extended and the toes are flexed).

JOINTS OF THE TRUNK, ETC.

Vertebral joints, see page 83.
Joints of the ribs, see page 86.

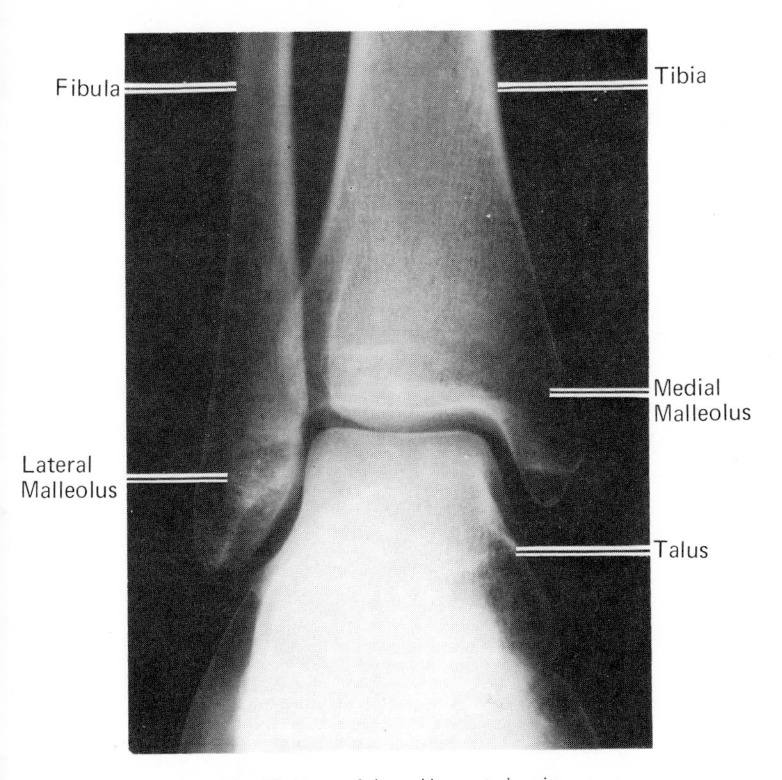

Fig. 77 X-ray of the ankle—anterior view.

PRACTICAL NOTES. Joints are liable to inflammation which is called arthritis. They may be involved in fractures of the bones in their vicinity. Ligaments may be torn and the bones may be forced out of their natural positions—dislocations.

QUESTIONS

1. Describe the hip joint. What are its movements and how are they brought about.
2. Describe the ankle joint and its movements.
3. Give an account of the types of joint found in the human body. Describe the shoulder joint.
4. Give a description of the knee joint. What are its movements and how are they brought about?
5. Describe and compare the shoulder joint and the hip joint.

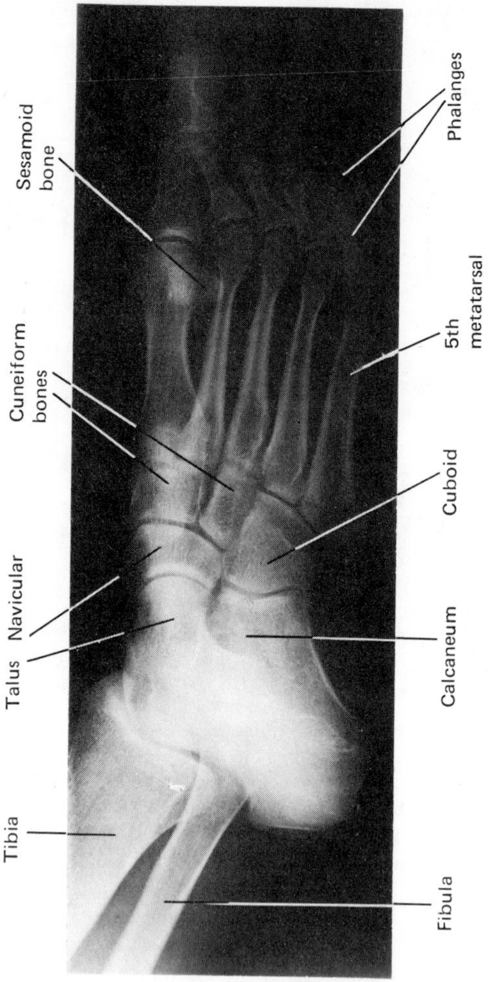

Fig. 78 X-ray of the foot from above.

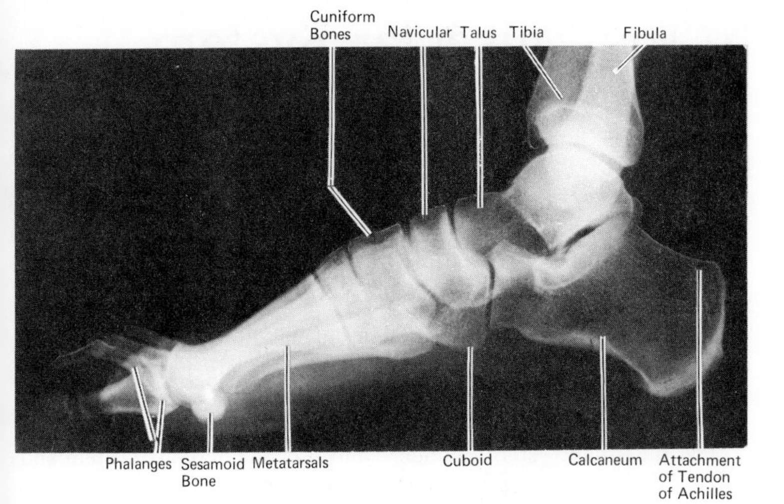

Fig. 79 X-ray of the foot—lateral view.

6 The Muscular System

One of the most important functions of the human body is motion, and it has been seen that the skeleton consists largely of a system of bony levers capable, by virtue of the joints and muscles, of moving upon one another.

The muscular system is under control of the nervous system and it has been noted (page 26) that it can be subdivided, according to the different types of fibres which it contains, into:

voluntary—controlled by the central nervous system;
involuntary—controlled by the involuntary or autonomic nervous system;
cardiac—found in the heart.

It is the voluntary or skeletal group consisting of striped muscle fibres which are now to be considered. The following are general anatomical points to be remembered:

1. The voluntary muscles constitute a considerable proportion of the total mass of the body (about 40 per cent).

2. Each muscle is made up of a large number of separate fibres bound together in bundles by a delicate sheath of connective tissue (the perimysium).

3. The whole muscle is closely covered by a sheet of fibrous tissue called **fascia.**

(*a*) Fascia not only covers individual muscles but also separates them into groups which are surrounded by it.

(*b*) Special sheets of fascia are found in different parts of the body, e.g. the **cervical fascia** enclosing the muscles and structures of the neck; the **pelvic fascia** lining the pelvis; the **fascia lata**, surrounding the muscles of the thigh and especially strong on the lateral side.

(*c*) The whole body is enclosed in an envelope of fascia which consists of two layers: (i) the **superficial** or subcutaneous fascia closely connected to the deep surface of the skin, and (ii) the **deep** fascia, including the fascial covering of the muscles, which encloses the individual parts of the body and helps to maintain their shape and to keep their contained structures in position.

There is a distinct plane between the superficial and the deep fascia. If the skin and subcutaneous fat are stripped from a part of the body, the line of cleavage is between these two layers. The superficial fascia comes away with the skin, leaving the rest of the structures covered by deep fascia.

(*d*) There are two important, greatly thickened portions of the deep fascia—the **palmar fascia** in the hand, and the **plantar fascia** in the sole of the foot, the main functions of which are protective.

4. **Muscles** are said to arise from, and to be inserted into, the various bones. (An exception is some of the muscles of facial expression which are attached to the soft tissues of the face.) Speaking generally, the **origin** is the fixed point and the **insertion** the movable point. For example, the deltoid muscle passes from its origin on the scapula and clavicle to its insertion into the deltoid tubercle of the humerus. Both the origin and insertion usually consist of fibrous tissue.

5. The main mass of the muscle is termed the **belly**. The fibrous ends taper off into tendons.

6. **Tendons**. These are the fibrous ends of the muscles by which they are attached to the bones. Tendons vary considerably.

(*a*) They may be hardly noticeable, so that the muscle appears to be directly attached to bone both at its origin and insertion.

(*b*) More usually, tendons are well developed at the insertion of the muscle.

(*c*) Frequently they are long, round or flattened bands, usually surrounded by sheaths of synovial membrane, especially in the neighbourhood of joints.

(*d*) Sometimes they form a broad, flat expansion termed an **aponeurosis**, e.g. the lumbar fascia or aponeurosis.

7. The intervals between various muscles are occupied by loose areolar tissue. In situations in which (*a*) a muscle or tendon comes in contact with or moves over a bony prominence, and when (*b*) the skin moves directly over bone, the areolar space is often enlarged and contains a small amount of synovial fluid. This is called a **bursa**; for example, there are several bursae around the shoulder joint, the prepatellar bursa, where the skin moves over the front of the patella, the olecranon bursa over the olecranon process of the ulna, and the bursa over the tuberosity of the ischium, on which we sit.

Bursae may become inflamed, usually as a result of repeated minor injury, e.g. prepatellar bursitis or 'housemaid's knee'; ischial bursitis or 'weaver's bottom'.

8. The nomenclature of the muscles is complicated. They may be named:

(*a*) According to function, e.g. flexors or extensors, adductors and abductors, etc.

(*b*) According to their attachments, e.g. sterno-mastoid.

(*c*) According to shape, e.g. deltoid (like Greek letter D or $\triangle$).

(*d*) According to their position or direction, e.g. pectoralis major (large breast muscle), rectus abdominis (straight abdominal muscle), the intercostal muscles (between the ribs); the oblique and straight muscles of the eye.

(*e*) According to their formation, e.g. biceps = two heads; triceps = three heads; quadriceps = four heads.

9. When the prime mover contracts there must be an equal and opposite action of the antagonist; for example, in order to flex the elbow the biceps contracts and the triceps relaxes.

THE MUSCLES

Almost all the skeletal muscles in the body are paired. They are very numerous and it is not to be expected that details of all of them will be learnt. There are, however, some individual muscles which require separate description, and it should be possible to appreciate the main muscle groups.

Muscles of the head and neck

The muscles of the head, face and neck are too numerous to consider in any detail. They can, however, be divided into several main groups.

1. Muscles of the scalp (**occipito-frontalis**) and ear (page 347, and of the eye (page 353).

2. The muscles of facial expression which are supplied by the VIIth cranial nerve. They are arranged:

(*a*) Around the eye (orbicularis oculi).

(*b*) Around the mouth and nose (orbicularis oris, buccinator).

The **orbicularis oculi** are circular muscles surrounding each eye which when contracted act as sphincters and close the eyes tightly. The **orbicularis oris** is situated in the lips between the skin and the mucous membrane of the mouth and has a similar type of action. The **buccinator** is the principal muscle of the cheek and forms the lateral wall of the mouth. It is used in chewing and sucking.

3. The muscles of mastication which move the lower jaw and are supplied by branches of the Vth cranial nerve, e.g. the temporal, masseter and pterygoid muscles. The **temporal** muscle arises from the temporal fossa of the skull. Its fibres converge into a strong tendon which is inserted into the coronoid process of the mandible. The **masseter** muscle is quadrilateral in shape and arises above from the zygomatic arch of the temporal bone. It is inserted into the outer surface of the lower jaw anterior to the angle.

4. Muscles in the neck attaching the head to the trunk. These are numerous and only a few will be mentioned.

Superficial

The **platysma** extends from the lower jaw in a thin flat sheet to the deep fascia on the front of the chest.

The **sterno-mastoid** is an important muscle extending upwards from the manubrium of the sternum and medial end of the clavicle to the mastoid process of the temporal bone. When operating singly each muscle rotates the head towards the opposite side; when acting together they flex the neck.

The **trapezius** is a large triangular muscle situated at the back of the neck. It arises from the occipital bone of the skull and from the spines of all the cervical and thoracic vertebrae. Its upper fibres are inserted into the lateral third of the clavicle, the middle fibres into the acromion process and the lower into the spine of the scapula. It may also be considered as one of the muscles of the shoulder girdle attaching the scapula to the trunk.

There are also a number of muscles in the front of the neck extending (i) from the lower jaw to the hyoid bone and (ii) from the hyoid bone and thyroid cartilage to the sternum. These are closely related to the trachea and the thyroid gland.

Deep

Examples are the **scalene** muscles extending from the cervical vertebrae to the first and second ribs.

5. Muscles of:

(*a*) The pharynx—the constrictor muscles which take part in the act of swallowing; also the muscles of the tongue and floor of the mouth.

(*b*) The larynx—external muscles which move the larynx as a whole and internal muscles which affect the tension of the vocal cords and are used in voice production.

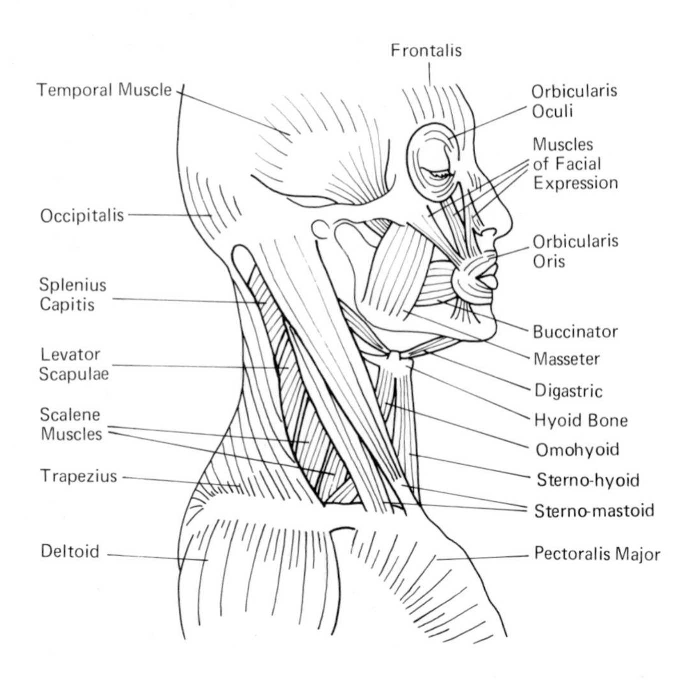

Fig. 80 Muscles of the face and neck.

Muscles of the shoulder girdle and upper limb

1. Muscles attaching the scapula to the trunk

Examples are: deep—**rhomboids**
superficial—**trapezius, serratus anterior**.

2. Muscles attaching the humerus:

(*a*) to the scapula, e.g. *supraspinatus, infraspinatus, subscapularis, deltoid*;

(*b*) to the chest wall, e.g. *pectoralis major* and *minor* and *latissimus dorsi*.

3. Muscles of the arm

The most important muscles of the arm are:

(*a*) The **biceps**, which has its origin by two heads, long and short, from the scapula. The long head arises from the top of the glenoid cavity; the short head from the tip of the coracoid process. It is inserted into the bicipital tubercle below the head of the radius. Passing as it does over two joints, it can produce movements at

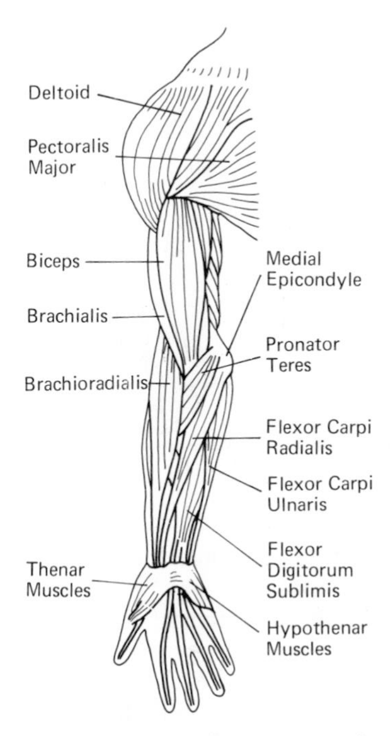

Fig. 81 Muscles of the right upper limb—anterior aspect.

both. Its main actions are to act as a powerful supinator of the forearm and to flex the elbow joint. It also helps in the forward movement at the shoulder joint.

(*b*) The **brachialis**, which arises from the front of the shaft of the humerus and is inserted into the ulna.

(*c*) The **triceps**, situated at the back of the arm, passes from the scapula to the olecranon process of the ulna. It therefore extends the elbow joint, and helps to support the shoulder joint and draw the arm backwards.

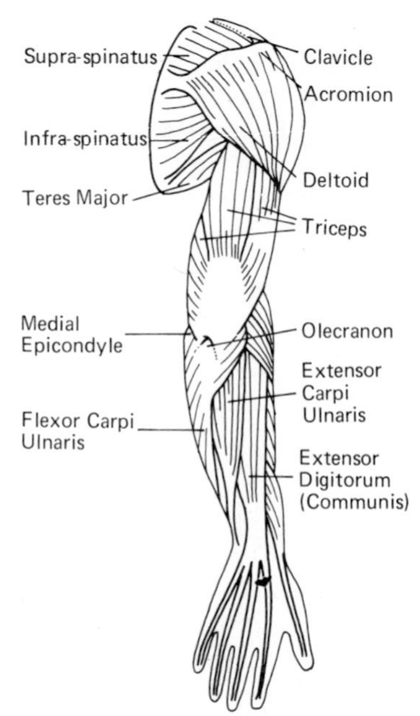

Fig. 82 Muscles of the right upper limb—posterior aspect.

4. Muscles of the forearm

On the anterior aspect. The muscles on the anterior aspect may be divided into three groups:

(*a*) The main group consisting of superficial and deep muscles. The superficial muscles are attached above to the medial epicondyle of the humerus and pass to the fingers; these include the **flexor digitorum sublimus,** which not only flexes the fingers but also the wrist joint and the elbow joint. The deep muscles include the **flexor digitorum profundus** arising from the ulna, which does not move the elbow joint but flexes the wrist and fingers.

(*b*) Muscles which flex the elbow and wrist only, passing from the humerus above to the wrist bones below (**flexor carpi radialis** and **flexor carpi ulnaris**).

(*c*) Muscles whose main action is concerned with pronation and supination (**pronator teres, pronator quadratus** and **brachio-radialis** or supinator longus).

On the posterior aspect. These may be divided into:

(*a*) The extensor muscles of the wrist and fingers (**extensor**

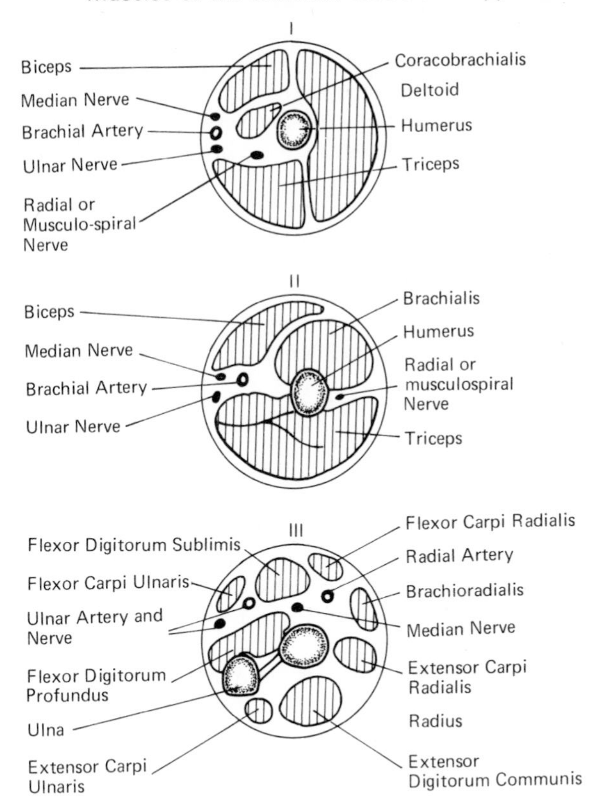

Fig. 83 Diagram illustrating position of most important structures in the upper limb. (*Top*) In upper third of arm. (*Middle*) In lower third of arm. (*Bottom*) middle of forearm.

digitorum communis) which, arising from the lateral epicondyle of the humerus, also extend the elbow joint.

(*b*) The extensors of the wrist (**extensor carpi radialis** and **extensor carpi ulnaris**).

5. Muscles of the hand and fingers

The tendons of the flexor muscles in front and extensor muscles behind are inserted into the bases of the terminal phalanges of the digits. Slips are also given to the other phalanges.

The thumb has separate muscles which, however, correspond to the main flexor and extensor groups in the forearm. Special short muscles situated in the palm of the hand also move the thumb and form the **thenar** eminence at the base of the thumb. The prominence on the ulnar side of the hand at the base of the little finger is called the **hypothenar eminence.** Arising between the metacarpal bones and inserted into the phalanges are the **lumbrical** and **interosseous** muscles.

Muscles of the trunk

1. Thorax
2. Abdomen—anterior abdominal wall
   —posterior abdominal wall

Muscles of the thoracic wall

The superficial muscles include:

(*a*) The **pectoralis major,** a large fan-shaped muscle on the upper anterior part of the chest, which also forms the anterior part of the axilla; and the **pectoralis minor,** a smaller triangular muscle lying deep to the pectoralis major. These muscles arise from the anterior aspect of the sternum, ribs and costal cartilages and are inserted into the upper end of the humerus and coracoid process of the scapula respectively.

(*b*) The **serratus anterior** arising from the ribs passes backwards to the vertebral border of the scapula.

(*c*) The intercostal muscles (eleven pairs) which pass from the lower border of one rib to the upper border of the rib below. They are formed by two distinct layers, the fibres of which pass in opposite directions. The **internal intercostals** are directed downwards and backwards; and the more superficial **external intercostals** pass downward and forwards. They are important muscles of respiration (see Fig. 61, page 88).

The diaphragm

The diaphragm is the large dome-shaped partition separating the cavity of the thorax from that of the abdomen. It consists partly of muscle (around the circumference) and partly of membrane or flattened tendon (in the centre). It is attached to the circumference of the thoracic cavity:

1. In front to the lower end of the sternum (sternal part).
2. On either side to the lower six ribs (costal part).
3. Posteriorly to the first two lumbar vertebrae by two slips called the crura (legs) of the diaphragm (lumbar part)—see Fig. 85)

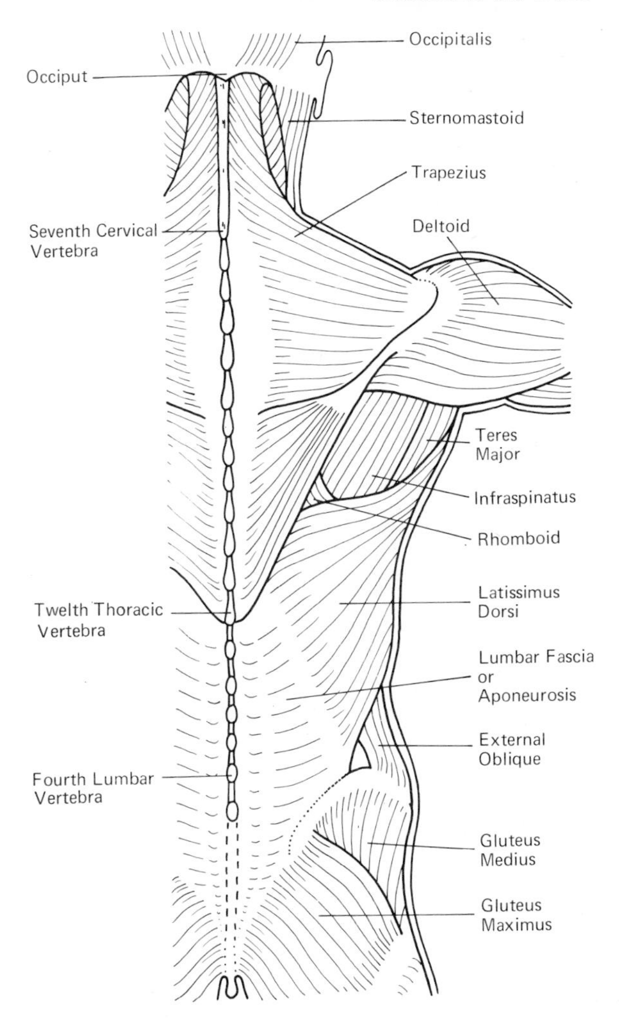

Fig. 84 Muscles of the back.

The heart and pericardium are related to the central portion of its **upper surface**. On either side it is covered by pleura and is related to the bases of the lungs. Its **lower** concave **surface**, largely covered by peritoneum, is related on the right side and centrally to the upper surface of the liver; on the left side are the fundus of the stomach and the spleen.

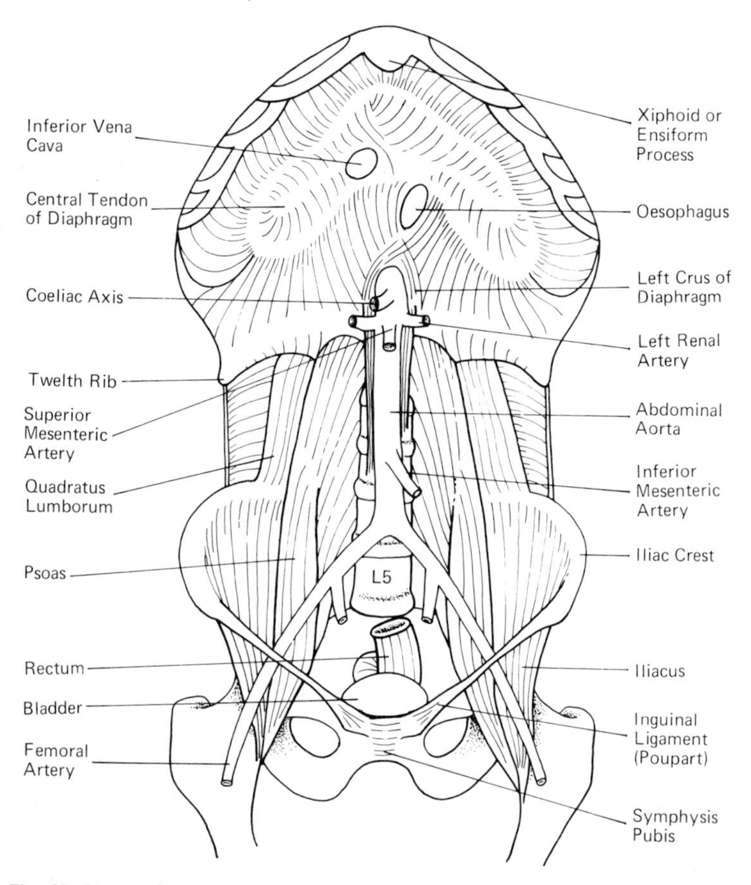

Fig. 85 Diagram showing: under surface of diaphragm; muscles of posterior abdominal wall; abdominal aorta dividing into common iliac arteries which give off internal and external iliac arteries.

It has **openings** in the posterior part close to its origin from the lumbar vertebrae for the passage of the following structures:

1. The aorta, in the mid-line.
2. The oesophagus, slightly to the left.
3. The inferior vena cava, slightly to the right.

The diaphragm is a very important muscle of respiration and is supplied by the **phrenic nerve** from the cervical plexus. During inspiration, the muscle of the diaphragm contracts so that the structure becomes flattened towards the abdomen, thereby helping to enlarge the size of the thoracic cavity. During expiration, it relaxes and resumes its dome-shaped appearance (see Fig. 173, page 267). Destructive lesions of the phrenic nerve cause paralysis of the diaphragm.

Sometimes abdominal contents herniate through the openings in the diaphragm, the commonest being 'hiatus hernia' in which a portion of the stomach enters the chest through the oesophageal opening.

Muscles of the abdomen

1. Anterior abdominal wall
2. Posterior abdominal wall

Anterior abdominal wall. (*a*) straight; (*b*) oblique and transverse.

(*a*) The **rectus abdominis** is the straight muscle of the anterior abdominal wall. It runs parallel with its fellow of the opposite side and is separated from it by a thin band of fibrous tissue, the **linea alba** (white line), which extends from the xiphoid process of the sternum to the upper border of the symphysis pubis. The muscle arises from the xiphoid process and the adjacent costal cartilages and is inserted into the upper border of the pubic bone. It is enclosed in a dense sheath of fibrous tissue which consists of anterior and posterior walls formed by fibrous expansions of the two oblique muscles (the **rectus sheath**). The fibres of the rectus muscle are interrupted in three places by bands of fibrous tissue which cross it transversely—the tendinous intersections (see Fig. 86).

(*b*) There are two oblique muscles which help to form the side and anterior walls of the abdominal cavity. The **external oblique,** the fibres of which pass downwards and forwards, arises from the lower eight ribs and is inserted in a fan-shaped manner: (i) into the rectus sheath; (ii) into the spine of the pubis; (iii) into the anterior half of the iliac crest; its lower border, between the anterior superior spine of the ilium and the pubic spine, consists of fibrous tissue and is called **inguinal ligament** (Poupart's ligament). The **internal oblique** arises mainly from the iliac crest and is inserted into the rectus sheath and the lower ribs. Its fibres, therefore, pass upwards and medially and cross those of the external oblique. The **transversus** is a muscle with fibres passing almost horizontally across the anterior abdominal wall and is the deepest of the three muscles. The parts of these muscles situated in the inguinal area are of importance in connection with the anatomy of hernia or rupture (see page 384).

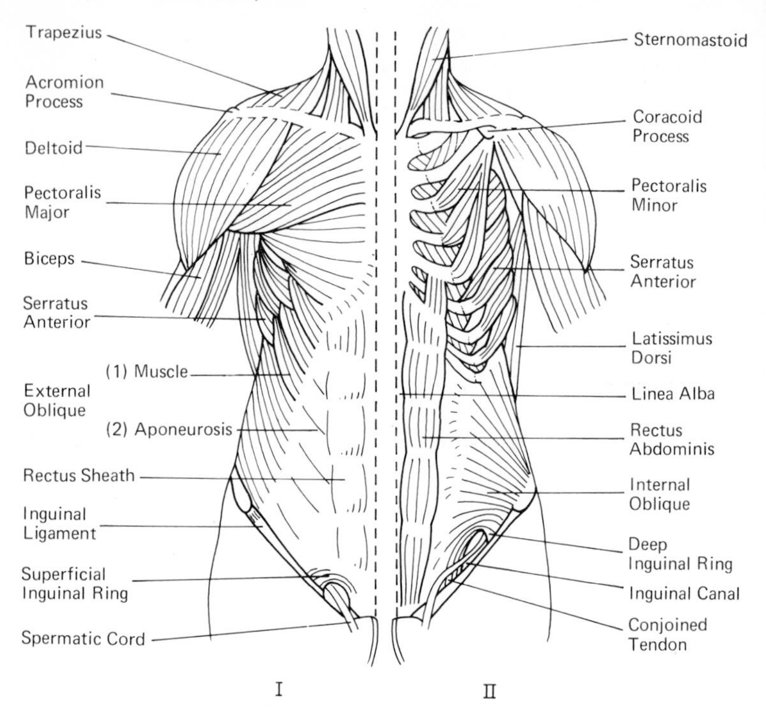

Fig. 86 Muscles of the front of the thorax and anterior abdominal wall. (*Left*) Superficial. (*Right*) Deep.

Posterior abdominal wall. The **psoas muscle** arises from the lumbar vertebrae. The **iliacus muscle** arises from the inner surface of the iliac portion of the innominate bone. They are inserted together into the lesser trochanter of the femur (Fig. 85).

The **quadratus lumborum** extends from the iliac crest upwards to the last (12th) rib. The posterior surface of the kidney is closely related to the psoas and quadratus lumborum muscles.

Muscles of the back

There are several groups of muscles which are placed on either side of the spine and extend for varying distances between the occiput above and the sacrum below. The upper ones extend the neck. The lower ones, including the **erector spinae**, straighten the spine. In the lumbar region they help to form a large aponeurosis of fibrous tissue—the lumbar fascia—and in this area, lesions cause 'lumbago'.

Muscles of the pelvis

Just as the diaphragm forms a roof to the abdominal cavity, muscles situated in the pelvis form its floor. The most important of these is

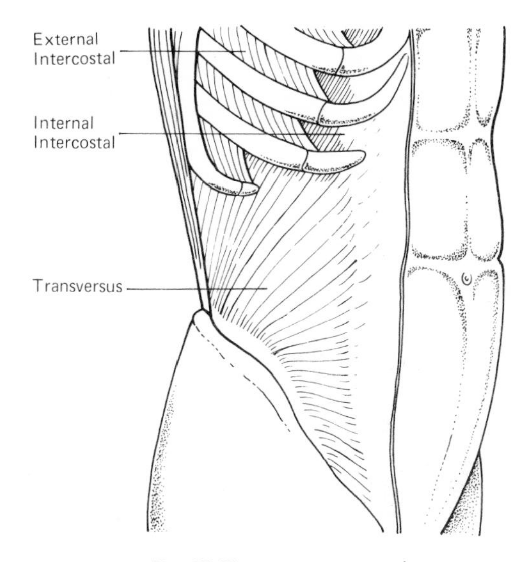

Fig. 87 The transversus muscle.

the **levator ani**, which also plays an important part in the act of defaecation. It is a muscular sheet extending across the outlet of the pelvis through which passes the rectum, urethra, and, in the female, the vagina.

Muscles of the lower extremity

Muscles of the buttock

Three muscles make up the rounded eminence of the buttock, the **gluteus maximus, gluteus medius** and **gluteus minimus** (the first named being the most superficial and the largest, the last the deepest and smallest). These muscles arise from the outer surface of the ilium (above, between and below the curved gluteal lines respectively) and are inserted into and around the greater trochanter of the femur. Their main action is to extend and abduct the thigh.

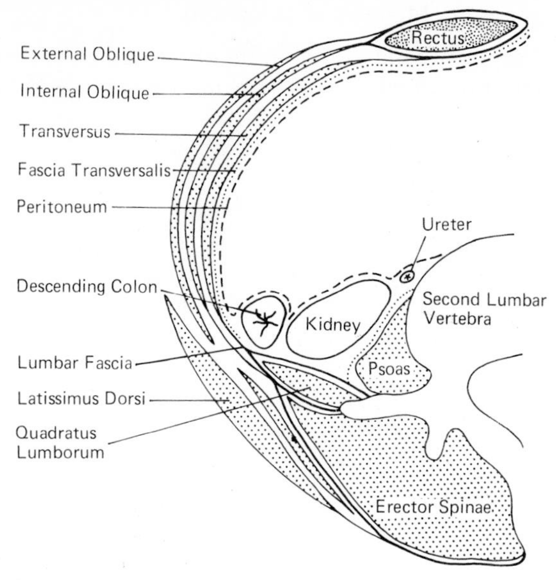

Fig. 88 Section of the abdominal wall through the lower part of the second lumbar vertebra.

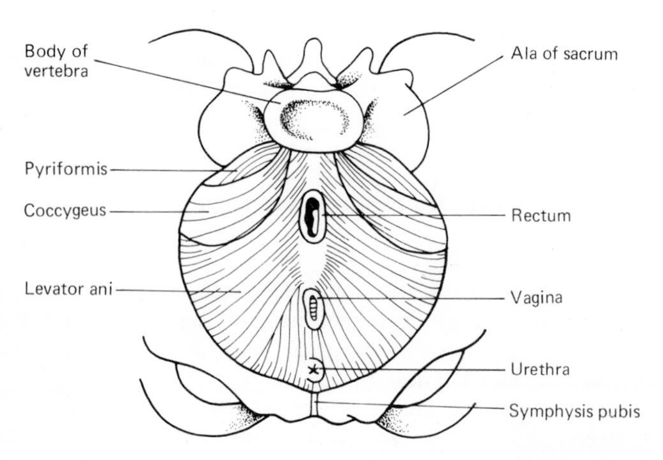

Fig. 89 Diagram of the muscles of the female pelvic floor.

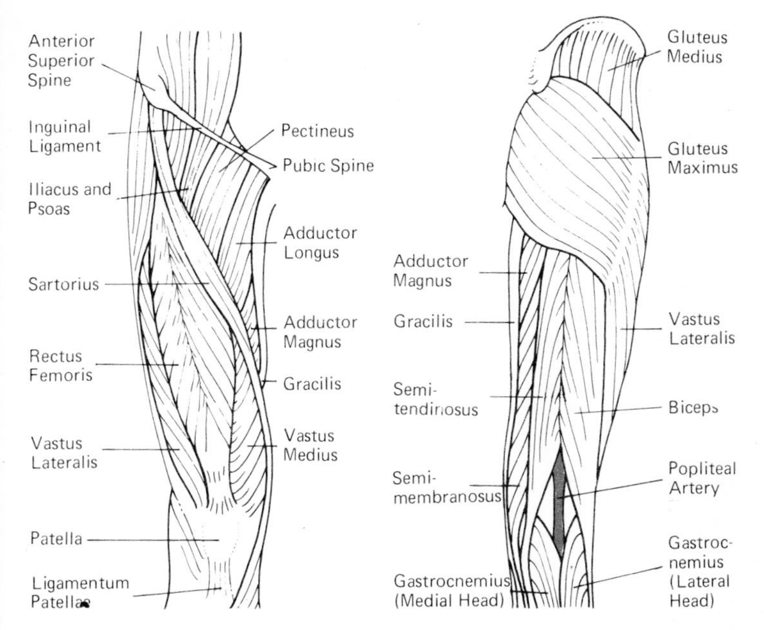

Fig. 90 Muscles of the front of the thigh.

Fig. 91 Muscles of the back of the thigh.

Muscles of the thigh

There are a number of muscles in the thigh which can be arranged in three groups:

1. The anterior group—in front of the femur.
2. The posterior group—behind the femur.
3. The medial group—to the inner side of the femur.

Anterior group (quadriceps). The main muscle of this group, the **quadriceps extensor** (made up of the **rectus femoris, vastus medialis, vastus lateralis** and **vastus intermedius** or crureus), terminates in a single tendon (the ligamentum patellae) in which the patella is developed as a sesamoid bone. It is inserted into the tuberosity of the tibia and its action is to extend or straighten the knee joint. The **sartorius** muscle passing obliquely from the anterior superior spine of the ilium to the medial side of the tuberosity of the tibia is also included in this group.

Posterior group (hamstrings). This group is formed by three muscles, **biceps, semi-membranosus** and **semi-tendinosus**, which arise from the tuberosity of the ischium and are inserted into the upper extremities of the leg bones. The biceps passes to the lateral side of the leg and also has a short head arising from the linea aspera of the femur. The other two muscles pass to the medial side of the leg. The biceps, therefore, forms the lateral boundary of the popliteal space, and the semimembranosus and semitendinosus the medial boundary. Having their origin above the hip joint and their insertion below the knee joint they are capable of producing movement at both, i.e. (i) the hamstrings straighten or extend the hip and (ii) they flex the knee.

Medial group (adductors). This consists of three muscles, the **adductors longus, brevis** and **magnus,** which arise from the pubic bone and are inserted mainly into the linea aspera of the femur. Their function is described by their name, viz. they adduct the thigh towards the mid-line.

Muscles of the leg

(*a*) Anterior group (*b*) Posterior group (*c*) Fibular group.

Anterior group. This is composed of those muscles which lie in front of the interosseous membrane between the tibia and fibula. It includes (i) the **tibialis anterior** passing from the tibia to the tarsal bones and, therefore, dorsi-flexes the ankle, (ii) the extensor muscles of the toes (**extensor digitorum longus**).

Posterior group. This consists of superficial and deep layers. The superficial muscles, i.e. the **gastrocnemius** and the **soleus,** form the back of the calf. The upper end of the former arises from the condyles of the femur, the soleus from the posterior aspects of the tibia and fibula. The former, therefore, in addition to its other action, helps to flex the knee joint. Both are inserted into the calcaneum by the tendo Achillis and plantar-flex the ankle joint. The deep muscles include the **tibialis posterior** arising from the tibia and fibula which also plantar-flexes the ankle joint as it passes to its insertion into the tarsal bones; and the flexor muscles of the toes (**flexor digitorum longus**).

It will be noted that, in contrast to the muscles of the forearm and hand, the flexor muscles of the toes are situated on the posterior aspect of the leg and the extensors of the toes on the anterior surface, but this will be clear if it is remembered that the palm of the hand corresponds with the sole of the foot.

Fibular group. These are also called the **peroneal** muscles. They arise from the lateral surface of the fibula and are inserted into the tarsal and metatarsal bones of the foot. Their action is to evert or

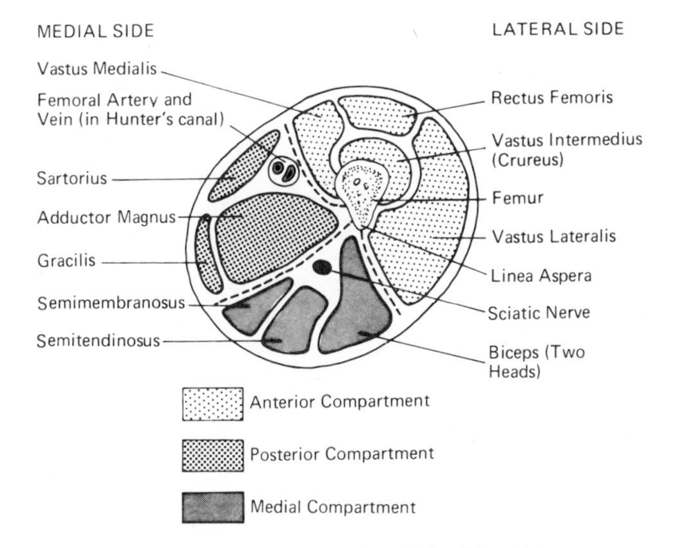

Fig. 92 Transverse section of the middle of the thigh.

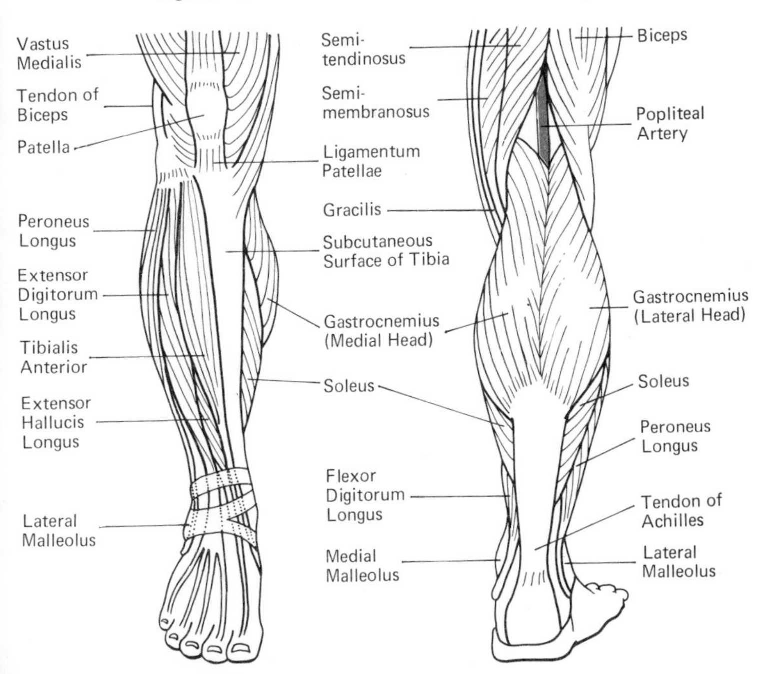

Fig. 93 Muscles of the front of the leg.

Fig. 94 Muscles of the back of the leg.

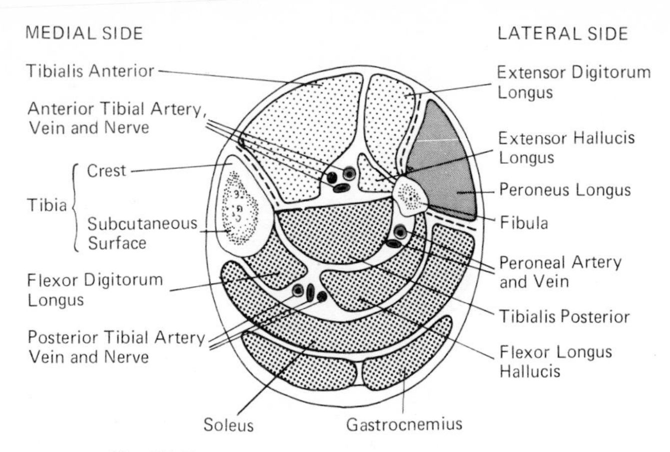

Fig. 95 Transverse section of the leg (middle of calf).

turn the foot outwards. There are a number of small muscles in the foot similar to those in the hands, viz. short muscles in the sole which are attached to the big toe, **interosseous** and **lumbrical** muscles for the toes.

QUESTIONS

1. Describe the diaphragm. What is its function?
2. Describe the chief muscles concerned in respiration.
3. Describe the types of muscle found in the body and their modes of action.

4. Give an account of (*a*) the abdominal wall and (*b*) the pelvic floor. Show how each fulfils its functions.

7 The Circulatory System

The life of every tissue and organ in the body depends on their receiving an adequate supply of nourishment and oxygen, and the removal of the waste products which result from their activities. These functions are carried out by the blood, and the heart and blood vessels are the mechanism by which a constant circulation of the blood throughout the body is maintained. The blood is pumped by the heart along the arteries to the capillaries and is returned by veins.

The capillaries form a dense and universal network throughout the body. They are microscopic channels which receive blood from the smaller arteries (**arterioles**) and deliver it into the smaller veins (**venules**). These canals consist of a single layer of cells partially joined together but with small spaces between them through which white blood cells (leucocytes) can pass by means of their power of amoeboid movement. Their diameter is about that of a red blood cell. The interchange of oxygen, nourishment and also waste pro-

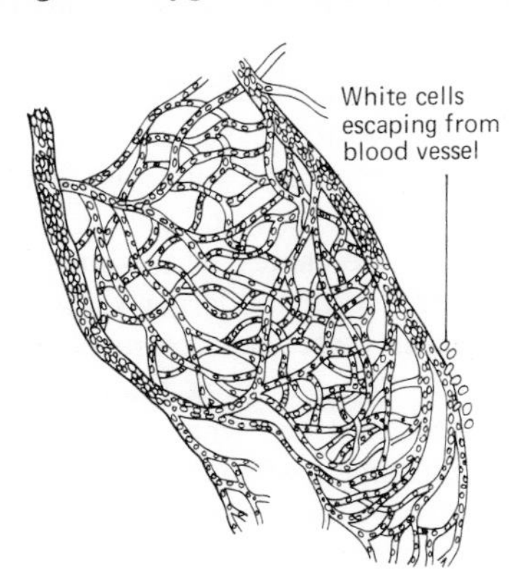

Fig. 96 Network of capillaries between an arteriole and a venule.

ducts takes place between the blood and the tissues through the walls of the capillaries.

The arteries which convey blood from the heart to the capillaries are thick-walled tubes consisting of three coats—outer, middle and inner. The outer coat (**tunica adventitia**) is composed of fibrous tissue and gives protection and strength to the vessel. The middle coat (**tunica media**) consists of plain (unstriped) muscle fibres with some yellow elastic fibres. The muscle fibres are arranged in a circular manner and by their contraction and relaxation the calibre of the vessel can be altered (page 162). This function is mainly under the control of the nervous system and plays an important part in determining the amount of blood supplied to an organ and in the

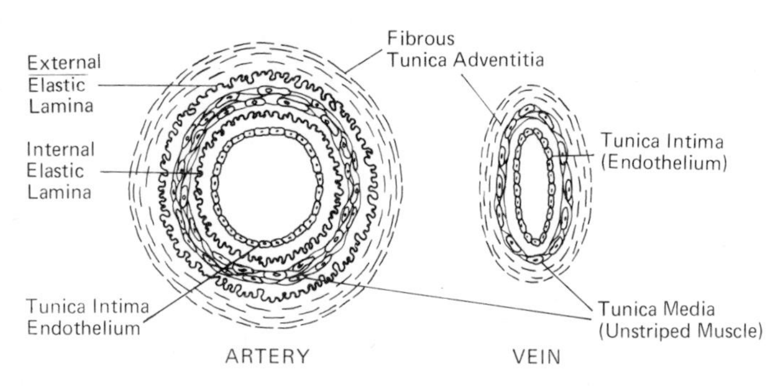

Fig. 97 Diagram illustrating structure of arteries and veins.

maintenance of the blood pressure. The inner coat (**tunica intima**) has two parts: the lining of the artery, consisting of flattened endothelial cells, and a layer of elastic fibres which separates the lining from the middle muscular coat. The amount of elastic tissue is greatest in the large arteries, especially the aorta, while in the smaller arteries and arterioles the muscular tissue predominates. This means that the blood supply to an organ or part is mainly controlled by variation in the calibre of the small arteries and arterioles, while the size of the aorta and larger vessels remains constant.

The veins also possess three coats corresponding to those found in the arteries, but they are all much thinner, hence the walls of a vein collapse and fall together when the vessel is opened. An artery, however, remains open when divided. Many veins of lower limbs and abdomen have **valves** in their interior, so arranged that they only allow blood to flow towards the heart and prevent any flow in the opposite direction. These valves consist of pouch-like folds with their openings directed towards the heart. They play a very important part in the return of the blood to the heart.

In the condition of varicose veins of the leg, the valves are defective and the veins are therefore over-distended with blood.

The heart

The heart is a hollow, muscular organ lying in the thorax between the lungs and in relation to the upper surface of the diaphragm. It is situated behind the sternum and extends outwards to the left for 9 cm ($3\frac{1}{2}$ in). Being conical in shape, it is described as having a base, directed upwards and to the right, and an apex directed downwards and to the left. The heart is divided by a partition or septum into right and left halves which do not communicate with each other. Each half consists of two chambers, an upper thin-walled atrium, and a lower thick-walled ventricle. The atria act as receiving chambers for the pump and the ventricles as distributors.

The opening between each atrium and ventricle is guarded by a valve which permits blood to flow only from the atrium to the ventricle and prevents any back-flow of blood in the opposite direction.

NOTE: It must be made clear that although the term 'auricle' has been in use for many years, the modern anatomical name for each of the receiving chambers of the heart is 'atrium', which is, therefore, now used in this book. In this new terminology 'auricle' refers to a small pouch projecting forwards from each atrium. In clinical medicine, however, an abnormality of cardiac rhythm is referred to either as 'auricular fibrillation' or 'atrial fibrillation'.

The course of the blood through the heart.

Blood from the veins of the head, neck and upper limbs enters the right atrium by the **superior vena cava**, and from the rest of the body and lower limbs by the **inferior vena cava**. Thence it passes through the right atrioventricular opening into the right ventricle. This opening is guarded by the **tricuspid valve**, so called because it consists of three folds or cusps. Blood leaves the right ventricle by the **pulmonary artery**. It passes to the capillaries of the lungs and is collected up in the **pulmonary veins** which pass to the left atrium. Four of these large pulmonary veins enter the left atrium. The left atrioventricular opening is guarded by the mitral valve. This has two cusps which resemble in shape a bishop's mitre. Blood leaves the left ventricle by the large, main artery of the body called the **aorta**. Both the openings of the pulmonary artery and the aorta are also guarded by valves, the aortic and pulmonary valves. Each consists of three half-moon-shaped (semilunar) cusps.

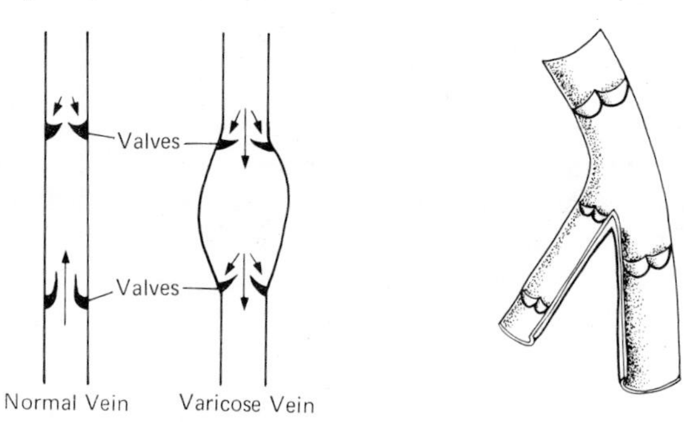

Fig. 98 Valves of veins.

Fig. 99 Vein cut to show valves.

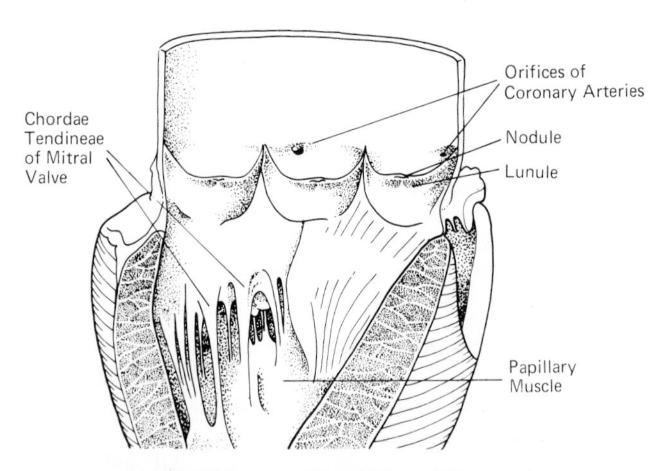

Fig. 100 The cusps of the aortic valve.

It will be remembered that blood from the veins is conveyed to the lungs in order to pick up oxygen and to become arterial blood. It follows, therefore, that the blood on the right side of the heart is venous blood deficient in oxygen. In contradistinction to other arteries of the body, the pulmonary artery contains de-oxygenated or venous blood and the pulmonary veins, unlike any other veins, contain oxygenated arterial blood.

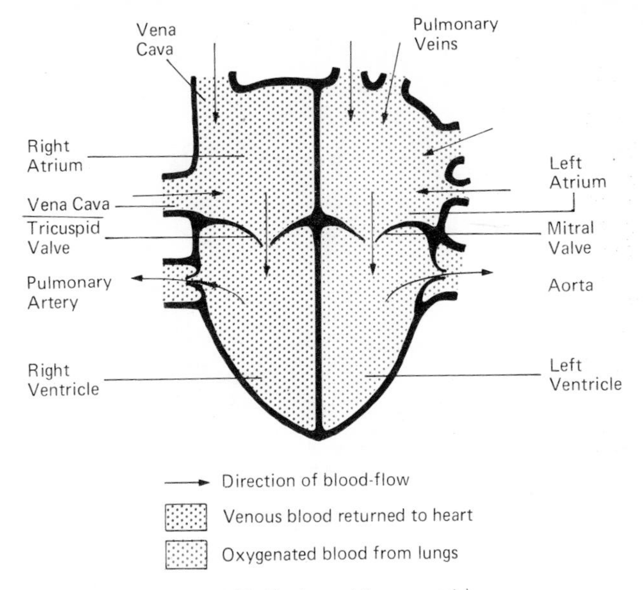

Fig. 101 The heart (diagrammatic).

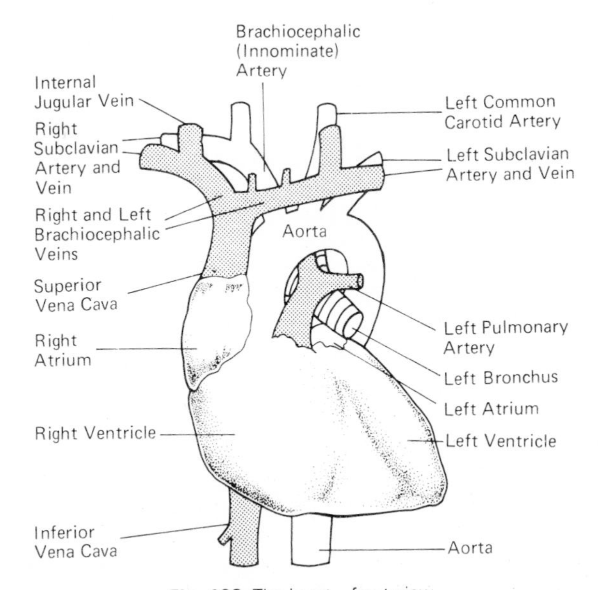

Fig. 102 The heart—front view

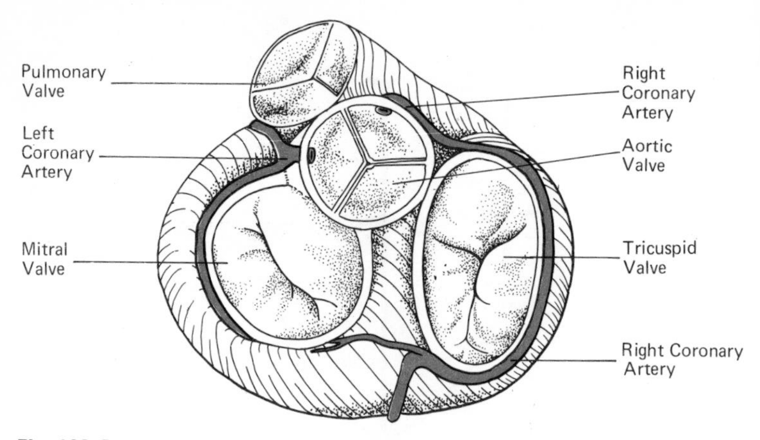

Fig. 103 Base of the ventricles seen from above, showing the relative positions of the valves and the courses of the coronary arteries. (Note that the atria have been removed.)

Summary

Chamber	*Openings*	*Type of blood*
right atrium	superior vena cava	venous
	inferior vena cava	venous
	right ventricle	venous
right ventricle	right atrium	venous
	pulmonary artery	venous
left atrium	pulmonary veins (4)	arterial
	left ventricle	arterial
left ventricle	left atrium	arterial
	aorta	arterial

Structure of the heart

From a structural point of view the heart consists of three layers:

The pericardium (outer covering).

The myocardium or heart muscle (middle layer).

The endocardium (inner lining).

The **pericardium** is a fibrous sac, having a serous lining, which surrounds the heart. Its upper part is attached to the great vessels leaving the heart and its lower part to the upper surface of the diaphragm. The serous lining consists of two layers; the outer or **parietal layer** lines the outer fibrous sac and the inner or **visceral** covers the heart muscle. A small quantity of serous fluid lubricates these two layers which are firmly attached to the fibrous sac and heart muscle respectively, so that when the heart beats the smooth serous layers glide over each other without friction.

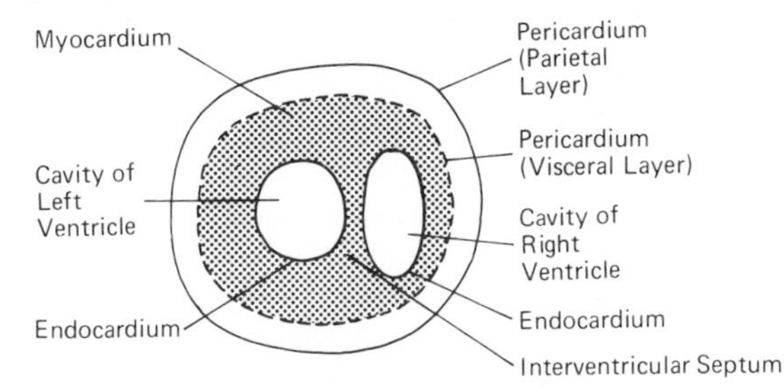

Fig. 104 Diagram illustrating general structure of the heart.

The outer fibrous sac is firm and inelastic and therefore prevents over-distension of the heart.

The **myocardium** or heart muscle. It has already been stated that the muscle of the heart consists of special fibres which are faintly striated but of the involuntary type and not under control

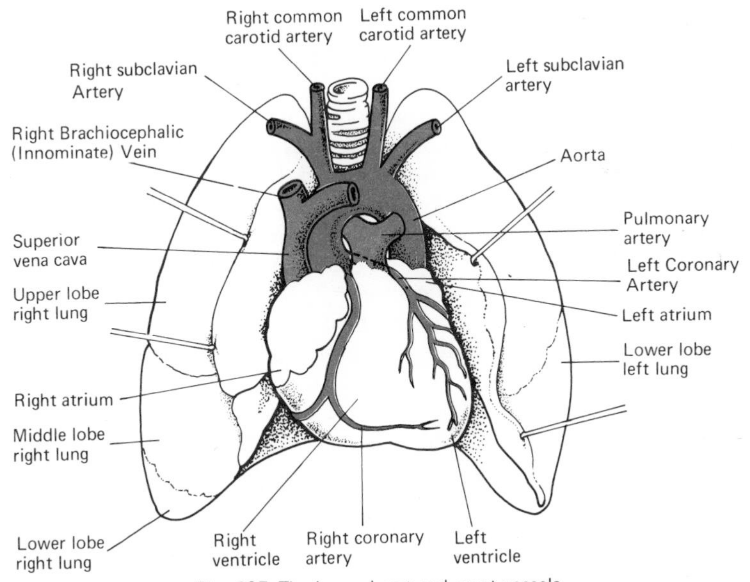

Fig. 105 The lungs, heart and great vessels.

of the will (page 28). These muscular fibres are arranged in a complex manner, but in such a way that when they contract they tend to squeeze the blood in a forward direction into the next opening through which the blood has to pass, i.e. the atria contract so as to force the blood downwards into the ventricles, and the ventricles so that they force the blood upwards into the pulmonary artery and aorta, respectively. At the same time some of the fibres of the two atria are continuous with each other and some of the right and left ventricular fibres are also continuous, so that the two atria contract simultaneously and the two ventricles also contract together.

The atria have only to pass on their contained blood through the comparatively wide atrioventricular valve openings into the ventricles. The amount of work they are called upon to do is, therefore, light and in consequence they have relatively thin walls. The right ventricle has more work to do, viz. to pump the blood through the lungs. Its wall is, therefore, thicker than the atria. The left ventricle having to force blood through the whole of the body must be stronger still and its walls are about three times as thick as those of the right ventricle.

The muscle of the atria is entirely separate from the muscle of the ventricles except at one point. This point of communication is situated in the septum of the heart and consists of special pale muscle fibres known as the atrioventricular bundle of His. Elsewhere the atria are separated from the ventricles by a band of fibrous tissue.

The walls of the atria are smooth, but a number of columns of muscular tissue project into the cavities of the ventricles. These are called the **papillary muscles** and to their free ends are attached a number of fine tendinous cords, the **chordae tendineae**, which extend to the lower borders of the atrioventricular valves. These attachments prevent the cusps of the valves being forced up into the atria when the ventricles contract.

The heart muscle receives its blood supply from the two (right and left) **coronary arteries**, the first branches of the aorta (page 137).

The **endocardium** or lining of the heart is a delicate membrane consisting of flat endothelial cells continuous with the endothelial lining of the arteries and veins. It lines the whole of the cavities of the heart and, at the valvular openings, is folded back on itself to form the cusps of the valves which, therefore, consist of two layers of endocardium separated by a little fibrous tissue to give them extra strength.

Each chamber of the heart has a capacity of about 120 ml (4 ounces). The weight of the adult heart is about 300 grams (10 ounces).

THE SYSTEMIC CIRCULATION

Blood vessels: (1) The arteries

The aorta

The aorta is the largest artery of the body. It arises from the left ventricle of the heart and passes through the thorax and abdomen. The portion situated within the thorax is, for convenience, termed the thoracic aorta and the rest of the vessel is known as the abdominal aorta.

The **thoracic aorta** is described in three parts:

The ascending aorta.
The arch of the aorta.
The descending (thoracic) aorta.

The ascending part, after leaving the left ventricle, passes upwards and to the right for about 5 cm (2 in). It then turns backwards and to the left to form the arch of the aorta, which lies behind the manubrium sterni, and becomes continuous with the descending aorta close to the left side of the 4th thoracic or dorsal vertebra. The descending aorta continues downwards in the thorax in front of the remaining thoracic vertebrae and passes through an opening in the diaphragm at the level of the 12th thoracic vertebra to become the abdominal aorta (see Fig. 106, page 138).

The **abdominal aorta** is related to the lumbar vertebrae and terminates at the level of the 4th by dividing into the two (right and left) common iliac arteries. On the surface of the body this division is represented by a point just below and to the left of the umbilicus and is approximately at the level of a line joining the highest points of the iliac crests (see Fig. 85, page 118).

As a result of disease, the wall of the aorta may be weakened and bulge outwards, forming an aneurysm. The thinning of the wall may result in rupture and fatal haemorrhage.

It will be noted that certain branches are paired and pass respectively to each side of the body, while others are single and arise from the front of the aorta.

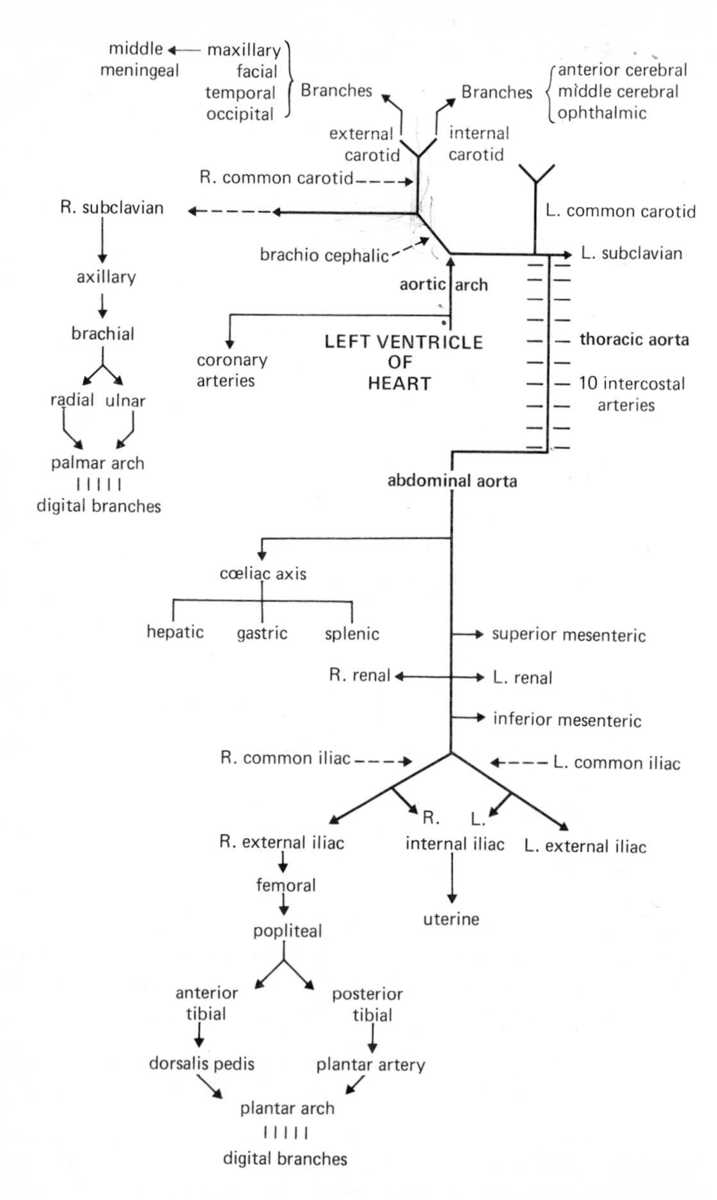

PLAN OF THE ARTERIAL SYSTEM

Branches of the aorta

Thoracic	(ascending)	right and left coronary arteries
	(arch)	brachiocephalic artery { right subclavian artery; right common carotid artery }
		left common carotid artery
		left subclavian artery
	(descending)	right and left intercostal arteries to each intercostal space
Abdominal		coeliac axis artery { hepatic artery; gastric artery; splenic artery }
		superior mesenteric artery
		right and left renal arteries
		inferior mesenteric artery
		right and left common iliac arteries

Coronary arteries. The right and left coronary arteries are the first branches of the aorta after it leaves the left ventricle. Each runs in the groove between its corresponding atrium and ventricle and gives branches to the heart muscle (see Fig. 103, page 132).

The coronary arteries supply the myocardium with oxygen and nourishment. They are, therefore, of great practical importance because disease of their walls may result either in narrowing of the artery or complete blockage of one of the branches. Both conditions cause cardiac pain (angina). If the block is complete, as in coronary artery thrombosis, (myocardial infarction), and if the patient survives, the part of the myocardium supplied by the affected branch will become deprived of its blood supply and the muscle will be replaced by a fibrous tissue scar. If this is extensive it will weaken the pumping power of the heart.

Brachiocephalic (innominate) artery. This important vessel arises from the arch of the aorta behind the right side of the manubrium sterni. It passes upwards and to the right and, after 5 cm (2 in), divides into the right subclavian and right common carotid arteries (see Fig. 106).

Common carotid artery. It must be remembered that the right common carotid arises from the brachiocephalic artery while the left common carotid springs directly from the arch of the aorta. Thereafter, their course and distribution are identical. The common carotid artery passes upwards in a sheath (the **carotid sheath**) along with the internal jugular vein and vagus nerve. It is surrounded

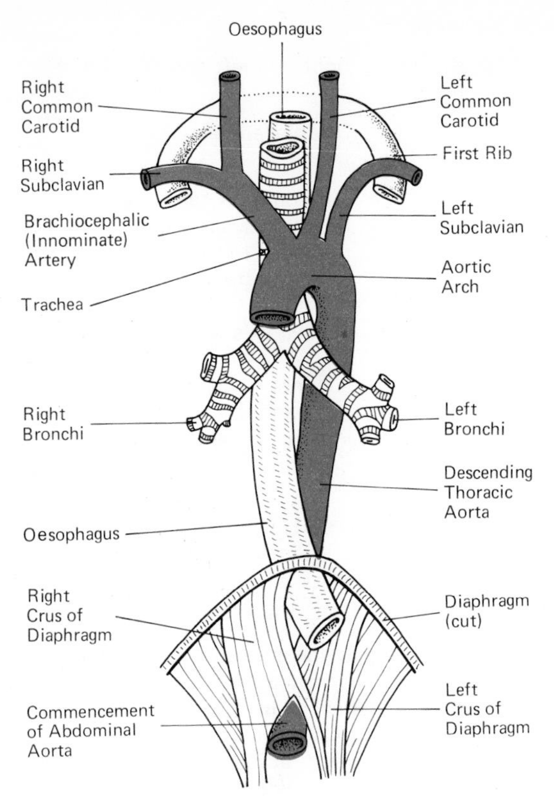

Fig. 106 Diagram of the thoracic aorta.

by the muscles of the neck and, at the level of the thyroid cartilage of the larynx, divides into two branches, the internal and external carotid arteries (see Fig. 107).

External carotid artery. This artery, as its name suggests, supplies the outer surface of the head and neck. It has four main branches:

1. The **facial artery**, which passes up over the outer surface of the lower jaw just in front of the angle and supplies the lower part of the face.
2. The **temporal artery** passing upwards in front of the ear to supply the frontal, temporal and parietal portions of the scalp.
3. The **occipital artery**, which passes behind the ear and supplies the occipital part of the scalp.

4. The **maxillary**, which supplies the structures around the jaws, and gives off the important **middle meningeal** artery to the interior of the skull.

It is of practical importance to remember that the pulse can be felt both in the facial artery as it crosses the lower jaw and in the temporal artery where it is placed immediately in front of the external auditory meatus.

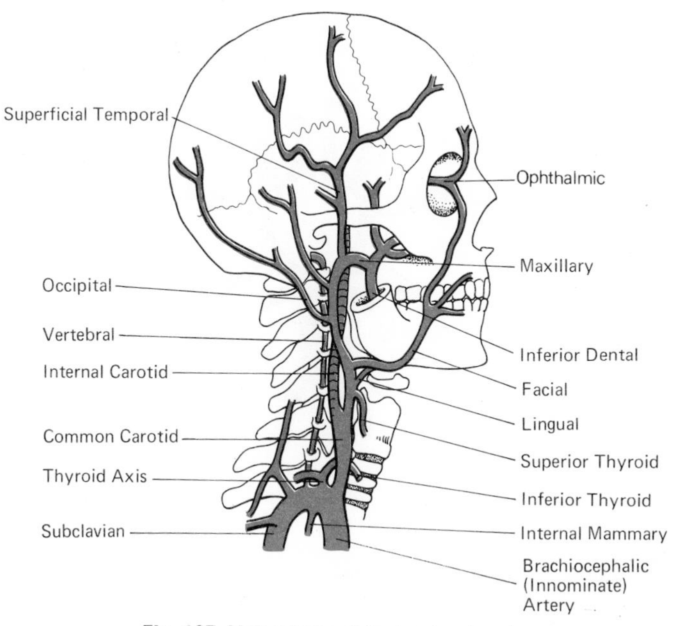

Fig. 107 Main arteries of the head and neck.

Internal carotid artery. This commences at the bifurcation of the common carotid and extends upwards to enter the interior of the skull through the carotid foramen in the petrous portion of the temporal bone. In the neck it lies deeply in the muscles. It enters the middle cranial fossa of the skull and terminates in two branches, the **anterior** and **middle cerebral arteries**, which supply the brain. It also sends a branch, the **ophthalmic artery**, to supply the eye.

The anterior and middle cerebral arteries communicate with each other and also with the **basilar artery**, a continuation of the vertebral arteries, thereby forming the **circle of Willis**, which ensures the even distribution of blood to the brain.

A congenital weakness may be present in the walls of one of these arterial junctions at which a small protrusion or aneurysm may develop. Later in life this may rupture causing haemorrhage into the subarachnoid space around the brain and spinal cord.

A stroke (hemiplegia, often with speech defect) may result from blockage of one of the cerebral arteries by a thrombus or an embolus or from haemorrhage. It may also result from narrowing (stenosis) or occlusion of the internal carotid artery by atheroma.

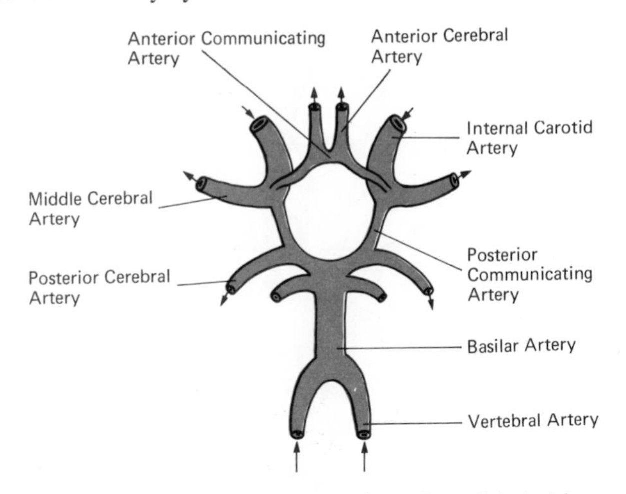

Fig. 108 The arterial circle of Willis (at the base of the brain).

Subclavian artery. The right subclavian artery arises from the brachiocephalic artery; the left directly from the arch of the aorta just beyond the origin of the left common carotid. It passes over the first rib, which it grooves, and behind the clavicle to enter the upper part of the axilla where it becomes the axillary artery. It is the main artery of supply to the upper limb, but has three branches:

1. The **vertebral artery**, which passes upwards through the special foramina in the transverse processes of the cervical vertebrae (page 81) to enter the skull by way of the foramen magnum. It is distributed to the posterior part of the brain and cerebellum.
2. Branches to the thyroid gland.
3. The internal mammary artery.

Axillary artery. This is the direct continuation of the subclavian artery and, at the lower boundary of the axilla, becomes the brachial artery (see Fig. 109).

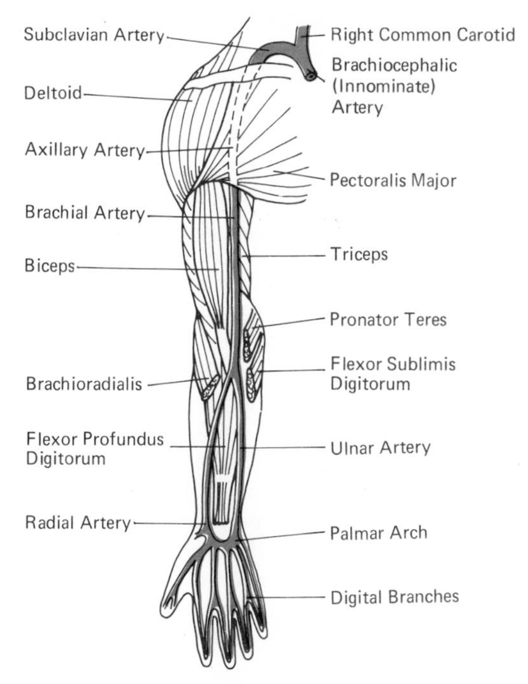

Fig. 109 Arteries of the upper limb.

Brachial artery. This continues from the lower border of the axilla to the bend of the elbow (cubital fossa), where it divides into the radial and ulnar arteries. In its course it runs close to the humerus on the medial side of the biceps muscle.

Radial artery. This follows the lateral bone of the forearm, the radius, and in the first part of its course is covered by muscles. In its lower part, just above the wrist, it is superficial and can be felt lying in front of the bone. It is in this position that the pulse is usually taken.

Ulnar artery. This runs down the medial side of the forearm close to the ulna.

The palmar arches. The ends of the radial and ulnar arteries pass in front of the wrist into the palm of the hand where they join in the form of two arches running transversely. One of the arches thus formed is superficial, the other deep, and from them branches to pass to the fingers.

Branches of the abdominal aorta. **Coeliac axis**. This is an artery which springs from the front aspect of the abdominal aorta immediately after it has passed through the diaphragm. It is a short trunk which divides into three branches:

The **gastric artery**, which helps to supply the stomach.

The **hepatic artery**, which supplies the liver.

The **splenic artery**, which passes to the left behind the stomach and along the upper border of the pancreas to the spleen (see Fig. 85).

Renal arteries. A little below the coeliac axis the right and left renal arteries arise from the sides of the abdominal aorta at the level of the second lumbar vertebra. Each passes to the corresponding kidney. Because the inferior vena cava is situated to the right of the aorta, the right renal artery has to pass behind this large vein in order to reach the kidney.

Superior mesenteric artery. Arising from the front of the aorta at almost the same level as the renal arteries is the superior mesenteric artery. This passes forward in front of the last part of the duodenum and reaches the fold of peritoneum called the mesentery, as its name suggests (page 204). Thence, it supplies the whole of the small intestine and the first portion of the large intestine.

Inferior mesenteric artery. This takes origin lower down from the front of the aorta and sends branches to the lower parts of the colon and to the rectum.

Common iliac arteries. These are the two branches into which the abdominal aorta divides opposite the fourth lumbar vertebra. Each passes downwards and laterally for about 5 cm (2 in) and then divides into the internal and external iliac arteries.

Internal iliac artery. This artery descends immediately into the cavity of the pelvis, the contents of which it supplies. In the female it gives origin to the important **uterine artery** which supplies the uterus.

External iliac artery. This is the direct continuation of the common iliac artery and runs downwards along the brim of the pelvis, passing under the inguinal (Poupart's) ligament to become the femoral artery, which is the main artery of the lower limb.

Femoral artery. Commencing at the inguinal (Poupart's) ligament, this is the continuation of the external iliac into the thigh. Its direction is indicated by a line drawn from a point midway between the anterior superior spine of the ilium and the symphysis pubis to a point just above the medial condyle of the femur (adductor tubercle). The upper half of the artery is comparatively superficial and lies in the space known as the **femoral** (Scarpa's) **triangle**

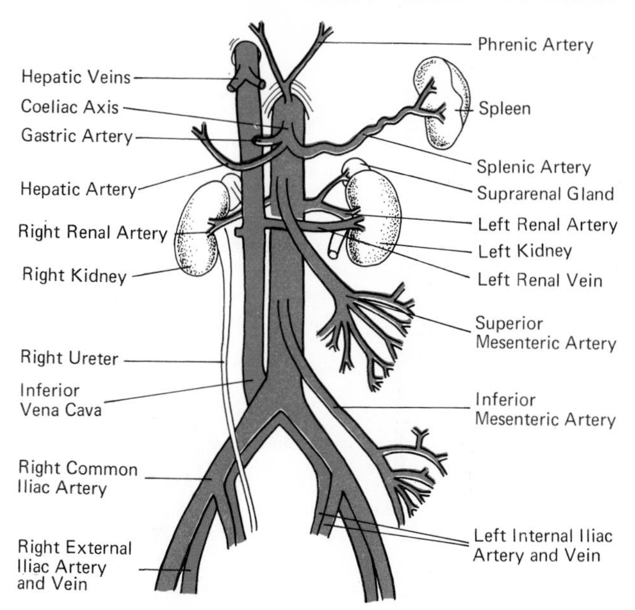

Fig. 110 Diagram showing abdominal aorta and inferior vena cava with important branches.

(see page 385). In the lower part it lies more deeply among muscles in a special tunnel called the **adductor** (Hunter's) **canal**. It ends by passing backwards behind the femur to enter the popliteal space where it becomes the popliteal artery. The femoral artery gives branches to the surrounding muscles and to the knee joint.

Popliteal artery. This is the continuation of the femoral in the popliteal space. At the lower end of the space it gives off the **anterior tibial artery** which passes forwards between the tibia and fibula to supply the front of the leg and is continued downwards on the dorsum of the foot as the **dorsalis pedis artery**. The **posterior tibial artery** is the direct downward continuation of the popliteal in the back of the leg, where it lies deeply in the muscles. It passes behind and below the medial malleolus to reach the sole of the foot as the plantar artery. Here it forms the plantar arch, similar to the palmar arch of the hand, from which digital branches pass to the toes.

It is sometimes useful to be able to identify and feel the pulse in the dorsalis pedis artery where it lies superficial to the metatarsal bone of the great toe.

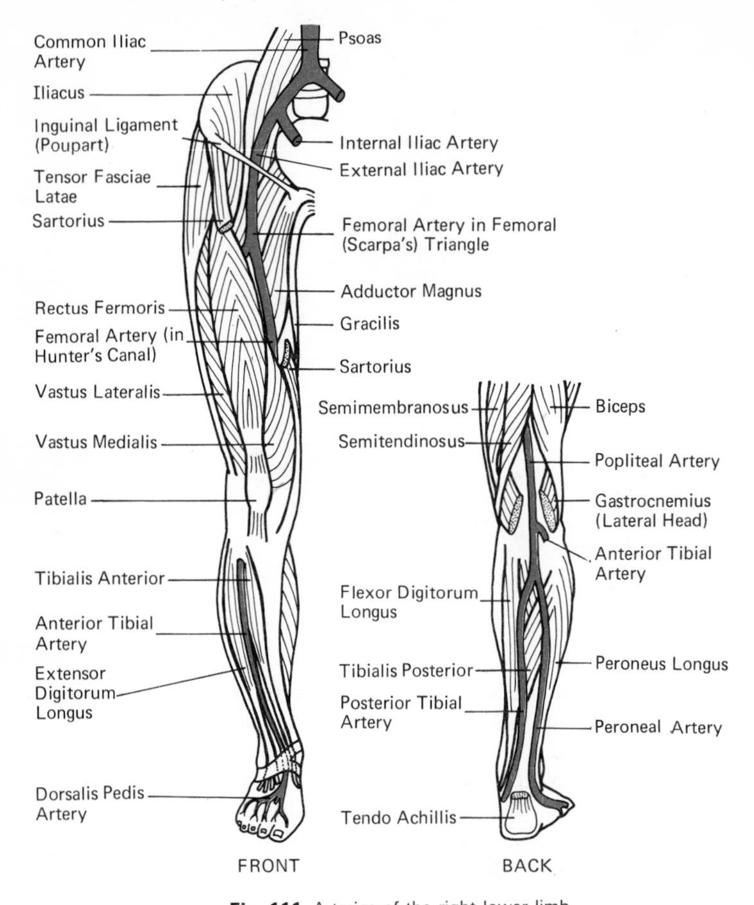

Fig. 111 Arteries of the right lower limb.

Pressure points

It is sometimes possible to arrest severe haemorrhage by compressing the artery which supplies the bleeding part. This can only be done in the case of those arteries which are near the surface and which can be compressed against some firm underlying structure, e.g.

Arteries of the head and neck

Facial: against the lower jaw.
Temporal: in front of the external auditory meatus.

Occipital: against the occipital bone 6.5 cm ($2\frac{1}{2}$ in) behind the ear.
Common carotid: against the cervical vertebrae to the side of the larynx.

Arteries of the upper limb

Subclavian: against the first rib in the hollow above the clavicle.
Brachial: against the medial aspect of the humerus in the middle of the arm.
Radial: at the lower end of the radius just above the wrist on its anterior surface.
Ulnar: against the anterior surface of the ulna.

Arteries of the lower limb

Femoral: against the pubic bone under the inguinal (Poupart's) ligament.
Posterior tibial: against the posterior surface of the medial malleolus.
Dorsalis pedis: against the upper surface of the navicular bone.

Blood vessels: (2) The veins

The veins commence at the termination of the capillaries, and convey blood back to the heart. The small venules unite with one another to form larger and larger vessels until two main trunks are formed which enter the right atrium of the heart. The upper of these trunks is the **superior vena cava** conveying the blood from the head and neck and the upper limbs. The lower trunk is the **inferior vena cava** which receives through its various branches blood from the rest of the body, including the abdominal cavity and its contents, and the lower limbs.

The veins are divided into two main groups:

(*a*) The **superficial**, which are situated on the surface of the body just under the skin, some of which can be easily seen.

(*b*) The **deep**, many of which accompany the main arteries of the body. It will therefore be unnecessary to describe them all in detail.

For convenience, the veins which go to form the inferior vena cava will be described first.

1. Veins going to form the inferior vena cava

Veins of the lower limb: (*a*) superficial
(*b*) deep

Veins of the abdomen: (*a*) outside the peritoneal cavity
(*b*) inside the peritoneal cavity (the portal circulation)

Veins of the lower limb. (*a*) There are two main superficial veins in the lower limb.

The **long saphenous vein**, which commences on the dorsum of the foot and passes up on the medial side of the leg, behind the medial aspect of the knee, to just below the medial end of the inguinal (Poupart's) ligament, where it passes deeper to enter the femoral vein. The point where it pierces the deep fascia of the limb is called the **saphenous opening**. The **short saphenous vein** commences on the lateral side of the foot and passes upwards along the centre of the back of the calf and pierces the deep fascia of the limb over the popliteal space at the back of the knee joint to reach the **popliteal vein**.

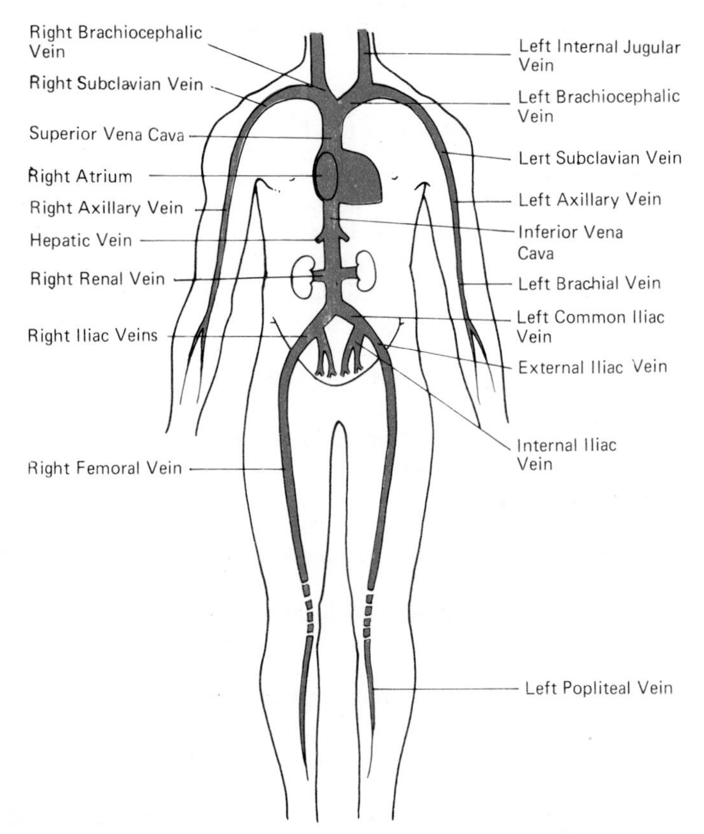

Fig. 112 Diagram of the main deep veins of the body.

(*b*) The main deep veins of the leg commence as the anterior and posterior tibial veins which accompany the corresponding arteries and unite to form the popliteal vein. This passes upwards in company with the artery and enters Hunter's canal as the **femoral vein.** In the femoral (Scarpa's) triangle it lies in a sheath of fibrous tissue, called the femoral sheath, together with the femoral artery. In the femoral triangle it receives the long saphenous vein and passes under the inguinal (Poupart's) ligament to become the **external iliac vein.**

Veins of the abdomen. *On the posterior abdominal wall.* The external iliac vein, joined by the internal iliac vein, becomes the common iliac vein and the right and left common iliac veins unite to form the inferior vena cava just below the bifurcation of the aorta into the common iliac arteries (5th lumbar vertebra). The **inferior vena cava** continues upwards in front of the bodies of the lumbar vertebrae lying to the right of the abdominal aorta. In the abdomen it receives the two important right and left **renal veins** from the kidneys and, just before it pierces the diaphragm to enter the thorax, two or three groups of **hepatic veins** pour the blood from the liver into it. In the upper part of the abdomen it lies behind the liver, the posterior aspect of which it grooves. Shortly after entering the thorax it reaches the right atrium, where it terminates.

The portal circulation (inside the abdominal cavity). The portal circulation concerns the blood which is supplied to and removed from the organs of digestion and its conveyance to the liver. It will be remembered that the blood to these organs comes from the following branches of the abdominal aorta:

The coeliac axis artery supplying the stomach, the spleen, the pancreas and the liver.

The superior mesenteric artery to the small intestine and first part of the large intestine.

The inferior mesenteric artery to the rest of the large intestine and the rectum.

These arteries all break up into capillaries in the organs which they supply and these capillaries unite in the usual way to form veins. The individual veins from the stomach, spleen and intestines in turn unite to form one large vein, the **portal vein**, which carries all their blood to the liver. Here, in the substance of the liver, the portal vein breaks up into smaller veins and finally into a second set of capillaries. These capillaries in the liver again unite to form the two groups of **hepatic veins** which leave the liver to join the inferior vena cava just before it pierces the diaphragm to enter the thorax

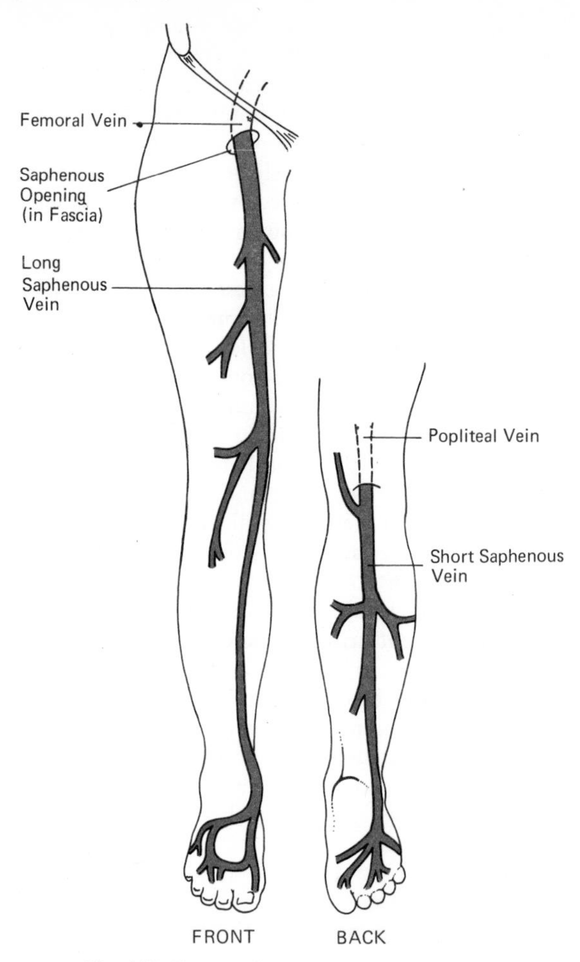

Fig. 113 The superficial veins of the lower limb.

(see also page 226). It will be noted that the portal vein, unlike other veins in the body, both commences and ends with capillaries. The importance of this is that the digested foodstuffs in the alimentary canal are absorbed into the capillaries which go to make up the portal vein and are carried by it to the liver. In order that these materials can come in contact with the individual liver cells for further chemical action or storage it is necessary for the blood to

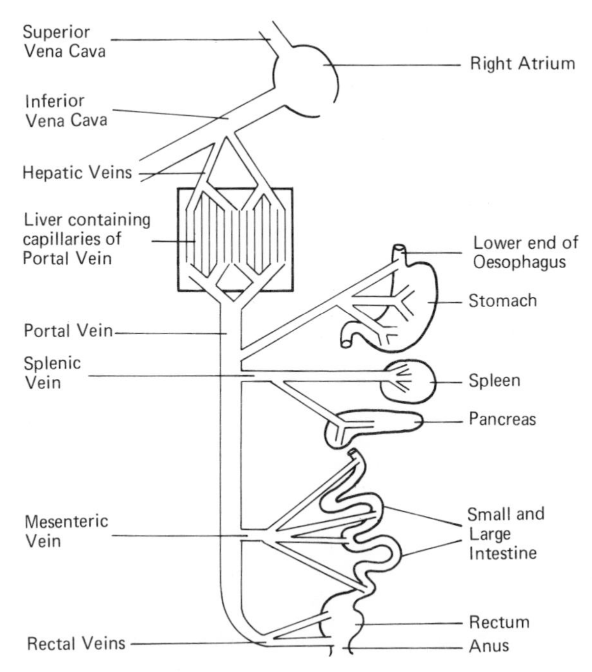

Fig. 114 Diagram illustrating the portal circulation.

pass through a second set of capillaries, for it will be remembered that interchange of substances between the blood and tissues can only take place through the capillaries. The portal vein itself is quite short (8 cm, 3 in) and is formed just behind the pancreas whence it passes upwards, behind the pylorus, to reach the portal fissure of the liver (page 225).

Obstruction of the portal vein or its branches (e.g. by thrombosis or by hepatic cirrhosis) causes a rise in blood pressure in the portal venous system, a condition known as portal hypertension. This results in enlargement of the spleen, oesophageal varices and haemorrhoids, etc.

In order to avoid any confusion, it must be pointed out that the hepatic artery supplies arterial blood to the liver in order to provide an adequate amount of oxygen to the organ and takes no part in the formation of the portal circulation.

2. Veins going to form the superior vena cava

(*a*) Veins of the upper limb: {superficial, deep}

(*b*) Veins of the head and neck: {superficial, deep}

Veins of the upper limb. The superficial veins commence mainly on the back of the hand and the more important of them converge on the cubital space in front of the elbow joint. Passing up the front of the forearm is the **cephalic vein**. At the end of the elbow this gives off the **median cubital vein** which goes to the medial side to join the **basilic vein**. The basilic vein runs up the medial side of the arm and the cephalic vein continues its course up the lateral side.

The veins at the bend of the elbow are of importance on account of their relatively large size and the fact that they are readily access-

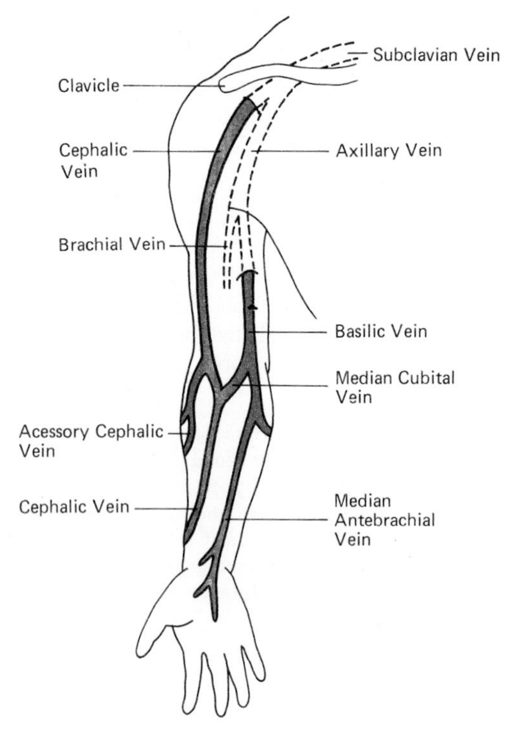

Fig. 115 Superficial veins of the right upper limb.

ible. They are the veins usually selected for the removal of blood or for the injection of fluids or drugs into the blood stream.

The cephalic and basilic veins pass upwards and pierce the deep fascia to accompany the axillary artery as the **axillary vein.** The deep veins of the arm accompany the radial, ulnar and brachial arteries and terminate in the axillary vein.

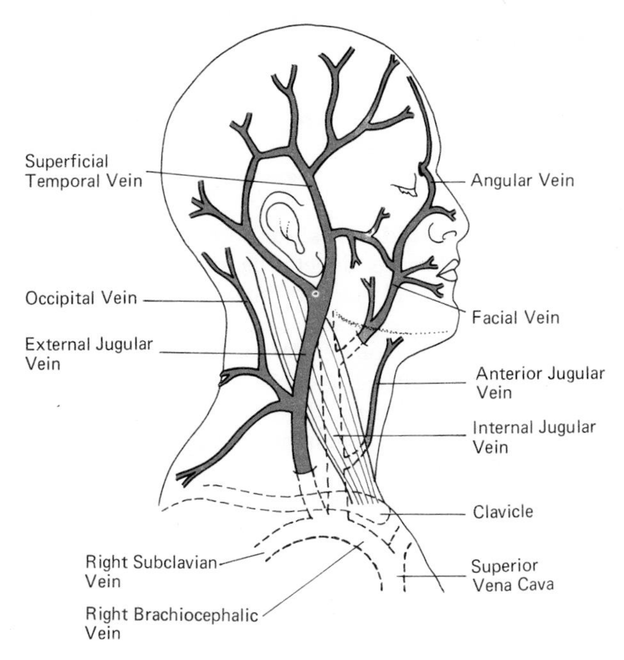

Fig. 116 Veins of the head and neck.

The axillary vein becomes the **subclavian,** which is finally joined by the **internal jugular vein** bringing blood from the head and neck to form the brachiocephalic vein. This unites with its fellow of the opposite side to become the superior vena cava. It will be noticed that although there is only one brachiocephalic artery, there are two (right and left) brachiocephalic veins.

Veins of the head and neck. The superficial veins of the scalp unite just behind the angle of the jaw to form the **external jugular vein.** This passes directly downwards, superficial to the sternomastoid muscle which it crosses obliquely, and enters the subclavian vein.

The veins of the face are, however, collected up into the main **facial vein** which crosses the lower jaw in front of the angle and then passes deeply to join the internal jugular vein. The facial vein actually commences as a small vein close to the medial angle of the eye (the **angular vein**). There is also a communication between the angular vein and the veins inside the skull which passes through the orbit. It is for this reason that boils and carbuncles on the face may be particularly dangerous, for infection may spread through this connection and cause inflammation within the skull (cavernous sinus thrombosis).

Apart from the veins of the brain, which pour their blood into the internal jugular vein, there are some other important venous channels within the skull. These are the **venous sinuses** which lie between layers of the dura mater (the outer covering of the brain).

The **superior longitudinal sinus** commences in the frontal region of the skull and runs directly backwards in the mid-line to the occipital region, in a fold of dura mater called the **falx cerebri.** This fold of dura mater separates the right from the left cerebral hemisphere. In the infant (under 18 months) before the bones of the skull are completely fused, there is a diamond-shaped opening, the anterior fontanelle, between the frontal and two parietal bones which is only covered by membrane (see Fig. 50, page 75). The superior longitudinal sinus passes directly beneath this opening in the mid-line, and it is possible to insert a needle into the sinus for purposes of removing blood. This method is sometimes used in infants as the limb veins are very small and difficult to find.

About the centre of the occipital bone the superior longitudinal sinus divides (at the torcular Herophili or confluence of sinuses) into right and left branches which become the transverse (lateral) sinuses. These pursue a curved course, in close relation to the inner aspect of the mastoid process, to reach the base of the skull which they leave by the jugular foramen to become the internal jugular vein. The lower end of the transverse sinus is sometimes called the sigmoid sinus.

The close association of the transverse (lateral) sinus with the mastoid process is of importance. The sinus may be exposed or opened during operations on the mastoid process and sometimes infection may spread from the mastoid into the sinus, a serious condition which may result in general blood infection.

The **cavernous sinus** is situated at the side of the sella turcica. It communicates with its fellow of the opposite side, with the angular vein on the face, and posteriorly with the transverse (lateral) sinuses.

The **internal jugular vein** commences at the jugular foramen in the base of the skull as the direct continuation of the sigmoid portion of the transverse (lateral) sinus. It passes downwards, deep to the sternomastoid muscle, in close association with the carotid artery with which it is included in the carotid sheath. Finally it unites with the subclavian vein to form the brachiocephalic vein (Fig. 116).

The **brachiocephalic vein**. Each brachiocephalic vein is formed behind the medial end of the clavicle by the junction of the internal jugular and subclavian veins. The left brachiocephalic vein passes obliquely behind the manubrium sterni to join the right brachiocephalic vein and by their union they form the superior vena cava.

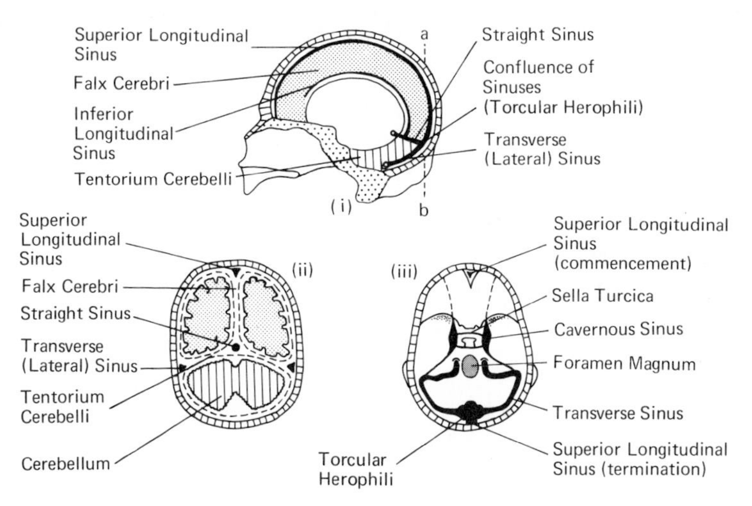

Fig. 117 Diagram of the venous sinuses within the skull. (i) Lateral view. (ii) Vertical section of skull—through *a–b* in (i). (iii) Interior of base of skull.

The **superior vena cava**, which returns the blood from the head and neck and upper extremities to the heart, is 8 cm (3 in) long and passes downwards behind the right margin of the sternum to end its course in the right atrium of the heart.

Cardiac catheterization. A very fine catheter can be passed into the **basilic vein** in the arm and thence onwards into the **superior vena cava** and right atrium. Its exact position can be located by X-rays. Intracardiac pressures can be measured and specimens of blood can be obtained and analysed to show the amount of oxygen they contain. Information can thus be obtained about the physiology of the heart in health and disease.

Venous thrombosis. Blood clots sometimes form in veins (thrombosis). It will be appreciated that if such a clot becomes dislodged (embolism) it will travel in the blood stream to the heart and thence via the pulmonary artery into a lung, cutting off some of its blood supply (pulmonary infarct).

The pulmonary circulation

By the pulmonary circulation is meant the passage of the venous blood from the right side of the heart through the lungs, where it becomes oxygenated, and its return to the left atrium. The **pulmonary artery** arises from the right ventricle, its origin being provided with semilunar valves which prevent back-flow of blood into the ventricle. At its commencement it is placed in front of the origin of the aorta from the left ventricle, but it passes backwards round

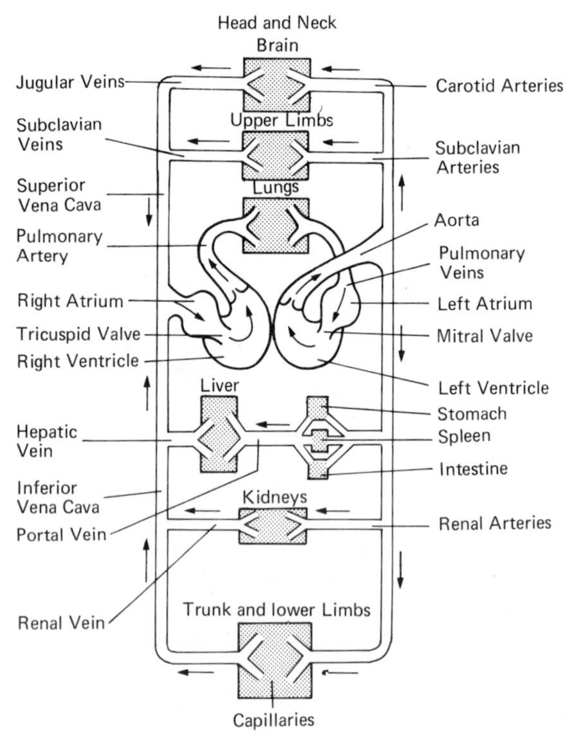

Fig. 118 Diagram of the circulation.

the left side of the ascending aorta and divides into right and left branches. This division takes place immediately below the arch of the aorta (which, it will be remembered, passes obliquely backwards and to the left from its origin in the ascending aorta to the commencement of the descending aorta) (Fig. 102).

Each branch of the pulmonary artery passes to its corresponding lung, breaks up first into smaller arteries which accompany the bronchi, and, finally, into capillaries surrounding the alveoli (air cells). The oxygenated blood from the capillaries is collected up into veins. Two main veins emerge from each lung and the four **pulmonary veins** thus formed enter the left atrium.

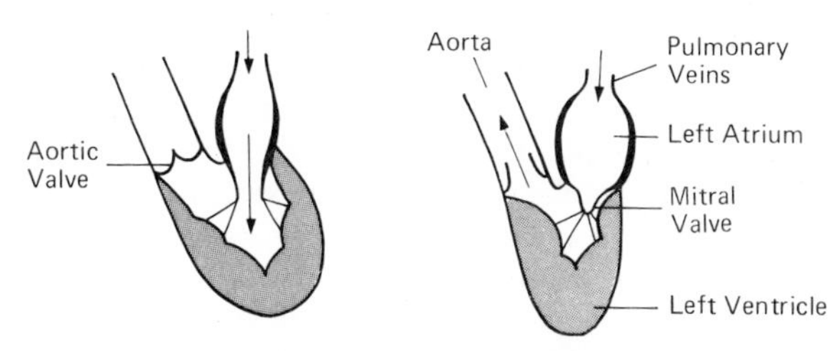

Fig. 119 (*Left*) Normal diastole—mitral valve open. (*Right*) Normal systole—mitral valve closed, aortic valve open.

PHYSIOLOGY OF THE CIRCULATORY SYSTEM

The circulatory system is so constructed anatomically and its functions specially adapted to the following requirements:

1. To maintain a constant blood supply to the brain and vital centres at all times. (Sudden temporary lack of blood supply will cause a fainting attack. Prolonged lack of oxygen may cause permanent damage or even death.)
2. To adjust the blood flow to other organs according to their immediate requirements: e.g. (*a*) the blood supply to the muscles is increased during exercise, (*b*) that to the abdominal organs during digestion, (*c*) that to the surface of the body is varied in order to regulate the body temperature.

It has been pointed out that the heart is mechanically a double pump maintaining two separate circulations, viz. the general and the pulmonary, in such a way that, in health, there is a complete balance between the two systems.

The heart-beat

The function of the heart is to maintain a constant circulation of blood throughout the body by regular rhythmic contractions of its muscle. The rate of contraction in the healthy adult at rest is approximately 72 beats per minute. In children it is a little faster. An increase in rate occurs with exercise, emotional disturbances and in many disorders of health.

The following sequence of events takes place during the process of cardiac contraction. First, blood pouring in from the great veins fills the atria (the superior and inferior venae cavae filling the right atrium and the pulmonary veins the left atrium). A wave of contraction then spreads over the muscle of the atria, which empty their blood into the corresponding ventricles. When the ventricles are full they contract, forcing their contained blood into the general and pulmonary circulations. At the same time, the atrioventricular valves are closed and the semilunar valves guarding the aorta and pulmonary arteries are opened. This process of muscular contraction of the heart is called **systole** and lasts approximately 0·4 (2/5) second. A period of rest follows during which there is no muscular contraction. The resting period is of the same duration as systole and is called **diastole**. This squence of events constitutes the cardiac cycle which, therefore, lasts 0·8 (4/5) second. It will be noticed also that the heart spends half of its time working and the other half resting.

The conducting system of the heart. A mechanism is present whereby rhythmic contractions and the proper sequence of atrial systole followed by ventricular systole is maintained. The impulse to contract starts in a specialized collection of pacemaker cells, lying close to the entrance of the superior vena cava into the right atrium,

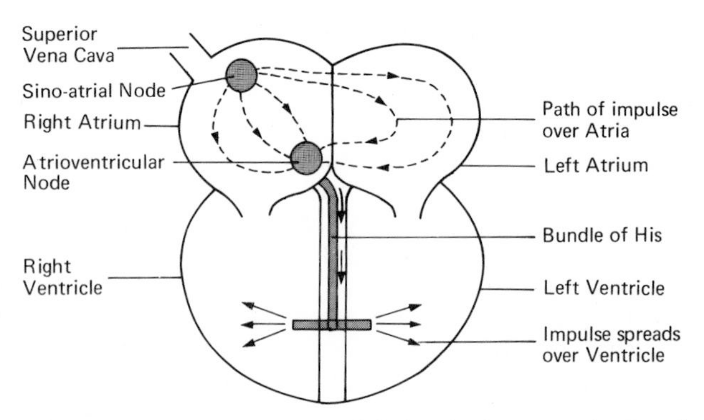

Fig. 120 Diagram illustrating the spread of the impulse for contraction from the sino-trial node over the atria to reach the atrioventricular node, whence it passes down to the bundle of His and is distributed to the ventricles.

known as the **sino-atrial node**. The impulse then passes in a wave-like manner over the atria and is, as it were, collected up at another similar node, called the **atrioventricular node**, situated near the right atrioventricular opening. It is at this node that the **atrioventricular bundle of His** (page 134), the only direct muscular communication between the atria and ventricles, commences. The bundle of His passes down the interventricular septum and divides into two branches, one of which is distributed to each ventricle.

This conducting mechanism is of great importance and may be damaged by disease, resulting in heart block.

Nervous control of the heart-beat. The rate of the heart-beat is controlled by the nervous system. Two sets of nerve fibres reach the heart.

1. The vagus nerve, which tends to slow the heart.
2. Sympathetic nerves, which tend to quicken its rate.

If both these nerves are acting equally the normal rate of the heart is maintained, but if one or other over-acts it produces its own effect on the rate of the heart-beat. Likewise, if the effect of one is diminished the other is able to act more strongly. Thus, if a drug known as atropine, which tends to paralyse the vagus, is given, the heart-rate is increased by the action of the sympathetic which is no longer held in check by the action of the vagus (see also page 338).

The heart-sounds. Normally an individual is not aware of his own heart-beat. However, if the ear is placed over the heart or a stethoscope is applied to the front of the chest, two sounds are heard with each beat of the heart which resemble the words 'lubb dup'. The first sound is made chiefly by the closure of the mitral and tricuspid valves, while closure of the aortic and pulmonary valves is responsible for the second sound. When the valves are diseased the character of these sounds is altered and murmurs may be heard.

The blood pressure

Blood pressure may be defined as the force or pressure which the blood exerts on the walls of the artery in which it is contained.

When an artery is cut across, the blood spurts out from the end nearest the heart with considerable force and is evidently under high pressure within the artery. It will further be observed that, in addition to the continuous stream, there will be regular spurts of increased pressure corresponding with each heart-beat. The continuous pressure is partly dependent on the elasticity of the arteries and is called the **diastolic pressure**. The increased pressure occurring with each beat of the heart is the **systolic blood pressure**. (It is to this maximum systolic pressure that we refer when we use the unqualified term 'blood pressure'.)

If a definite amount of fluid is pumped through a large tube with a certain amount of force it will flow out at the far end at a definite rate and pressure. If the same amount of fluid is pumped through a narrower tube with the same force the pressure with which it leaves the tube will be increased. The narrower the opening of the tube the greater will be the friction of the fluid on the walls of the tube.

An illustration of this can be seen in the use of the fire or ordinary garden hose. If no nozzle is attached and water is turned on at moderate pressure a steady stream will emerge from the end of the hose and project for about a foot. If a narrow nozzle is then attached without altering the supply of water this will leave the nozzle in a fine jet projected for a number of yards, and which is obviously producing much friction against the walls of the nozzle as it leaves. Compression of any part of the hose without the nozzle will be easy, in other words, the pressure within is relatively low. The application of the nozzle, by narrowing the opening and causing resistance to the outflow, will cause an increase in the pressure within the whole of the rest of the hose which can no longer be so easily compressed.

The same mechanical factors occur in the production of blood pressure within the arteries. The heart pumps with a constant force. The aorta is a relatively wide channel but the arteries gradually become smaller and therefore by their resistance to the flow of blood maintain a high pressure in the whole of the arterial system. If the heart beats more strongly the pressure will be increased, but it will fail if the force of the heart-beat is reduced. Again, if the normal size of the arteries is reduced the resistance will be increased and the blood pressure will rise.

Lastly, if the total amount of circulating blood is diminished, e.g. after severe bleeding, the blood pressure will also fall.

Blood pressure is therefore maintained by the following factors:

1. The force of the heart-beat.
2. The resistance to the blood flow in the narrowing peripheral arteries (peripheral resistance).
3. The elasticity of the arteries.
4. The total amount of circulating blood.

Measurement of blood pressure. Blood pressure is measured by an instrument called the **sphygmomanometer**. This consists of a rubber bag which is placed round the arm (the position in which it is customary and convenient to take the blood pressure, i.e. in the brachial artery). The interior of the bag is connected by a rubber tube to a mercury pressure-gauge. When the pressure in the bag equals the pressure in the artery, the latter is compressed flat and the flow through it temporarily arrested. The pulse, therefore, disappears at the wrist. This may be appreciated either by the finger or by listening over the brachial artery at the bend of the elbow with a stethoscope. The height to which the column of mercury has been forced at this moment is therefore the systolic blood pressure. The height is measured in millimetres and the blood pressure is referred to as being so many millimetres of mercury (mm Hg).

The blood pressure varies with the age of the individual. With certain reservations, the systolic blood pressure may be reckoned as 100 plus the age, in millimetres of mercury. Thus, the normal systolic blood pressure of a person of 20 is 120 mm Hg, and for a person of 30 it is approximately 130 mm Hg. However, this increase is less marked as age advances and the maximum normal blood pressure does not exceed 145 to 150 mm Hg. In the adult female the blood pressure may be 10–20 mm less than the male of the same age.

NORMAL SYSTOLIC BLOOD PRESSURE

	mm Hg
children under 10	100
young adults, 20–30	100–120
middle age, 30–60	124–140
old age, over 65 to 70	145–150

The diastolic blood pressure is about 80 mm Hg and the difference between the diastolic and systolic pressures is called the **pulse pressure.** This is usually expressed as:

$$BP = 120/80 \text{ mm Hg}$$

The maintenance of a normal systolic blood pressure is very necessary to health. A fall below 100 mm (hypotension) in an adult is an indication of a very feeble circulation, below 80 mm it usually means impending death. A low blood pressure occurs in cases of shock and severe haemorrhage. It is also found in some cases of heart failure and in Addison's disease of the adrenal gland. At the same time, abnormally high pressures e.g. 220/120 mm Hg (the figures refer respectively to the systolic and diastolic pressures), in addition to causing symptoms, may be dangerous, and there is risk of a blood vessel bursting under the strain especially in the brain, giving rise to an apoplectic stroke. High blood pressure (or hypertension) may be associated with hardening of the arteries (arteriosclerosis) and some cases of kidney disease.

Venous pressure and venous return to the heart. The pressure both in the capillaries and veins is quite low, so that if either are cut blood flows out in a slow steady stream.

It has been seen that the force of the heart-beat is sufficient to drive the blood round the arteries and into the capillaries. This force, however, is insufficient to ensure the **return of blood** from the veins **to the heart**, especially in the distal parts of the body. Other mechanisms are therefore necessary.

1. Depending on the position of the body at the time, the force of gravity aids in the return of venous blood from those parts which are at a higher level than the heart, e.g. from the head and neck when the body is erect.
2. Muscular movement has a constant squeezing effect on the veins. Since the veins are provided with valves the effect must force the blood onward towards the heart.
3. In the act of breathing the expansion of the chest produces a suction or negative pressure within the thorax. In addition to inhaling air into the lungs this suction also acts on the large veins near the heart and helps to suck blood upwards to the heart.

The venous pressure may be raised in cases of congestive heart failure.

The pulse. Each time the left ventricle contracts it forces its contained blood into the aorta. The aorta being elastic expands in order to accommodate this additional amount of blood. At the same time, the blood which was present before the ventricular contraction is now pushed on into the next section which, in turn, also expands, and so on. Thus a wave of expansion is created which starts at the root of the aorta and spreads over the whole of the arterial system, gradually dying away as it reaches the capillaries. This wave of expansion constitutes the pulse and it travels rapidly over the arteries, much more rapidly, in fact, than the velocity of the blood stream.

The pulse can be felt and often seen in the superficial arteries of the body, but it is customary to study it in the radial artery. (The other arteries in which it can be felt with ease are the temporal in front of the ear, the facial as it passes over the lower jaw, the carotid artery in the neck and the dorsalis pedis on the dorsum of the foot.)

Observations to be made are:

1. The rate.
2. The rhythm.
3. The quality:

(*a*) volume; (*b*) tension; (*c*) state of the artery.

1. The rate. The normal rate of the pulse is that of the heartbeat. It is usually about 72 beats per minute in the adult at rest, but is increased by exercise, emotional disturbances and in disease. Undue slowness may also be caused by abnormal conditions e.g. myxoedema.

Normal rates

	beats per minute
adults at rest	60–80
infants	100–120
children 6–10	80–100

2. The rhythm. Under normal conditions this is regular but in diseases of the heart may become irregular.

3. The quality. (*a*) By **volume** is meant the degree of expansion of the artery. If the heart is beating strongly it imparts a big pulse-wave to the arterial system and the vessel is felt to undergo considerable increase in size. A weakly beating heart, on the other hand, produces only a small pulse-wave which can scarcely be felt.

(*b*) The **tension** of the pulse is dependent on the blood pressure. If the blood pressure is low the pulse is easily compressed, thereby stopping the circulation. With a high blood pressure, on the other hand, more force is required to compress the pulse. With practice it is possible to say whether a pulse is of normal, low or high tension.

(*c*) The **state of the artery**. In health the artery is a soft elastic structure which, when the pulse is stopped above, can hardly be felt by the fingers. In the condition known as arteriosclerosis, lime salts are deposited in the wall of the artery which becomes hard and tortuous.

All these features of the radial pulse can be determined by practice, and it is very important to acquire an accurate knowledge of the normal pulse in order that any abnormal feature may be readily recognized when they are present.

Cardiac output. The cardiac output (CO) is the volume of blood pumped out of the left ventricle in one minute. Clearly it is the amount pumped in one stroke multiplied by the number of beats or strokes per minute:

i.e. CO = stroke volume × heart rate.

The normal resting cardiac output is about 5 litres per minute. Exercise, emotion and fever increase the cardiac output.

Some diseases such as thyrotoxicosis and anaemia increase the cardiac output, while others, e.g. myocardial infarction and aortic stenosis may cause a low output.

Nervous control of the blood vessels

Most of the arteries of the body are directly under the control of the autonomic nervous system. Special centres, known as the **vaso-motor centres**, exist in the hypothalamus and medulla oblongata to exercise this control. Two sets of nerves are present.

1. Vasoconstrictor.
2. Vasodilator.

These nerves may affect the whole of the circulatory system at a time, or their action may be limited to a localized organ or part.

Vasoconstrictor nerves. As their name suggests, these nerves narrow the lumen of a blood vessel and thereby diminish the amount of blood to the part or organ which it supplies. If all the vaso-constrictor nerves are sending out impulses, a general effect will be produced, the whole of the arterial system will be narrowed and therefore the blood pressure will be raised (see page 158).

Vasodilator nerves. These act by dilating the blood vessels and allowing a greater blood supply to the organ. Thus, the blood vessels of the alimentary tract are dilated by this action during the process of digestion; also during muscular exercise the blood vessels to the muscles are dilated so that they are able to carry more blood.

In the serious condition known as 'shock' the blood vessels as a whole become dilated and, in consequence, the blood pressure falls. One of the methods of treatment is to give drugs which stimulate the α-adrenergic receptors causing the arteries to narrow, thereby raising the blood pressure.

NOTE: Most sympathetic impulses stimulate vasoconstriction and parasympathetic impulses result in vasodilatation.

An example of the nervous control of blood vessels which every-one has either felt or observed is the phenomenon of **blushing**. This is a purely local modification of the circulation and variation in the amount of blood in the skin of the face, and surrounding parts if the blush is extensive. An emotion, pleasure, embarrassment, disgust, or offended modesty perchance, possesses the mind and with inconsiderate haste the skin grows red and a hot flush is felt. This is due to the conscious nervous system affecting the vasomotor centre over which the individual has no control. Vasodilator impulses are sent out and the small arteries in the skin of the affected part dilate, bringing an excess of blood to the surface.

Conversely, in extreme terror or rage the face may become very pale and cold ('white with rage'). In this case the vasoconstrictor nerves are limiting the flow of blood through the arteries.

QUESTIONS

1. Trace the course of the blood from the left ventricle, through the vessels of the stomach and back to the left atrium.
2. Trace the course of the circulation of the blood from the foot to the heart and back to the foot again.
3. Give the names and positions of the chief blood vessels of the limbs.
4. Compare and contrast the structure of a vein and an artery. How is venous blood returned to the heart from the limbs?
5. Give an account of the structure of the heart. How is arterial blood supplied to the upper limb?
6. Name the main branches of the abdominal aorta. How is blood returned from the intestines to the heart?
7. Describe the aorta and enumerate its main branches.
8. Give an account of the principal blood vessels entering and leaving the heart.
9. Give an account of the position and structure of the heart. What factors influence the heart rate?
10. Give an account of the cardiac cycle. How does exercise affect the heart and circulation?

8 The Blood (Haemopoietic system)

The blood plays a very important part in the maintenance of life. It flows throughout the body and, when in the capillaries, it is in intimate relationship with the tissues, taking oxygen and other nutritive substances to them and at the same time removing their waste products.

The total volume of blood in the body is about 6 litres (10 pints) and constitutes about one-twentieth of the total body weight.

Composition of blood

When blood is examined under the microscope it is seen to consist of cells or corpuscles floating in a yellowish fluid, the blood plasma Normal blood contains approximately:

	per cubic millimetre
red corpuscles	5 000 000
white cells	8 000
platelets	250 000

Red corpuscles or erythrocytes

These are the most numerous cells in the blood and contain a substance called haemoglobin, which gives them their red colour. It is actually correct to refer to them as corpuscles rather than cells, for the term cell implies the presence of a nucleus. Red corpuscles, however, have no nucleus. They are flattened, circular discs consisting of a thin membrane or envelope, the centre of which is depressed so that their two outer surfaces are concave. (They are, therefore, circular, bi-concave, non-nucleated discs.) They are very small, having a diameter of 7·5 thousandths of a millimetre (7·5μ) or 1/3200 inch. They are, therefore, able to pass through the capillaries. Their function is to carry oxygen.

Haemoglobin. This is a complex protein which gives the red colour to the erythrocytes. It consists of a protein, globin, combined with an iron-containing pigment called haem. Iron, therefore, is essential for the formation of haemoglobin, and this is of great importance in the treatment of anaemia. The normal red corpuscle is said to contain 100 per cent of haemoglobin (approximately 15 grams per 100 ml of blood). If deficient in haemoglobin, the percentage is lower.

Haemoglobin has a very strong affinity for oxygen, and when they come into contact the oxygen is absorbed, forming oxyhaemoglobin. This process normally takes place in the lungs where the

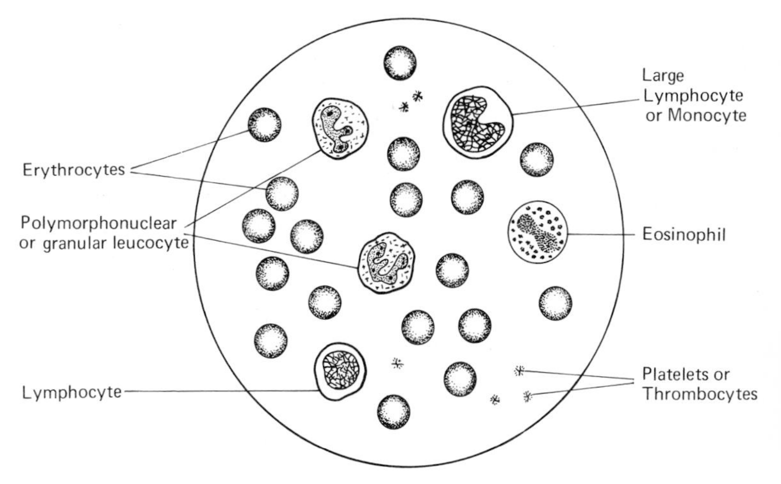

Fig. 121 Blood cells.

venous blood, deficient in oxygen, is able to absorb oxygen from the air in the alveoli and leave the lungs by the pulmonary veins as arterial blood.

Oxygenated arterial blood has a bright red colour, while that in the veins, having lost its oxygen, has a bluish-purple hue.

Haemoglobin also has a strong affinity for the poisonous gas, carbon monoxide, with which it forms carboxyhaemoglobin. Haemoglobin bound this way cannot carry oxygen and the victim may die of anoxia. Carboxyhaemoglobin causes the individual to develop a cherry-red hue in contrast to the cyanosis of anoxia.

Development. The erythrocytes are formed in the red bone marrow. In the infant the cavities of all the bones contain red marrow. In adults much of this is replaced by yellow marrow and erythrocytes are mainly produced in the red bone marrow of the sternum, ribs, vertebrae, cranial bones and the proximal epiphyses of the femur and humerus. They commence as large cells having nuclei (called proerythroblasts). Their next stage is a smaller cell, the **normoblast**, which also contains a nucleus and is red in colour, because haemoglobin is present in its substance. The normoblast loses its

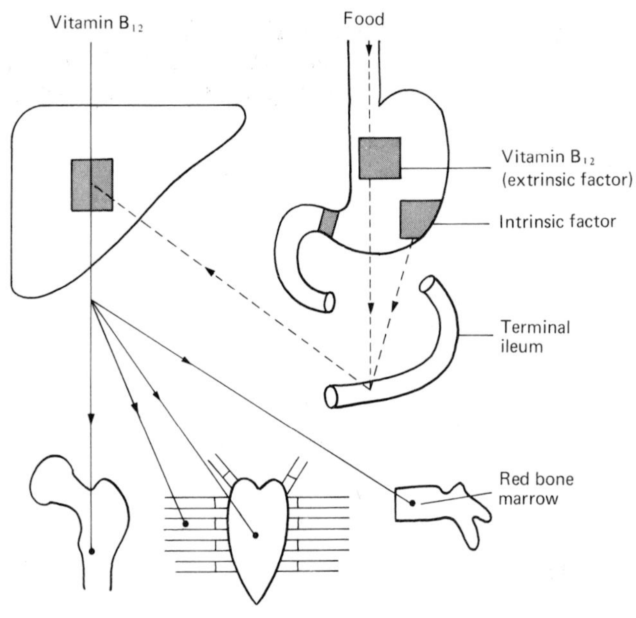

Fig. 122 The absorption of vitamin B_{12}, its storage in the liver, and its distribution to red bone marrow.

nucleus and passes into the circulation as the mature erythrocyte. In order that these changes may take place:

(*a*) Vitamin B_{12} (the erythrocyte-maturing factor cyanocobalamin) and folic acid are necessary for the development of the proerythroblast into the normoblast (see page 217).

(*b*) Iron is necessary for the provision of haemoglobin to fill the normoblasts before they can become the mature red corpuscle. Lack of iron in the diet or loss of iron in bleeding causes iron-deficiency anaemia.

(*c*) In addition to a normal diet containing protein and iron, small amounts of cobalt, copper, vitamin B (nicotinic acid and riboflavine) and vitamin C are also necessary.

The life of a red cell is limited, probably to about four months.

When worn out, it is destroyed by the spleen, which, however, saves and stores the iron for use in the preparation of fresh haemoglobin.

Cyanocobalamin, vitamin B_{12}

This is a substance, occurring mainly in animal proteins and bacteria but absent from vegetables, which is necessary for the development of the proerythroblast into the mature erythrocyte. Its absence leads to the development of a type of anaemia known as **pernicious anaemia**, the study of which has helped in the understanding of red cell formation.

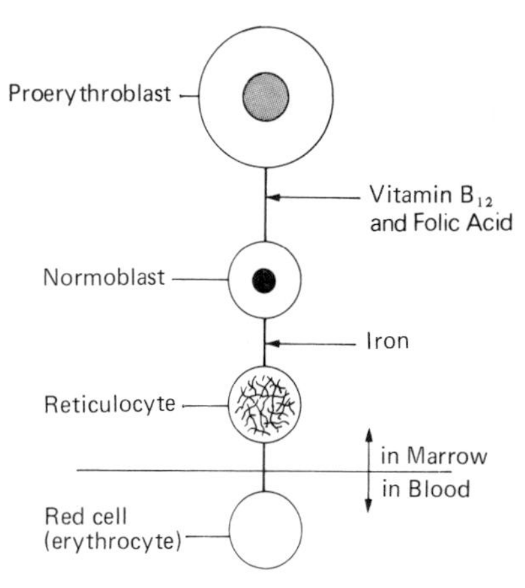

Fig. 123 Diagram illustrating the development of the red cell (erythrocyte) from the proerythroblast.

Although it was over 100 years ago that Addison first described this type of anaemia, it was not until Minot and Murphy in 1926 showed that it could be cured by giving uncooked liver that its cause began to be understood. Ten years later Castle showed that a special blood-forming factor was necessary for the maturing of the proerythroblast into the normoblast. He further suggested that before this could be used another factor was necessary and that this was secreted by the mucous membrane of the stomach. He, therefore, called vitamin B_{12} the **extrinsic factor** because it was present in the diet and that secreted by the stomach the **intrinsic factor**.

It is now known that vitamin B_{12} cannot be absorbed from the lowest part of the small intestine in the absence of the intrinsic factor with which it probably forms temporarily some type of compound.

In 1948 it was found that from over a ton of liver about 1 gram of minute red crystals could be obtained which had the same effect as the whole liver. Later is was discovered that this substance could be prepared commercially by other methods and it was identified as vitamin B_{12} and shown to contain traces of the metal cobalt. Regular injections of vitamin B_{12} are now the recognized treatment of pernicious anaemia.

It is clear that in pernicious anaemia the absence of intrinsic factor from the gastric juice leads to failure of vitamin B_{12} to be absorbed and stored in the liver whence it is conveyed to the bone marrow. Its absence results in defective formation of the red cells in the bone marrow and the proerythroblasts fail to develop into normoblasts, so that anaemia results. In other words, the number of red cells is considerably reduced, often to between 1,000,000 and 2,000,000 per cubic millimetre, while the haemoglobin content of each cell remains relatively high.

Folic acid is also necessary for normal red cell production. It is found in green vegetables, liver, kidney and yeast. Its deficiency results (like cyanocobalamin deficiency) in a megaloblastic anaemia.

White cells

These are colourless cells containing nuclei and are a little larger in size than the red cells (10μ, 1/2500 inch) but much less numerous (1 WBC to 500 RBC). Their actual number varies between 4000 and 11000, but 8000 per cubic millimetre is a normal average.

Two main varieties are found in the blood:

1. The polymorphonuclear leucocyte (75 per cent).
2. The lymphocyte (25 per cent).

The polymorphonuclear or granular leucocyte. This is so called because its nucleus is irregular and variable in shape and its cytoplasm contains granules. It forms about 75 per cent of the white cells. The distinguishing feature of these cells is their power of independent movement; in this respect they resemble the amoeba.

They have two important functions:

1. To protect the body against the invasion of bacteria.
2. To remove dead or injured tissue.

If bacteria enter the tissues, the granular leucocytes immediately attempt to surround the organisms and destroy them by taking them into their own bodies. Sometimes the toxins of the bacteria are powerful enough to kill the leucocytes and it is the accumulation of leucocytes destroyed in this manner, together with liquefied dead tissue, which forms pus.

After the bacteria have been killed it is the leucocytes which remove the tissues which have been damaged or destroyed by the action of toxins and, thus, they play their part in the first stages of the process of healing.

It is because of this power of eating or ingesting bacteria and dead matter by means of various enzymes that the granular leucocytes are sometimes called **phagocytes**, and the process they carry out, **phagocytosis**. This is similar to the way in which the amoeba takes in food (Fig. 6, page 12).

Development. The polymorphonuclear leucocytes or granulocytes are derived from special cells in the bone marrow (myeloblasts).

Varieties. If special dyes are used it is possible to show several varieties according to the actual stains which the granules in each take. Some stain evenly and are called **neutrophile** cells; others take a red, acid stain and are called **eosinophile** cells; a few pick up an alkaline dye and are referred to as **basophile** cells. The number of eosinophiles is sometimes increased in allergic states such as asthma, certain skin diseases and infections with intestinal parasites.

The lymphocytes. These are non-granular cells and constitute about 25 per cent of the total number of white cells. They are distinguished from the polymorphonuclear cells by having a large round nucleus. Two types, large and small, are recognized. They play some part in the defence of the body against infection, but do not show the same power of phagocytosis as the polymorphonuclear cells, nor is their function completely understood.

Lymphocytes are concerned in the immunity mechanisms of the body. Hence the use of anti-lymphocytic serum (ALS) to prevent rejection of the foreign tissues introduced in an organ-transplant operation.

Development. Lymphocytes are derived from the lymph glands, the thymus and the spleen and other masses of lymphoid tissue in the body. Their total number is relatively increased in infancy.

Leucocytosis. In almost every infection involving the body Nature responds by increasing the number of granular leucocytes circulating in the blood. This is called leucocytosis, and the number of cells may rise to 20,000 or 30,000 per cubic millimetre, especially if an abcess (pus) is present.

In some conditions, such as pneumonia, if the body fails to produce a leucocytosis the outlook becomes serious. Virus infections do not often cause leucocytosis but often cause a leucopoenia.

If the leucocytes are diminished in number the term leucopoenia is used. This may occur when the bone marrow is depressed by the action of some drugs, toxins and also radiation.

In leukaemia there is a proliferation of abnormal leucocytes and a deficiency of normal blood cells.

The blood platelets (thrombocytes). These are minute spherical structures found in the blood, numbering about 250,000 per cubic millimetre. They are produced by large cells (megakaryocytes) present in the bone marrow. They are smaller than the red corpuscles and their main function is concerned with the clotting of blood. In certain diseases their number may be seriously decreased (thrombocytopoenia), and the patient shows a tendency to bleeding into the skin (purpura); producing either small or even larger bruises. Often bleeding from the mucous membranes also occurs.

The blood plasma

The yellowish; slightly alkaline fluid in which the blood corpuscles float is called the plasma. It has the following composition:

Proteins (7 grams per 100 ml) serum albumin (4·5 G) serum globulin (2·5 G), and fibrinogen (0·3 G).
Salts { including chlorides, sulphates, phosphates of sodium, potassium and calcium.
Urea (20–40 mg per 100 ml).
Glucose (70–120 mg per 100 ml).
Water (90%).
Other substances such as prothrombin, vitamins, enzymes and antibodies to disease are also present.

It will be noticed that plasma contains three types of protein:

(*a*) Serum albumin (4·5 per cent) and serum globulin (2·5 per cent).
(*b*) Fibrinogen (0·3 per cent).

By various means the fibrinogen can be removed from the plasma. The fluid then remaining is called serum, i.e.

$$\text{plasma} = \text{serum} + \text{fibrinogen}$$
or
$$\text{serum} = \text{plasma} - \text{fibrinogen}.$$

It is important to understand what serum is, because specially prepared types may be used in the prevention and treatment of disease. It will be noticed that, since it consists of plasma from which fibrinogen only has been removed, it will still contain the valuable antibodies to disease which are mainly carried in a special protein called immuno-globulin.

The total amount of protein may be reduced in certain diseases. This is specially likely to occur in chronic nephritis when large quantities of albumin are lost in the urine. In this instance the serum albumin is particularly lowered. This loss of protein from the blood is an important factor in the production of oedema in this condition because it results in a lowering of the osmotic pressure of the blood.

Blood has a slightly salt taste, which a patient may mention if there has been any haemorrhage into the mouth.

The clotting of blood

Since man is liable to injury and the shedding of blood, a mechanism is provided within the body whereby there is a spontaneous tendency for the loss of blood to be limited.

The actual mechanism is a complicated one, but the general principles are simple and important. Shortly after being shed, the blood becomes sticky and sets into a jelly-like mass. After a few hours this mass contracts and from it is squeezed a yellowish fluid, the serum, in which the mass or clot floats.

The essential change in the clotting of blood is the conversion of the protein fibrinogen into a substance called fibrin. This forms fine threads which entangle the blood cells and then contract. It is this contraction of fibrin which expresses the serum and binds the clot into a firm mass.

Since blood consists of cells and plasma (serum + fibrinogen), it follows that, if serum is expressed, the remaining clot must be formed of cells + fibrin (fibrinogen having been converted into fibrin during the process of clotting).

This may be expressed simply in the following way:

$$\text{blood} = \begin{cases} \text{corpuscles (red and white)} \\ + \\ \text{plasma} = \begin{cases} \text{fibrin} \\ + \\ \text{serum} \end{cases} \end{cases} \quad \text{corpuscles (red and white)} + \text{fibrin} = \text{clot}$$

The formation of this firm mass in a wound acts as a plug and effectively seals off the opened blood vessels, thereby preventing further loss of blood.

Certain conditions hasten the clotting of blood, while others retard it. These often have a practical importance.

Factors hastening clotting

1. The presence of calcium salts. These salts are sometimes given in conditions where there is a tendency to excessive bleeding.
2. Injury to the tissues or vessel wall. A clean cut with a sharp knife or razor bleeds much more freely than a crushed wound

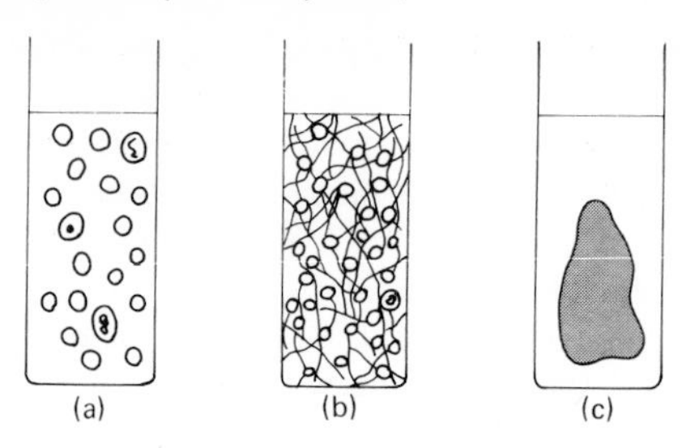

Fig. 124 Diagram illustrating clotting of blood. (*a*) Cells floating freely in plasma. (*b*) Cells entangled in threads of fibrin. (*c*) Clot floating in serum (clot retraction).

in which there is considerable bruising and damage to the surrounding tissues.

3. Contact with a foreign body. The application of a surgical dressing such as gauze aids very considerably in the speedy formation of a clot and arrest of haemorrhage.
4. Temperature slightly higher than that of the body. Use of this fact is often made during surgical operations when bleeding surfaces are packed with swabs soaked in hot saline 49°C (120°F).

Factors retarding clotting

1. The addition of sodium or potassium citrate. This acts by removing the activity of calcium salts. It is very frequently used to prevent blood clotting in blood transfusions. The blood from the donor is taken directly into a solution of potassium citrate (3·8 per cent) and may be kept for very long periods without risk of clotting.
2. Contact with oil, grease or paraffin wax. This means that ointments and greasy materials should not be applied wounds until bleeding has ceased. An old-fashioned method of blood transfusion was to take the blood into a bottle which had been previously coated with paraffin wax.
3. Local cold.

Certain other substances are used to alter the coagulability of the blood. A substance which can be obtained from the liver, called heparin, prevents coagulation, and drugs such as phenindione (Dindevan) produce a similar effect. The leech secretes a substance (hirudin) that delays clotting, which explains why bleeding continues for so long after a leech bite. The poison of certain snakes (especially viper venom) hastens clotting and is sometimes used for this purpose.

It may have been noticed that a substance called adrenaline is sometimes applied to bleeding surfaces in order to arrest haemorrhage. This, however, has no actual effect on the mechanism of clotting, but acts by causing the blood vessels to contract, thereby diminishing the actual escape of blood (see page 297).

Certain factors are necessary for the conversion of fibrinogen into fibrin.

1. The presence of platelets (thrombocytes).
2. The presence of calcium salts which are normally found in the blood.
3. Two substances:
   (*a*) prothrombin found in the blood,
   (*b*) thrombokinase (thromboplastin) formed by platelets and also by damaged tissues.

The main details of clotting may be summarized in the following way and may be expressed diagrammatically. In the presence of calcium salts the substance prothrombin which is present in the blood is converted into thrombin by the action of thromboplastin. The latter substance is derived

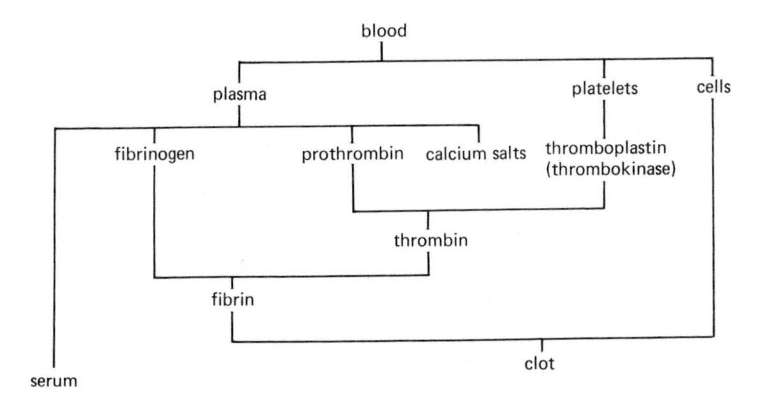

from the blood platelets and the damaged tissues. Vitamin K is necessary for the formation of prothrombin. Thrombin then acts on the fibrinogen of the plasma and converts it into fibrin, which traps the red cells and platelets in its meshes, forming a clot from which serum is expressed by contraction of the fibrin threads.

SUMMARY OF THE FUNCTIONS OF THE BLOOD

1. To convey oxygen to the tissues by means of the haemoglobin in the red cells.
2. To remove waste products from the tissues and convey them to the appropriate organs for excretion, e.g.:
   (*a*) carbon dioxide is carried to the lungs,
   (*b*) urea is carried from the liver to the kidneys for excretion, and
   (*c*) water is carried to the kidneys, lungs and skin where any excess is removed.
3. To carry nourishment to all parts of the body. This includes not only digested foodstuffs but also vitamins.
4. To carry the hormones or chemical messengers of the body, i.e. the internal secretions of the ductless glands (e.g. insulin from the pancreas).
5. To carry antibodies to disease in the immuno- (gamma) globulin of the serum.
6. To aid in the defence of the body by the phagocytic action of the white cells.

Blood groups

In connection with the important procedure of blood transfusion, individuals may be divided into four main (A, B, O) groups and it is essential that the correct type of blood is employed in each transfusion. The table that follows shows the groups and the names given to them.

The division of persons into these four groups depends on the following facts:

1. Human **blood serum** may contain substances called **agglutinins**; that, is substances which have the power of causing the red cells of persons belonging to another group to run together in clumps or agglutinate if they are mixed with this serum.

Name	*Percentage of persons*
AB	5
A	40
B	10
O	45

2. Human **red cells** may contain substances called **agglutinogens**; that is, substances which have the power of stimulating the production of agglutinins.

   These substances are designated in the following way:

The agglutinins are called α (alpha) and β (beta).
The corresponding agglutinogens are called A and B.
Human serum may contain α or β or both (α + β) or no agglutinins.
Human red cells may contain A or B or both (A + B) or no agglutinogens (O).

It follows that if a person possesses red cells belonging to Group AB his serum cannot contain α or β agglutinins, otherwise his own red cells would agglutinate and life would be impossible. In the case of an individual of Group A, his red cells will contain the agglutinogen A and his serum the agglutinin β which will not affect them. If both the agglutinins α and β are present in the serum the red cells can contain no agglutinogen and the individual will belong to Group O.

In other words:

Group	*Cells contain*	*Serum contains*
AB	A and B agglutinogens	no agglutinins
A	A agglutinogens	β agglutinins
B	B agglutinogens	α agglutinins
O	No agglutinogens	α and β agglutinins

The main point in blood transfusion is that the cells of the donor must be compatible with the serum of the recipient who receives them. (For practical purposes, the effect of the serum of the donor on the cells of the recipient may be ignored as it becomes so well diluted in the act of transfusion that, at the worst, only minor ill-effects can be produced.) In the following table which sets out the principles on which blood transfusions are based:

+ = agglutination of donor's cells by recipient's serum.
– = no agglutination of donor's cells by recipient's serum.

It will be seen that individuals of Group AB have serum containing no agglutinins and so they can receive cells from any other group by transfusion. They are therefore sometimes called universal recipients.

Recipient's serum		*Donor's cells (agglutinogens)*			
Group	*Agglutinins*	AB	A	B	O
AB	nil	–	–	–	–
A	β	+	–	+	–
B	α	+	+	–	–
O	α and β	+	+	+	–

On the other hand, those of Group O have red cells which contain no agglutinins so that their cells can produce no reactions in members of the other groups. Persons having Group O blood are called universal donors.

Group A and B can only receive blood from their own groups or Group O. Group O patients can only receive Group O blood, since A and B agglutinogens in donor blood would be agglutinated by the recipient's α and β agglutinins.

In practice, the group to which an individual belongs is ascertained by testing their cells against the serum of a known group. Both the receipient and the donor having been appropriately labelled, it is customary to take the added precaution of testing the serum of the recipient against the corpuscles of the donor to confirm that the mixture is compatible (cross-matching).

If incompatible blood is transfused, very serious symptoms may follow. The foreign red cells are first of all agglutinated into clumps which may block capillaries in various parts of the body. Later, these clumps are broken down and free haemoglobin is liberated into the blood stream (haemolysis). This is partly converted into bilirubin by the liver and will produce jaundice. Part, however, is excreted by the kidneys and may block the renal tubules, especially if the urine is acid, leading to suppression of urine and kidney failure which may prove fatal. Pain may be complained of early in the transfusion, and this should be an indication to discontinue the procedure before too much damage has been done.

Another interesting feature is that blood groups are inherited, the important factors being the A and B agglutinogens. This is sometimes of importance in proving that a child could not be the offspring of an alleged father.

The Rhesus (Rh) factor

This substance was first discovered in experiments on the rhesus monkey and was consequently given the name of 'the rhesus factor', abbreviated to Rh.

Further experiment showed that this agglutinogen was also present in the red cells of 85 per cent of human beings, who can therefore be divided into two groups:

Rh positive = 85 per cent
Rh negative = 15 per cent

Under certain circumstances antibodies may develop in the blood of an individual which are capable of causing the agglutination and destruction of Rh positive cells. These antibodies are called anti-Rh.

If two parents are both Rh positive their offspring will be Rh positive. If one of the parents is Rh positive their offspring will probably be Rh positive but, if the mother is Rh negative and her child is Rh positive, in a certain number of cases the mother then becomes sensitive to the positive factor in the child's blood and she develops the anti-Rh bodies mentioned above.

father Rh + ve plus mother Rh − ve → child Rh + ve
mother rh − ve plus child Rh + ve → mother Rh − ve with anti-Rh.

In any future pregnancies this anti-Rh may affect the Rh + ve red cells of the foetus, causing their agglutination and destruction, if it passes through the placenta and enters the foetal circulation. This condition is known as erythroblastosis foetalis and results in severe and often fatal jaundice (icterus gravis neonatorum). It is also responsible for hydrops foetalis and some cases of repeated stillbirths or miscarriages.

Father	Mother	Child	
Rh + ve	Rh + ve	Rh + ve	= normal
Rh − ve	Rh + ve	Rh + ve	= normal
Rh − ve	Rh − ve	Rh − ve	= normal
Rh + ve	Rh − ve	Rh + ve	= 4% risk of abnormality in first child.

A similar state of affairs can be produced in the maternal blood by transfusion. If a woman who is Rh negative is transfused with blood of the correct ordinary group but which is Rh positive she may also develop anti-Rh bodies in her blood. If she subsequently has an Rh positive child (from an Rh positive father), the child may be affected by erythroblastosis fetalis. In other words, the commonest cause of haemolytic disease in the new born is when Rh negative women form antibodies against the Rh positive foetal blood cells.

It follows that all Rh negative women of the childbearing age should only be transfused with Rh negative blood if this risk is to be avoided. This is especially important during pregnancy and the puerperium.

These are the general principles which apply to the subject of the Rh factor, but the subject is complicated by the existence of a number of sub-groups.

Immunity

One of the functions of the blood is to carry antibodies to some infections in the immunoglobulin of the serum.

Immunity may be:

1. natural
2. acquired

1. Natural immunity is inborn and is conveyed during pregnancy from the mother to the child.
2. Acquired immunity results from recovery from attacks of some diseases during which the body of the individual produces anti-bodies in excess of those required to overcome a particular infection and which remain in the serum and tissues. They are then available to combat immediately any subsequent invasion by the original organism. Against some organisms such resistance is lifelong; in other instances the immunity is of shorter duration.

In many cases an active immunity can be produced by artificial means by the use of a vaccine consisting either of dead germs or their toxins. Such immunity lasts for a relatively short time and further doses of vaccine are required at regular intervals.

In a few instances, antibodies can be injected in either human or horse serum which contains the appropriate antibodies (passive immunity).

NOTE: Further details of this important subject on the prevention and treatment of infections are available in textbooks of Medicine etc.

QUESTIONS

1. What do you understand by clotting of blood? What conditions (*a*) hasten, (*b*) hinder, clotting?
2. Describe the composition and functions of the blood.
3. Give an account of the composition of blood. What substances in the diet are necessary for the formation of blood?

9 The Lymphatic System

It has been seen that all the tissues of the body are permeated by a vast network of capillaries containing blood. The walls of the capillaries consist of a single layer of cells and, except for the white cells which at certain times are able to make their way through these walls, the blood does not actually come into direct contact with the tissues.

The whole of the tissues, however, are bathed in fluid called **tissue fluid**. This may be regarded as a sort of 'middle man' between the blood on one hand and the tissues on the other, and all interchange of nourishment and waste products between them takes place through the medium of the tissue fluid.

The lymphatic system is, therefore, a subsidiary or second circulatory system, which drains the tissue fluids. From the tissue spaces the tissue fluid passes into narrow vessels (lymphatic vessels) which consist of extremely fine endothelial cells. These unite to form channels which ultimately, as will be seen, rejoin the general circulation.

Strictly speaking, therefore, lymph is the name given to the tissue fluid when it has entered the lymphatic vessels. The larger lym-

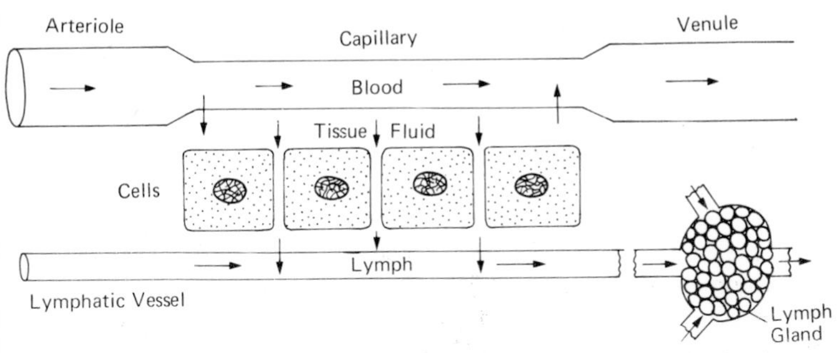

Fig. 125 Circulation of tissue fluid and lymph.

phatic vessels resemble small veins in structure and are provided with valves to prevent back flow.

Lymph itself may be looked upon as part of the plasma which has passed by diffusion from the blood through the walls of the capillaries.

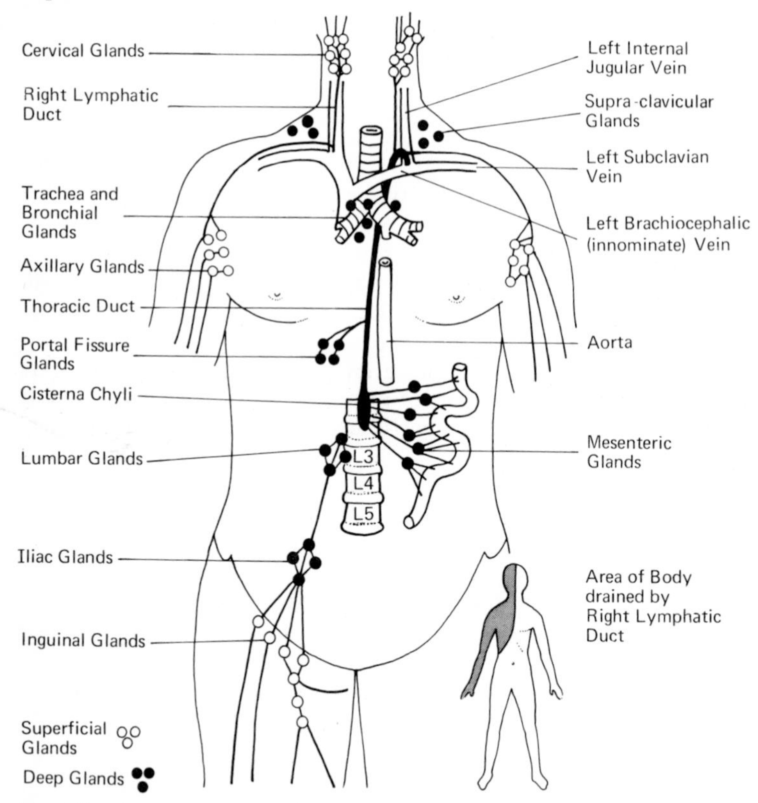

Fig. 126 Diagram of the thoracic duct showing position of main groups of lymphatic glands.

The lymphatic vessels

The smaller channels situated within the tissues unite to form larger vessels which converge into two large ducts. The most important of these is the **thoracic duct**. This duct is 40 cm (15 in) long and commences as a dilated sac, the **cisterna** (receptaculum) **chyli**, into which the lymph vessels from the lower limbs and the contents of the abdominal cavity (particularly the intestines) empty their

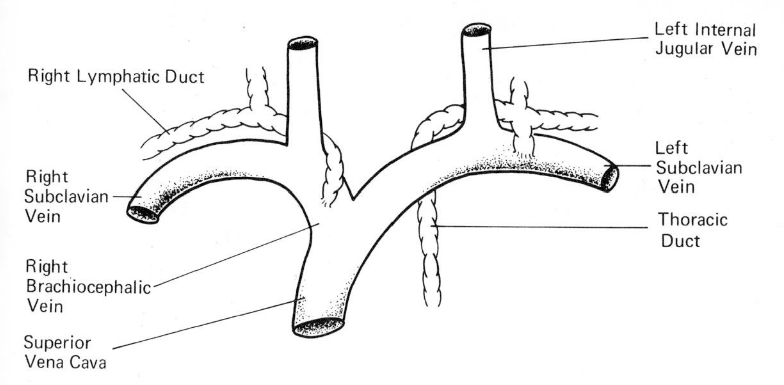

Fig. 127 Diagram illustrating the return of lymph to the great veins.

lymph. The cisterna chyli is situated in front of the upper lumbar vertebrae on the right of the abdominal aorta and lies deep in the epigastrium. From it, the thoracic duct passes upwards in front of the left side of the thoracic vertebrae in the thorax and ends by joining the left brachiocephalic vein in the root of the neck. Before its termination it is joined by the lymphatics from the left arm and left side of the head and neck.

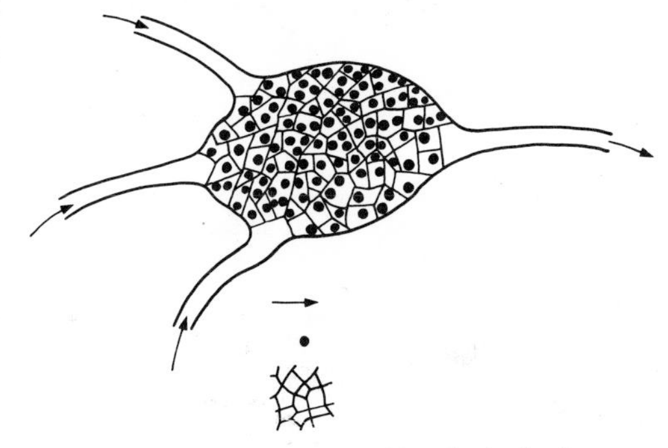

Fig. 128 Diagram of lymphatic gland.

The lymph from the right side of the head and neck, the right arm and the right side of the chest wall enter the right innominate vein by the **right lymphatic duct**.

By means of the thoracic duct and the right lymphatic duct, therefore, the whole of the lymph of the body returns to the blood. There is thus a constant circulation of lymph from the capillaries into the tissue spaces and back again into the blood stream.

The lymphatics of the intestine

These are of special importance because all the digested fat is absorbed through them. It will be seen later (page 210) that the minute projections from the mucous membrane of the intestine called villi each contain a small lymphatic vessel into which the digested fat is absorbed. These channels are called **lacteals** and ultimately all pass to the cisterna chyli. The lymph reaching the thoracic duct from the intestines differs from the lymph from other parts of the body in that it contains an excess of fat absorbed from the intestines, giving it a milky appearance. It is called **chyle,** the term cisterna chyli meaning 'the receiver of the chyle'.

The lymphatic glands

Situated on the course of the lymphatic vessels and generally occuring in groups are the lymphatic glands. These consist of a fibrous framework containing round cells which appear to be identical with the lymphocytes found in the blood, but when collected together in this way are described as lymphoid tissue.

The lymph glands act as a filter for the lymph. The vessel bringing lymph to the gland is called the **afferent vessel.** The lymph passes through the substance of the gland and comes in contact with the lymphoid tissue; it is then collected up and leaves the gland by an **efferent vessel.** Lymph may pass through several groups of glands before it is actually returned to the blood stream.

There are many groups of lymphatic glands in the body. Among the most important are:

Glands	*Where situated*
SUPERFICIAL	
cervical	in neck
axillary	in axilla
inguinal	in groin
DEEP	
iliac	in iliac fossa
lumbar	close to lumbar vertebrae
thoracic	at root of lung
mesenteric	in mesentery of intestines
portal	in portal fissure of liver

Lymphoid tissue elsewhere in the body

Although a great deal of the lymphoid tissue is situated in the lymphatic glands, there are a number of other collections of this tissue.

1. The tonsils (page 185).
2. A mass on the posterior wall of the pharynx which when enlarged constitutes 'adenoids'.
3. Scattered patches in the small intestine called Peyer's patches (page 210). (These become involved in typhoid fever.)
4. The spleen.
5. The appendix.

PRACTICAL IMPORTANCE. The lymphatic vessels and glands are of great importance in medicine and surgery. Owing to the intimate contact of the lymph with the tissues it is clear that products of inflammation of the tissues including also bacteria will be able to enter the lymph-stream. These products of inflammation are carried to the glands which act as filters and prevent them from reaching the general circulation. It is for this reason that a septic finger, for example, produces tender swelling of the glands in the axilla. Sometimes the bacteria are very active and powerful and are able to destroy the gland, producing an abscess. Inflammation of the lymphatic vessels is called lymphangitis, and in this condition they may be seen as fine red streaks in the skin, running from an infected area to the nearest group of lymph glands.

Cancer cells may also be carried by the lymphatics to the glands where they are held up, but unfortunately they are able to grow within the glands and produce a secondary growth of cancer in them.

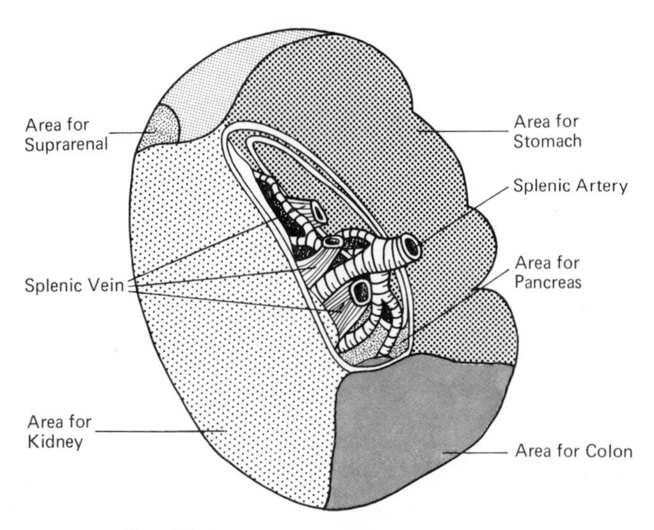

Fig. 129 The medial or visceral aspect of the spleen.

The spleen

The spleen is a dark purple-coloured organ lying in the left hypochondriac region of the abdomen behind the lower ribs and costal cartilages. It is about 13 cm (5 in) long, 8 cm (3 in) broad and weighs about 200 g (7 oz). Its upper surface is in contact with the under surface of the diaphragm. Its posterior surface is related to the anterior part of the left kidney. On its medial aspect it is in contact with the stomach and at its tip or lower pole is the splenic flexure of the colon.

On the medial aspect is the **hilum**, where the splenic artery, a branch of the coeliac axis, enters. The splenic vein also leaves the spleen at the hilum and helps to form the portal vein. The tail of the pancreas extends to the hilum of the spleen.

Structure. The spleen consists of a supporting structure of fibrous tissue while the mass of the gland, the splenic pulp, is made up of lymphoid tissue. The organ is surrounded by a fibrous capsule.

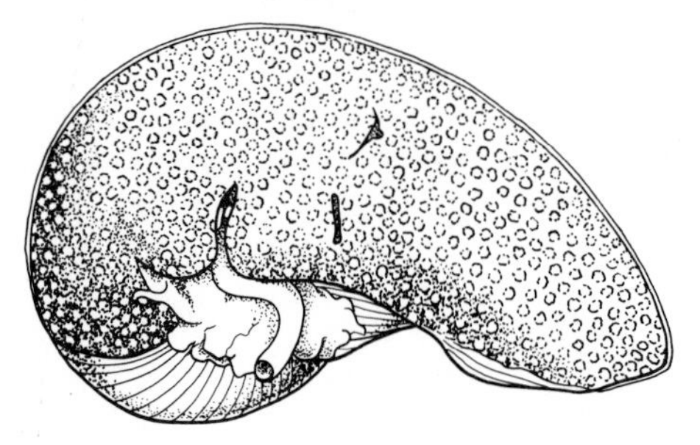

Fig. 130 The naked-eye appearance of a section of the spleen.

Functions. The spleen is sometimes described as a gland, but as it is not known to produce any important internal secretion it is better considered as a lymphatic organ.

1. It is not essential to life, for the spleen can be removed without any permanent deterioration in health.
2. The spleen destroys worn-out red blood corpuscles, removes and probably stores iron from their haemoglobin. From the remaining pigment it forms some of the bilirubin, which the liver also helps to form and secrete in the bile.
3. It plays some part in the production of antibodies and the defences of the body against infection and is often found to be enlarged in disease.
4. It provides some of the lymphocytes of the blood.

5. In some animals it acts as a reservoir for red cells. This function is probably of minor importance in man.
6. It may sometimes destroy platelets.

The spleen may be ruptured by injury, which may be trivial if it is enlarged by disease. Rupture may be difficult to diagnose but will result in severe internal haemorrhage and shock.

The reticulo-endothelial system. The term reticulo-endothelium denotes a system of special connective tissue cells which are present in the spleen, liver, lymphoid tissue and bone marrow and also other parts of the body. They act as phagocytes and can ingest particles of foreign material, some have the power of movement and they play an important part in the defences of the body against infection. It is the reticulo-endothelial cells in the spleen which remove the worn-out erythrocytes. Those in the liver convert haemoglobin into bile pigment.

The cells of this system may be affected by disease, e.g. in leukaemia and Hodgkin's disease.

The tonsils

The tonsils are two oval masses of lymphoid tissue lying in their bed in the side walls of the oral pharynx, between the anterior and posterior pillars of the fauces. Their size varies, but they tend to be relatively larger in children than in adults. The lower pole of each tonsil is continuous with lymphoid tissue situated in the base of the tongue. On the surface of the organ, small openings can be seen which are the mouths of small pits in its substance called **crypts**.

The lymphatic vessels of the tonsils drain into glands situated below the angle of the jaw and into the deep cervical glands. Their blood supply is derived from branches of the external carotid artery.

The tonsils form part of a ring of lymphoid tissue guarding the entrance of the alimentary and respiratory tracts against bacterial invasion, viz. the tonsils, the lymphoid tissue in the pharynx (adenoids) and that in the base of the tongue. They are frequently the site of inflammation (tonsillitis). The formation of an abcess in the bed of the tonsil is called a quinsy.

SUMMARY OF FUNCTIONS OF LYMPHATIC SYSTEM

1. Lymph, a product of the tissue fluid, is the 'middle man' conveying nourishment and oxygen from the blood to the cells and removing their waste products to the blood.

2. The lacteals in the intestine absorb digested fat and carry it via the cisterna chyli to the thoracic duct.
3. The lymphatic glands act as filters, and remove bacteria from the lymph in cases of infection.
4. Special collections of lymphoid tissue constitute the spleen, tonsils and Peyer's patches.
5. It supplies the lymphocytes to the blood.

QUESTIONS

1. What are lymph glands? Where are they found in the body?
2. Describe the spleen and discuss its functions.
3. Give a description of the thoracic duct. What are (*a*) lymph, (*b*) serum, (*c*) plasma?
4. Describe the lymphatic system and outline its functions.

10 Anatomy of the Digestive System

The purpose of digestion is to alter the foodstuffs by chemical action and to convert them into such simple forms that they can be absorbed into the blood and utilized by the various tissues of the body according to their requirements. The process of digestion takes place in the alimentary canal and is aided by certain accessory organs of digestion, the salivary glands, the liver and pancreas.

Foodstuffs are dealt with by the body in four stages:

1. Ingestion, mastication and swallowing.
2. Digestion.
3. Absorption.
4. Excretion (egestion) of residue and waste products.

1. **Ingestion**, or the taking in of food, and **mastication** are functions performed by the mouth and teeth, aided by the tongue. The pharynx and oesophagus are concerned with swallowing.
2. **Digestion**, although commencing in the mouth, is mainly carried out in the stomach and upper part of the small intestine.
3. Absorption can occur from any part of the alimentary tract, but is dependent on the nature and state of the substance concerned. The ordinary foodstuffs are mainly absorbed in the small intestine.
4. The large intestine is responsible for absorption of water and transit of the residue for **excretion** (egestion) in the form of faeces (page 220).

Enzymes

As already stated, the process of digestion is the splitting up of the complicated foodstuffs into simple forms which can easily be absorbed. This involves a number of chemical processes which are performed by means of substances secreted by the glands of the alimentary tract and called enzymes.

Definition. An enzyme is a substance secreted by the glands of

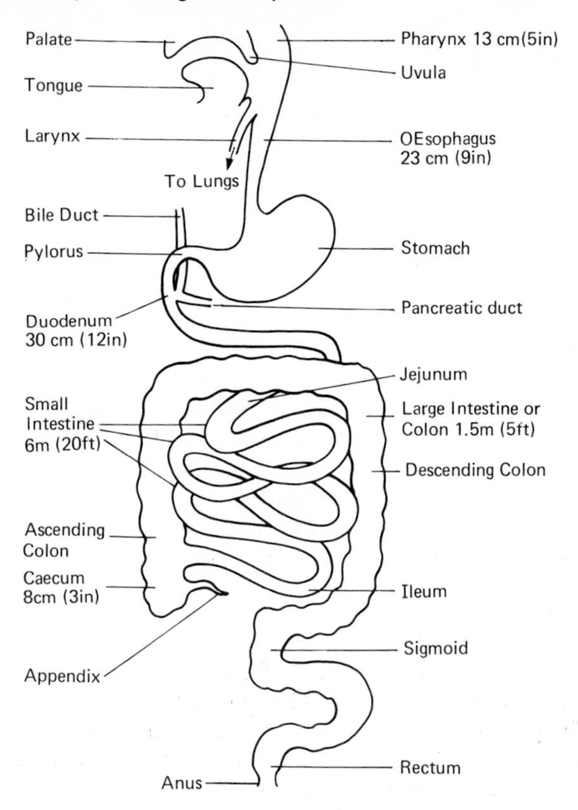

Fig. 131 The alimentary tract (diagrammatic).

the alimentary tract (including the salivary glands and pancreas) which has the power of acting on a foodstuff and converting it into a simple form in which it can be absorbed. There are a number of such enzymes and each has a specific action, i.e. each enzyme acts on one particular type of foodstuff. The important enzymes are set out on page 191.

The alimentary system can conveniently be described in two main sections:

1. The organs concerned with ingestion of food, viz. the mouth, pharynx, oesophagus.
2. The organs concerned with digestion, absorption and excretion of foodstuffs which are situated in the abdominal cavity, viz. the stomach, small and large intestines.

SCHEME ILLUSTRATING THE ALIMENTARY CANAL

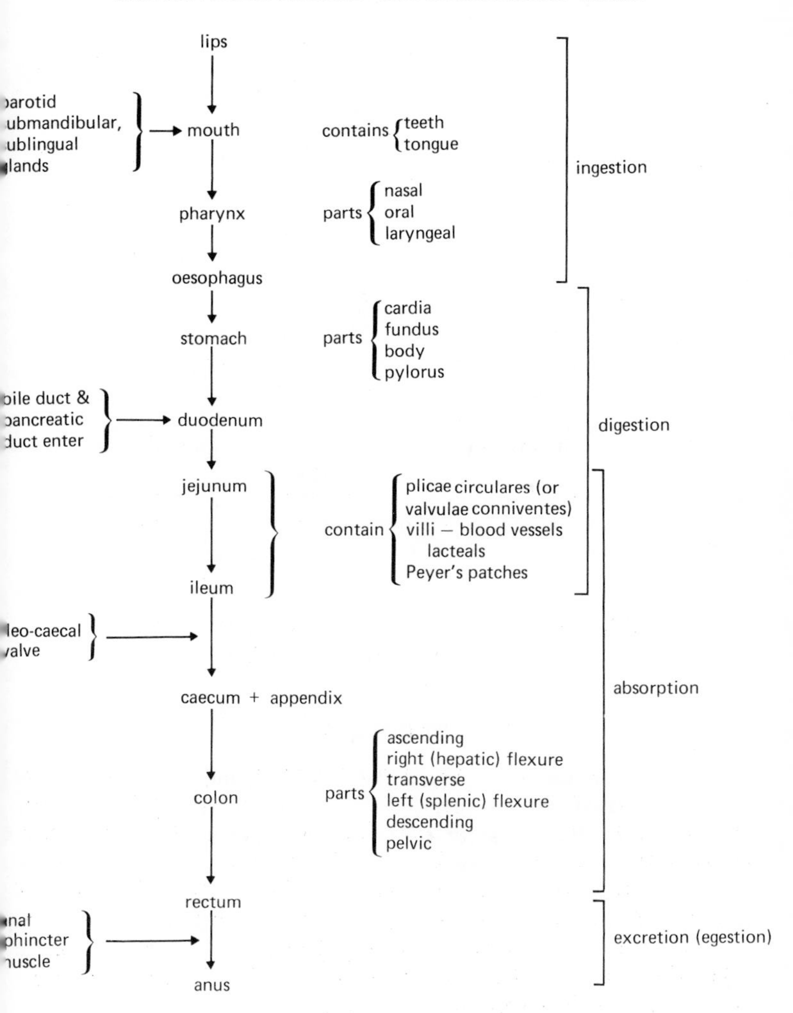

Summary of the important enzymes in digestion (see page 188)

Name	*Site*	*Function*
ptyalin	saliva	converts starch into maltose
pepsin	gastric juice	converts proteins into proteoses and peptones in presence of hydrochloric acid
rennin	gastric juice	curdles milk, converting caseinogen into casein
trypsin* chymotrypsin	pancreatic juice	converts proteoses and peptones into peptides
amylase	pancreatic juice	converts starches into maltose
lipase	pancreatic juice	converts fats into fatty acids and glycerol
erepsin	intestinal juice	converts peptides into amino-acids
maltase	intestinal juice	converts maltose into glucose
lactase	intestinal juice	converts lactose into glucose and galactose
sucrase	intestinal juice	converts sucrose into glucose and fructose

NOTE: Trypsin is formed by the enterokinase from the intestinal juice acting on trypsinogen.

ANATOMY AND PHYSIOLOGY OF UPPER PART OF ALIMENTARY TRACT

The mouth

The mouth, or buccal cavity, is the upper expanded portion of the alimentary canal. It is a cavity divided into two parts. The central or main part of the cavity contains the **tongue**. The outer part, or **vestibule**, is the recess between the teeth and the inner surface of the cheeks. The anterior opening of the mouth is bounded by the lips which contain muscle fibres and are lined on their inner surface by mucous membrane, giving them their natural red colour. The whole of the interior of the mouth is also lined with mucous membrane which is kept moist by saliva.

The roof of the mouth is formed by the bony **hard palate** in front, and the fleshy **soft palate** behind. The soft palate has a cone-shaped prolongation at its posterior end which hangs down in the back of the mouth and is called the **uvula**. On either side of the

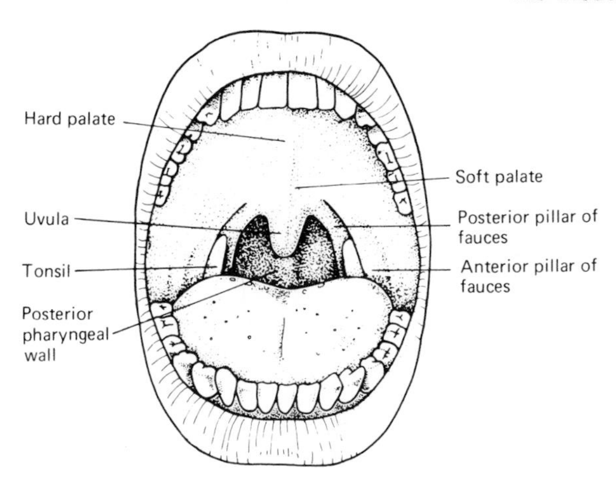

Fig. 132 The back of the mouth.

uvula are two folds of mucous membrane, the **anterior** and **posterior pillars of the fauces** between which lie the **tonsils** (page 185). These pillars form the boundary of the posterior opening of the mouth into the pharynx.

The tonsils (see also page 185) are masses of lymphoid tissue, consisting largely of lymphocytes. Small orifices on the surface of each tonsil lead down to deep pits, or crypts. In these, cheesy plugs may form from lymphocytes, desquamated epithelial cells and bacteria. The tonsil receives most of its blood supply from the tonsillar artery. The function of the tonsils is thought to be a protective one, in some way guarding against infection. However, the tonsils are themselves frequently infected by haemolytic streptococci or viruses, resulting in acute tonsillitis.

The floor of the mouth is mainly formed by the tongue, but the anterior portion under the tongue consists of a sheet of mucous membrane covering various muscles.

The ducts of the salivary glands open into the mouth. The duct of the **parotid gland** (Stenson's duct) enters the vestibule on each side opposite the second upper molar tooth. Under the tongue, close to the mid-line, are two small raised papillae which are the openings of the ducts of the **submandibular** (submaxillary) **glands** (Wharton's duct). The **sublingual glands** have a number of small ducts which open independently under the front of the tongue, close to the entrances of the submandibular ducts.

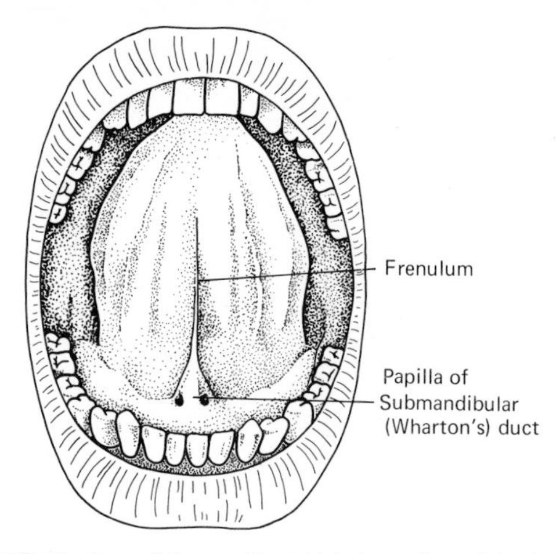

Fig. 133 The floor of the mouth and inferior surface of the tongue.

The tongue

This is a muscular organ covered on its free surface by a special type of mucous membrane. Unlike the mucous membrane of the rest of the mouth, which is smooth, that of the tongue is rough. The roughness is produced by numerous minute elevations called **papillae.**

If the mucous membrane from the sides of the tongue is examined under a microscope it is seen to contain a number of specialized collections of epithelial cells which are called **taste buds** (see also page 342).

The hindmost part of the tongue is attached to the hyoid bone.

Three types of papillae are found on the tongue:

(*a*) Eight or nine large papillae arranged in a V-shaped manner at the base of the tongue—the **circumvallate papillae** which contain numerous taste buds.

(*b*) **Fungiform** or flattened papillae.

(*c*) **Filiform papillae** which are small pointed elevations.

The teeth

Each individual has two sets of teeth, the temporary or deciduous (milk) teeth of childhood and the permanent teeth of adult life. The arrangement of these sets is somewhat different, for the temporary set consists of 20 teeth and the permanent of 32.

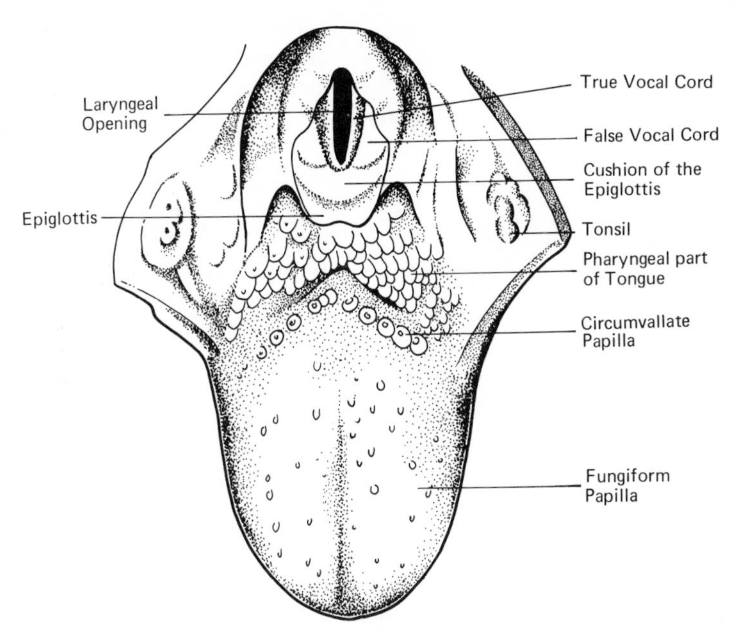

Fig. 134 The tongue and opening of the larynx.

Permanent teeth. These are thirty-two in number, sixteen above and sixteen below, or eight in each half of both upper and lower jaws. According to their shape and position they can be classified into four groups:

Class of teeth	*Numbers on each side of both jaws*
incisors	2
canines	1
premolars or bicuspids	2
molars	3

The **incisors** are the sharp cutting teeth situated in front and are described as central and lateral according to their position. To the lateral side of the incisors are the **canine** or grasping teeth (eye teeth). Behind the canines are the **premolars** and the **molars** or grinding teeth. The teeth are set in each jaw in the shape of a horseshoe, which is sometimes referred to as the dental arch.

Temporary teeth (deciduous or milk teeth). These are twenty in number, ten in each jaw, viz. two incisors, one canine and two molars in each half of the jaw. Compared with the permanent set, therefore, there are no premolars and only two molars instead of three.

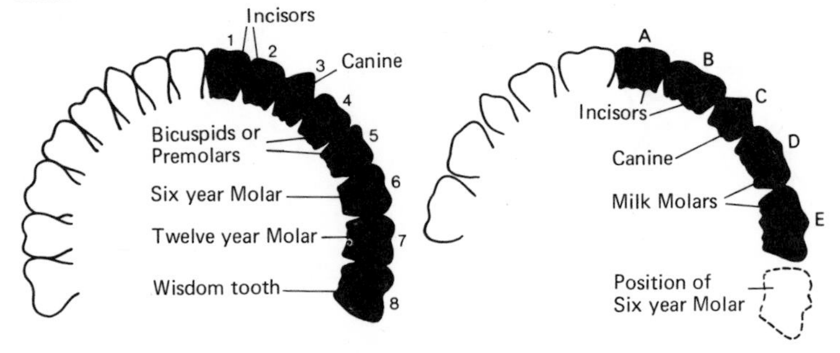

Fig. 135 Arrangement of the teeth—permanent (*left*) and temporary (*right*).

Eruption of the teeth

The eruption of both sets of teeth follows a fairly regular order, but as a general rule those of the lower jaw appear before those of the upper jaw.

At birth all the temporary teeth and most of the permanent teeth are buried in the alveoli of the jaws. Several of the latter are, however, formed after birth.

The infant usually cuts its first teeth, the lower central incisors, about the 6th or 7th month. Others continue to appear and the full set of milk teeth is completed about the age of 2 years.

The earliest permanent teeth to appear are the first molars, which project from the gums behind the second temporary molars. Others erupt in turn and the permanent set is complete about the age of 12, except for the third molars or wisdom teeth, which appear between the 17th and 25th years, or may never erupt but remain buried in the gums. Sometimes unerupted third molars cause trouble and have to be removed.

Structure and appearance of teeth

A tooth consists of a crown, a neck and a root. The **crown** is the portion projecting above the gum. The slight constriction where it is surrounded by the gum margin is the **neck** and the part buried in the alveolus of the jaw is the **root**.

The main mass of the tooth consists of a substance called **dentine**,

Dentition table

Temporary teeth	*Months to appear*
incisors	6–12
first molars	12–14
canines	14–20
second molars	20–24
Permanent teeth	*Year*
first molars	6th
central incisors	7th
lateral incisors	8th
premolars	9th and 10th
canines	11th
second molars	12th
third molars	17th–25th

For convenience a dental surgeon may use the following formula (see Fig. 135):—

LEFT (upper)	RIGHT (*upper*)
87654321	12345678
87654321	12345678
(*lower*)	(*lower*)

a very hard material which resembles bone in composition but contains no Haversian canals. The exposed portion of the tooth is covered with a thin layer of **enamel**, which is even stronger than dentine.

The interior of a tooth is hollow and is called the **pulp cavity**. This contains soft connective tissue, small blood vessels and nerves which enter the root of the tooth through a fine canal at its apex.

The teeth are firmly fixed in the alveoli of the jaw by a special cement substance and a strong layer of connective tissue called the periodontal membrane.

The incisor teeth have a crown consisting of a sharp cutting edge and a single, pointed root. The canines have a crown which terminates in a point, and a single root. The premolars have a single root and a crown consisting of two elevations or cusps, which explains why they are also called bicuspids. The molars have square-shaped crown with four cusps. The upper molars possess three roots, the lower two.

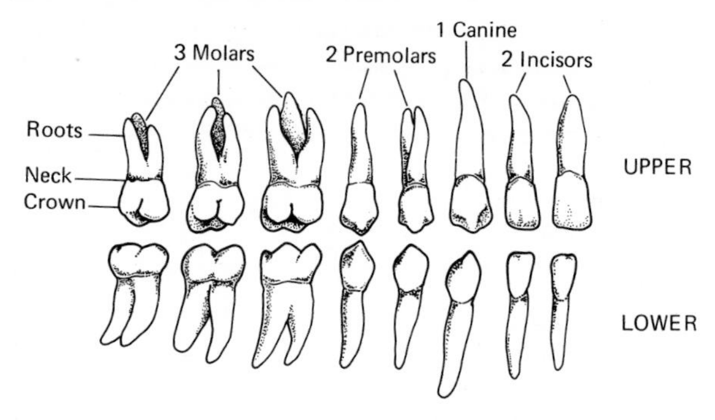

Fig. 136 Shapes of human teeth.

Functions of the mouth, tongue and teeth

The main functions of the **mouth** are the intake of food, its mastication by the teeth and mixture with the saliva. In this way food is prepared for the act of swallowing. Another important function is in the production of speech, and it can also be used for breathing.

NOTE: Some drugs (e.g. glyceryl trinitrate, isoprenaline) can be absorbed through the mucous membrane of the mouth if tablets are allowed to dissolve under the tongue.

The **tongue** moves the food about in the mouth and thus helps proper mastication. By passing the food back into the pharynx it performs the first part of the act of swallowing. It contains the

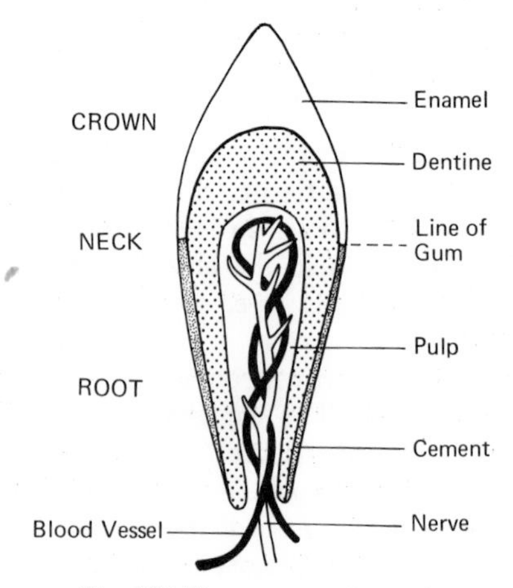

Fig. 137 The structure of a tooth.

organ of taste and also plays an important part in the mechanism of speech. (See page 273).

The **teeth** are essential for efficient mastication. The incisors are the cutting teeth and permit the individual to bite off small pieces of food. The canines are really grasping teeth and, in animals such as dogs, are much larger and longer than in man. They prevent the food slipping while the incisors are cutting it off. The molars and premolars are the grinding teeth and it will be noticed that mastication is a side-to-side movement of the lower jaw so that the food is ground between them.

A film known as 'plaque' normally covers teeth and is difficult to remove, even with a good toothbrush. It contains bacteria which break down carbohydrates into acids which attack the enamel and eventually cause dental decay (caries). Dental caries is undoubtedly reduced when the concentration of fluoride in drinking water is about 1 part per million, which makes enamel more acid-resistant.

The salivary glands

There are three pairs of salivary glands, the **parotid**, **submandibular** (submaxillary) and **sublingual** glands. They secrete saliva which enters the mouth by means of their ducts.

Parotid glands. These are situated at some distance from the mouth partly in front of and partly below each ear. The deepart part of the gland lies between the upper part of the ramus of the lower jaw and the mastoid process. Each has a duct, **Stenson's** (parotid) **duct**, which runs diagonally in the substance of the cheek and enters the vestibule of the mouth at a small papilla which can be seen opposite the second upper molar tooth. (It is these glands which become enlarged in mumps.)

Submandibular (submaxillary) glands. These are smaller than the parotids. Each is situated just under cover of the angle of the jaw. The submandibular or **Wharton's duct** passes forwards in the floor of the mouth and enters by a papilla close to the mid-line under the tongue (Figs 133, 138).

Sublingual glands. These lie in the front part of the floor of the mouth behind that part of the lower jaw which forms the chin. They are quite close together and pour their secretion directly into the mouth by a number of small ducts.

Functions of saliva

The saliva is the mixed secretion of the three pairs of salivary glands. It is a fluid consisting of 90 per cent water, and is, in contrast with the secretion of the stomach, alkaline in reaction. It contains the important enzyme **ptyalin** (salivary amylase) which acts on the cooked starches in the mouth and converts them into a sugar called maltose. In addition, it contains mucin and a small amount of calcium salts which may become deposited on the teeth in the form of tartar.

Another important function of saliva is to moisten the food, thereby acting as a lubricant aiding the act of swallowing and the passage of food down the oesophagus. The mouth is kept constantly

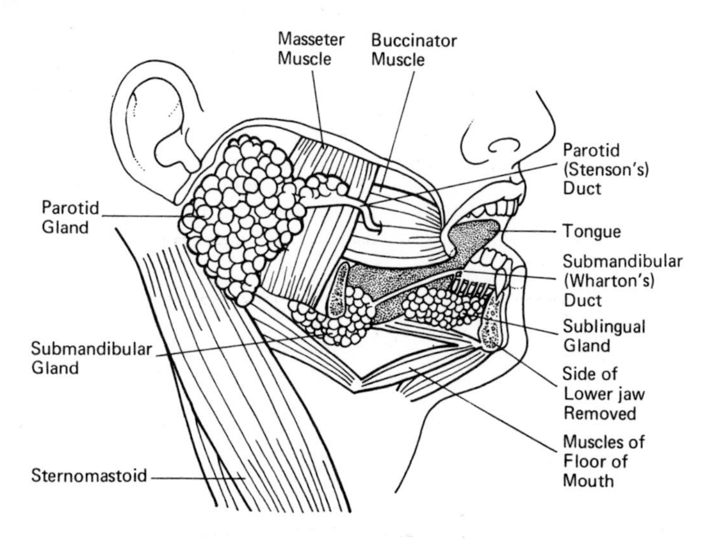

Fig. 138 Diagram of the salivary glands.

moist by the saliva, but when food is taken the output of saliva from the glands is greatly increased. This is because the glands are supplied with nerves connected to special salivary nuclei in the medulla of the brain. On the intake of food a nervous message is sent to this centre which transmits it to the secretory nerves of the glands, causing an increase in their output (the salivary reflex). This centre is also influenced by impulses from the higher or psychological centres of the brain, and it is well known that the sight or smell of appetizing food will cause an increase of saliva, in fact it 'makes the mouth water'.

Dryness of the mouth may be caused by mouth breathing, dehydration and drugs such as atropine which decrease the amount of saliva.

Inflammation of the mouth, due to infection or vitamin deficiency, is known as stomatitis.

The pharynx

After leaving the mouth food enters the pharynx. The pharynx is an expanded portion of the alimentary canal about 13 cm (5 in) long, and is divided into three parts.

1. The nasopharynx.
2. The oropharynx.
3. The laryngopharynx.

The nasopharynx is not functionally part of the alimentary tract but belongs to the respiratory system. The oral pharynx is shared as a passage by both the alimentary and the respiratory systems, while the laryngeal portion is concerned only with the transport of food.

The **nasopharynx** is situated at the back of the nasal cavity and extends from the base of the skull to the level of the soft palate, where it becomes continuous with the upper part of the oral pharynx. In addition to these openings it has, on each side, the orifice of the Eustachian tube which passes to the middle ear (page 345).

The **oropharynx** lies behind the mouth, between the soft palate above and the upper opening of the larynx below. Its anterior aspect

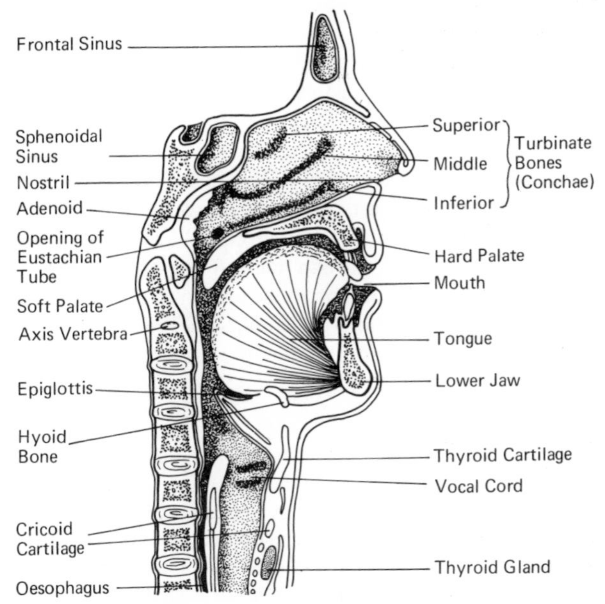

Fig. 139 Section through the nose and throat.

is open to the mouth and is bounded by the anterior pillars of the fauces (Fig. 132). The tonsils and posterior pillars of the fauces, therefore, lie in its side walls. The posterior wall can be seen through the mouth and is situated in front of the second (axis) and third cervical vertebrae.

The **laryngopharynx** lies behind the cartilages of the larynx. Above it is continuous with the oral pharynx; below it merges into the oesophagus at the level of the cricoid cartilage.

Structure. The pharynx is a muscular bag, suspended from the base of the skull and lined by mucous membrane. The chief muscles are the 'constrictor muscles' which contract in the act of swallowing and thereby pass the food on into the oesophagus. They are supplied by the accessory nerve (page 322).

The oesophagus

The oesophagus or gullet is a muscular tube, lined by mucous membrane, which extends from the pharynx to the cardiac orifice of the stomach. It is 25 cm (10 in) long and, in its course, lies in turn in the neck, the thorax and the abdominal cavity. **In the neck** it lies between the trachea in front and the cervical vertebrae behind. The carotid arteries are situated a short distance away on either side (Fig. 182). **In the thorax** it continues downwards in front of the thoracic part of the vertebral column. The trachea, until it divides into the two main bronchi, remains as an anterior relation. In the rest of its course the oesophagus lies behind the pericardium. In its lower half the thoracic aorta is situated to its left side (Fig. 106). **In the abdomen**, the oesophagus leaves the thorax by piercing the diaphragm and enters the abdominal cavity for a short distance before terminating in the stomach at the cardia (Fig. 106).

The oesophagus has sphincter muscles at its upper (cricopharyngeal) and lower (cardiac sphincter) ends, which relax when swallowing takes place. The cardiac sphincter prevents regurgitation of stomach contents into the oesophagus.

Diaphragmatic herniation. The stomach may herniate through the hiatus (L. opening) in the diaphragm which transmits the oesophagus. The stomach may either herniate directly upwards into the mediastinum (oesophagogastric or 'sliding' hernia) or a portion of the stomach may herniate upwards and come to lie alongside the oesophagus (para-oesophageal or 'rolling' hernia). These hiatal herniae can be diagnosed by radiography (barium meal).

The act of swallowing or deglutition

When a mouthful of food has been masticated and well mixed with saliva the movements of the tongue and cheeks convert it into a soft rounded mass called a **bolus.**

The act of swallowing takes place in three stages:

1. The lips are closed. The tongue is raised against the palate and forces the bolus of food past the pillars of the fauces into the oral pharynx. The soft palate rises and shuts off the nasopharynx. At the same time the larynx is raised and the epiglottis acts as a guard to its upper opening.

2. The bolus is grasped in the pharynx by the contraction of the constrictor muscles and forced into the oesophagus.

So far the act of swallowing, though more or less automatic, is a deliberate and voluntary one. Subsequently, however, the individual has no further control and it becomes involuntary (a reflex action). It will be noticed also that during the first two stages the breath is held. This is a protective reflex preventing the aspiration of food into the respiratory tract. This too is an involuntary act not under control of the will. In fact, the only motor functions of the alimentary tract under the direct control of the voluntary nervous system are the acts of mastication, swallowing and defaecation.

3. The bolus of food is carried down the oesophagus by the contraction of its muscular walls, and finally reaches the stomach.

The whole act of swallowing occupies about ten to twenty seconds, depending on whether liquids or solids are taken, the passage of the latter being slower. It can be watched by means of X-rays if the patient is given a barium mixture to take.

In this way any obstruction can be demonstrated and, if necessary, further investigated by oesophagoscopy. Difficulty in swallowing is called dysphagia.

Peristalsis The special method of muscular contraction by which the bolus of food is carried down the oesophagus is called peristalsis. This form of muscular action is of great importance and occurs throughout the whole of the alimentary tract for the purpose of passing on its contents. It is an involuntary movement outside the conscious control of the individual, and is dependent upon the muscular walls of the tube being stretched by their contents.

Consider the bolus of food in the oesophagus or the contents of any segment of the alimentary canal. The circular muscle of the tube immediately behind the bolus contracts and that directly in front relaxes. This results in the bolus being forced into the relaxed

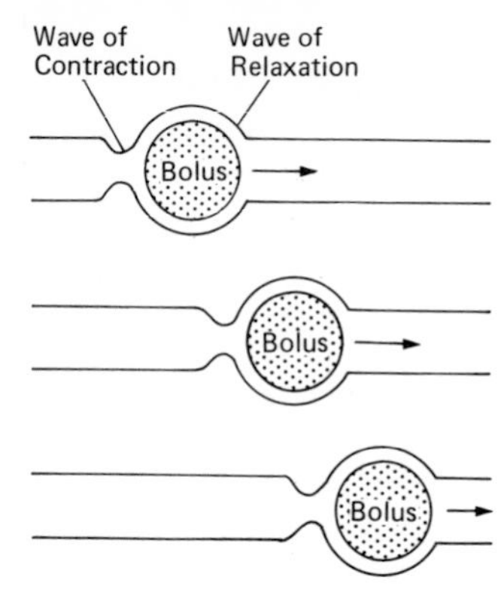

Fig. 140 Diagram illustrating peristalsis, showing wave of contraction preceded by wave of relaxation of muscle, resulting in forward movement of bolus of food.

portion. The contraction of muscle follows closely behind the bolus and further relaxation occurs in front of it, thus it is passed steadily forwards.

Peristalsis may be defined as a wave of muscular contraction preceded by a wave of relaxation which causes the contents of a hollow tube to be passed onwards.

ANATOMY OF ABDOMINAL ORGANS OF DIGESTION

The regions of the abdomen

For convenience of description the abdomen is divided into regions by imaginary lines drawn through certain fixed points. Reference to the diagram (Fig. 141) makes these divisions quite clear.

Two horizontal lines are taken:

1. The upper line joining the lowest point on the ribs of each side.
2. The lower line joining the highest points of the iliac crests.

There is a vertical line on each side which passes vertically upwards from a point midway between the anterior superior iliac spine and the symphysis pubis. This line passes approximately through the nipple and middle of the clavicle.

Nine regions are thus marked out.

In the middle of the abdomen from above downwards are: the

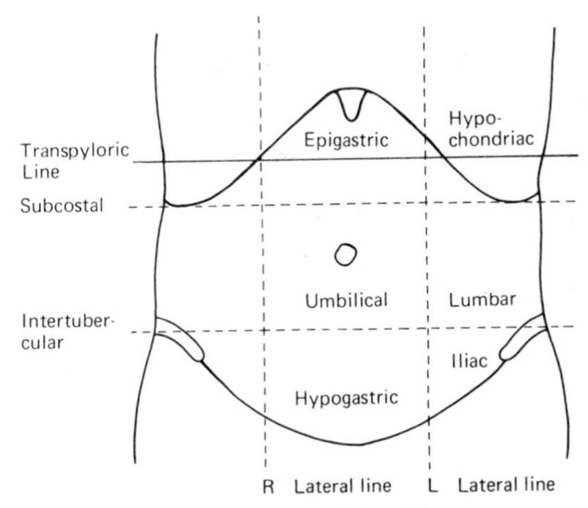

Fig. 141 The regions of the abdomen.

epigastric region or epigastrium, the umbilical region, the hypogastric region or hypogastrium.

On each side from above downwards: the hypochondriac region or hypochondrium, the lumbar region, the iliac region.

The greater part of the alimentary canal and the organs concerned with digestion lie within the abdominal cavity, and are wholly or in part covered by the peritoneum. The boundaries of the cavity are described on page 37.

The peritoneum

The peritoneum is a serous membrane and, like the other important serous membranes of the body, the pleura and the pericardium, consists of two separate layers. The **parietal** layer lines the walls of the abdominal cavity and the **visceral** layer covers the organs.

The smooth, slightly moist membrane facilitates the movements of the viscera so that the stomach and intestines are free to move without friction within the abdomen.

The detailed arrangement of the peritoneum and its various folds is very complicated and difficult to follow, but a general conception of its disposition can be obtained in the following way: Imagine it to be a closed bag having front and back surfaces. Then consider the effect if an organ were pushed into the middle of the back surface in such a way that this surface completely surrounded the organ.

It will then be understood that one organ may be completely surrounded by peritoneum while structures like the pancreas, lying on the posterior abdominal wall, will be covered by the membrane only on their anterior surfaces.

Those organs completely surrounded by peritoneum will, therefore, be suspended from the posterior abdominal wall by a double fold of the membrane. It is in this way that a **mesentery** or fold of peritoneum by which the intestine is attached to the posterior abdominal wall is formed. It is between these two layers that the blood vessels reach the organs, for the abdominal aorta and its branches lie outside the peritoneal cavity.

The stomach, intestines (except for the duodenum and rectum), the liver and spleen are almost completely surrounded by peritoneum, which therefore forms an outer coat for these structures. The duodenum, rectum and pancreas are only covered on their anterior surfaces.

Peritoneal ligaments. The liver, uterus and other organs are maintained in position partly by means of double folds of peritoneum which form suspensory ligaments.

The omenta. These are folds of peritoneum connected to the stomach. The **great omentum** hangs from the lower border of the stomach like an apron in front of the small intestines; in its posterior portion lies the transverse colon. The **lesser** or **gastrohepatic omentum** stretches from the lower border of the liver to the upper or lesser curvature of the stomach.

The mesentery. This is the fold of peritoneum which encloses the small intestine and anchors it to the posterior abdominal wall. The attachment to the abdominal wall is relatively short, whereas the intestinal part is many feet long, so that the mesentery can be described as a fan-shaped structure.

The pelvic peritoneum

The peritoneum in the pelvis is continuous with that of the rest of the abdominal cavity. It covers the front aspect of the rectum. In the male it passes forwards over the posterior and upper surfaces of the bladder to become continuous with that on the anterior abdominal wall. In the female it passes from the rectum over the posterior and anterior surfaces of the uterus before reaching the bladder. The sac between the rectum and uterus in the female is called the recto-uterine pouch (*of Douglas*). On either side, in the female, the membrane covers the uterine (*Fallopian*) tubes. These tubes have an opening into the peritoneal cavity where their mucous membrane is directly continuous with the peritoneum. It is in this way that the ova are able to pass from the ovaries, which lie inside the peritoneal cavity, into the uterus. In the male the peritoneum is a completely closed sac.

Functions of the peritoneum

1. It is a serous membrane which enables the abdominal contents to glide over each other without friction.

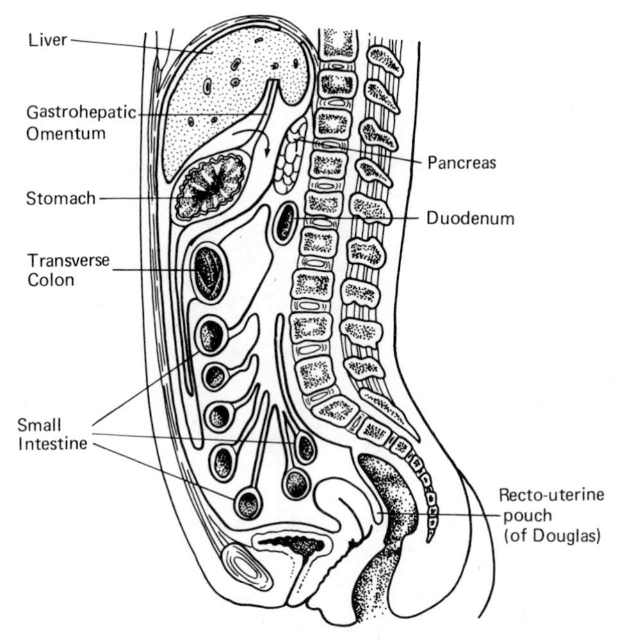

Fig. 142 Sagittal section of the abdomen showing disposition of the peritoneum. (The arrow passes through the foramen of Winslow into the lesser sac.)

2. It forms a partial or complete covering for the abdominal organs.
3. It forms ligaments and mesenteries which help to keep the organs in position.
4. The omentum and mesentery contain a considerable amount of fat and act as important fat stores for the body.
5. The omentum can move about inside the cavity and in the event of inflammation occurring tends to wrap itself round the affected part of the alimentary tract and prevent the infection

from spreading to the rest of the peritoneum. (It is very common in cases of acute appendicitis to find the appendix completely surrounded by omentum. On account of its mobility it has been called 'the abdominal policeman'. It is shorter and less well-developed in infancy and early childhood, so that appendicitis in these cases may be very serious.

6. It has the power to absorb fluids in large quantities (hence the former value of intraperitoneal saline in infants).

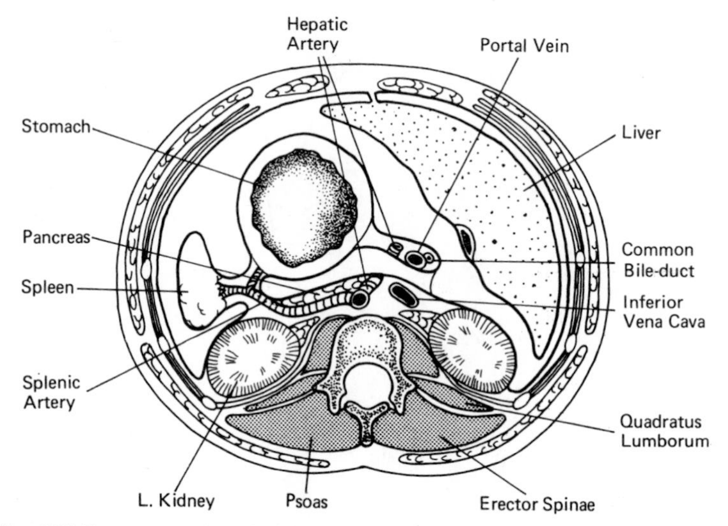

Fig. 143 Tranverse section of abdomen, showing the arrangement of the peritoneum (at the level of the foramen of Winslow).

Peritoneal dialysis. The peritoneum is a membrane through which some electrolytes and simple substances can be exchanged with others in the blood. For example, if a suitable glucose–electrolyte solution is introduced into the peritoneal cavity its strength will equalize with that in the blood, i.e. urea will pass from the blood into the dialysis fluid which lacks this substance. If the fluid is then removed from the peritoneal cavity the blood urea, which accumulates in some cases of kidney disease, will be lowered.

The stomach

The stomach is a dilated portion of the alimentary canal which forms a receptacle for the food after its passage down the oesophagus. It is a J-shaped structure when seen by X-rays with the individual in the upright position. It lies in the upper part of the

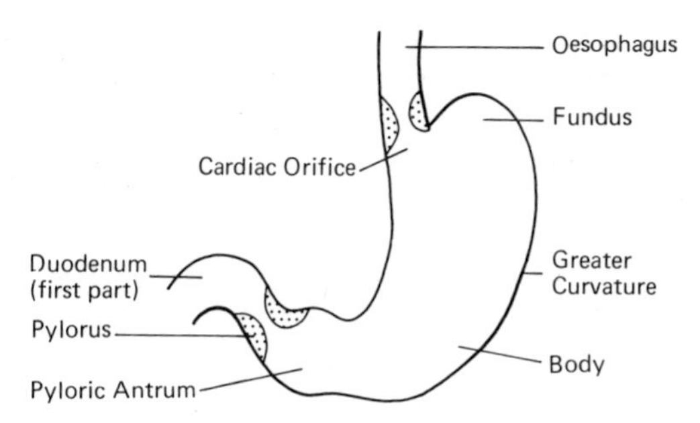

Fig. 144 Diagram illustrating parts of stomach.

abdominal cavity, mainly in the epigastrium but partly in the left hypochondrium and umbilical regions.

The stomach has anterior and posterior surfaces and upper and lower curved borders. The upper border is called the lesser curvature, and the lower, the greater curvature.

It consists of the following parts (see Figs 144, 159):

The **cardia** or cardiac orifice at its upper end which is the entrance of the oesophagus.

To the left of the cardia is the dome-shaped upper part called the **fundus**, which in life generally contains a bubble of air. This lies immediately below the left part of the diaphragm. Sometimes

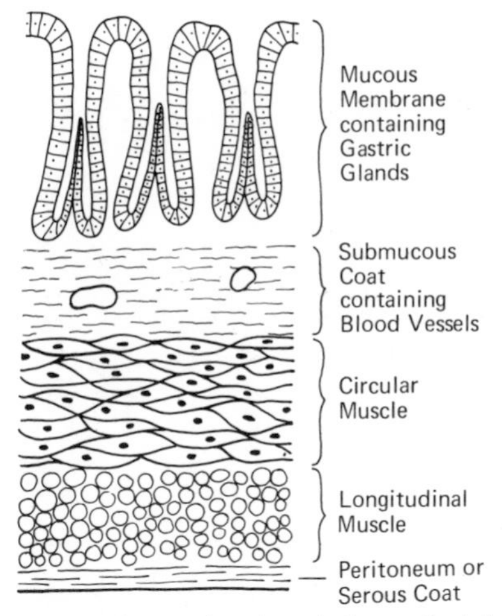

Fig. 145 Magnified diagram of a section through the wall of the stomach, showing the various layers (oblique muscles omitted).

a part of the fundus slips back through the oesophageal opening in the diaphragm and enters the thorax (hiatus hernia).

The **body** constitutes the main part of the stomach.

The **pyloric antrum** forms the lower part of the J and leads to the pylorus, a canal surrounded by a sphincter muscle, which separates the stomach from the duodenum.

Relations. The anterior surface lies partly behind the left lobe of the liver and under-surface of the diaphragm, and partly in contact with the anterior abdominal wall. The posterior surface lies in the 'stomach bed' which is formed by the left kidney, the pancreas and the spleen.

The stomach receives its arterial blood supply from the coeliac axis artery, the first branch of the abdominal aorta. The blood from the stomach passes into the portal vein and thence to the liver.

Structure. The stomach consists of four coats:

1. The peritoneum or serous coat.
2. The muscular layer, having from without inwards:
   (*a*) longitudinal fibres,
   (*b*) circular fibres, and
   (*c*) oblique fibres, fewer in number than (*a*) or (*b*).
3. The submucous layer.

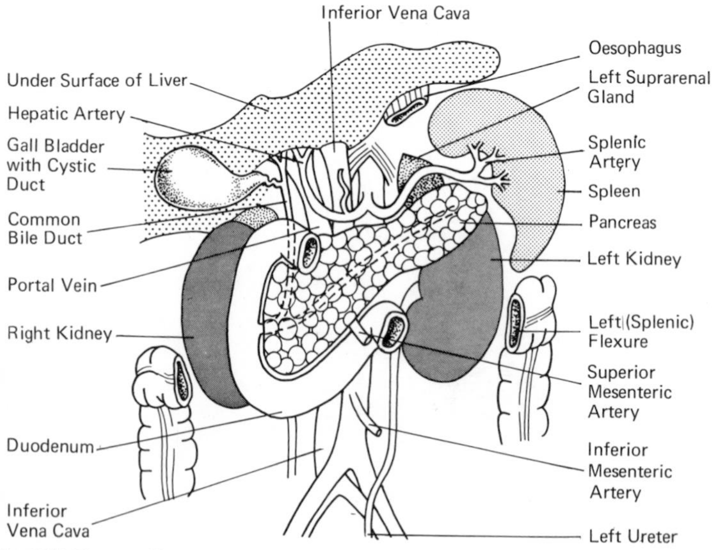

Fig. 146 Diagram illustrating the structure of the posterior abdominal wall and showing the 'bed of the stomach'.

4. The mucous membrane.

The peritoneum forms the outer covering of the stomach. The muscle is arranged in layers beneath the peritoneal coat and an increase in the number of circular fibres forms the pyloric sphincter muscle. The submucous layer consists of fibrous tissue and contains the blood vessels. The inner lining is the mucous membrane. This contains the glands which secrete the gastric juice and is thrown into folds, thereby increasing the total surface area of the stomach from which secretion can take place.

The small intestine

The small intestine is the portion of the alimentary tract extending from the pyloric sphincter of the stomach to its termination in the first part of the large instestine called the caecum. It is about 6 metres (20 feet) long and consists of three parts:

duodenum (25–30 cm; 10–12 in long).
jejunum (about two-fifths of small intestine).
ileum (about three-fifths of small intestine).

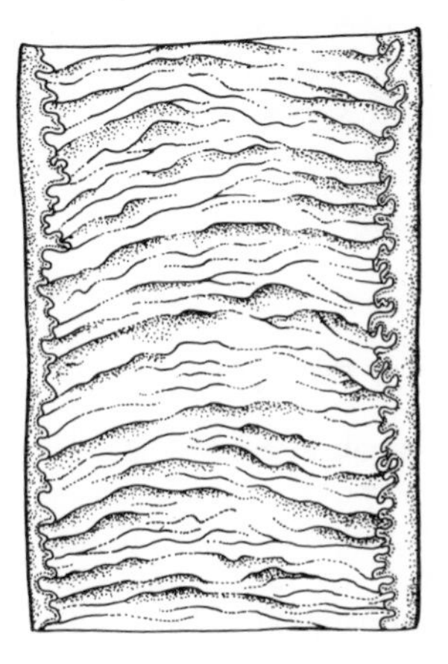

Fig. 147 Interior of jejunum, showing the plicae circulares.

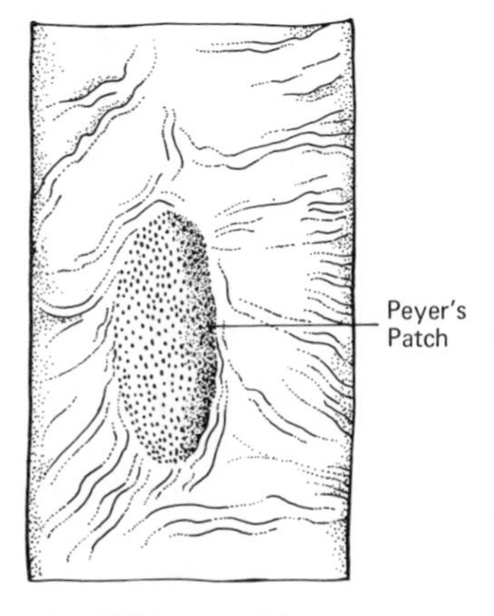

Fig. 148 Interior of ileum, showing a Peyer's patch.

The duodenum

The duodenum, or first part of the small intestine, differs from the other parts by being fixed to the posterior abdominal wall by peritoneum which covers only its anterior surface. It is C-shaped and lying in its concavity is the head of the pancreas. About halfway along the concave surface of the duodenum the bile duct and the pancreatic duct enter together at a small papilla called the **ampulla of Vater.**

The jejunum and ileum

These are partially movable within the abdominal cavity and are attached to the posterior abdominal wall by a mesentery about 10 cm long. Although given separate names, the former passes imperceptibly into the latter about two-fifths of the way down the course of the small intestine. Some variation of appearance and structure constitutes the differences between them.

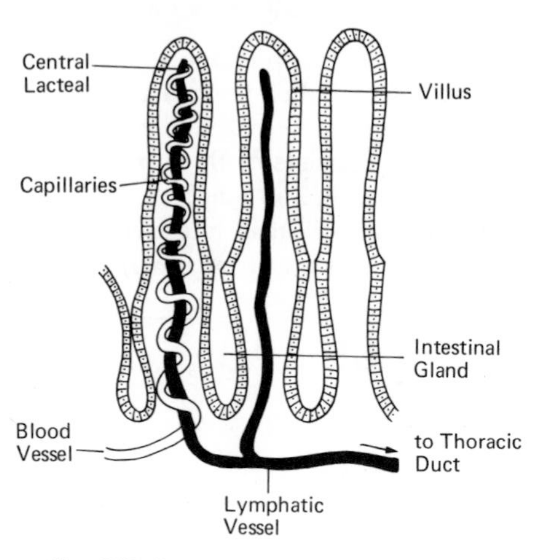

Fig. 149 Structure of villi in small intestine.

Structure of small intestine. The small intestine, like the stomach, has four coats, viz. peritoneal or serous; muscular consisting of longitudinal and circular fibres only; submucous; and mucous. As in the stomach, the mucous membrane is arranged in folds, here called **plicae circulares** (valvulae conniventes). Their effect is greatly to increase the surface from which secretion and absorption can take place. The surface of the mucous membrane is slightly roughened by minute projections from it, called **villi.** Each villus contains capillaries into which are absorbed the products of carbohydrate and protein digestion (glucose and amino-acids), and a central lymphatic vessel or lacteal into which fats are absorbed.

The mucous membrane of the small intestine contains glands secreting the intestinal juices (succus entericus). Lymphatic tissue is found in the submucous layer either in solitary nodes or in collected masses (often 2·5–8 cm (1–3 in) long) called ***Peyer's patches*** which are present in the lower part of the ileum. It is important to remember Peyer's patches because they are affected in typhoid fever.

The large intestine

The large intestine is 1·5 metres (5 ft) long and has the following parts:

caecum and vermiform appendix
ascending colon
transverse colon
descending colon
pelvic or sigmoid colon
rectum

It commences at the entrance of the terminal part of the ileum into the caecum and ends at the anus, which is the opening of the rectum to the exterior.

The caecum

This is a dilated sac situated in the right iliac fossa into which the ileum enters at the **ileocaecal valve**. The mucous membrane is so arranged at this point that, together with a sphincter muscle, it acts as a valve permitting the contents of the ileum to enter the caecum but preventing their return into the ileum.

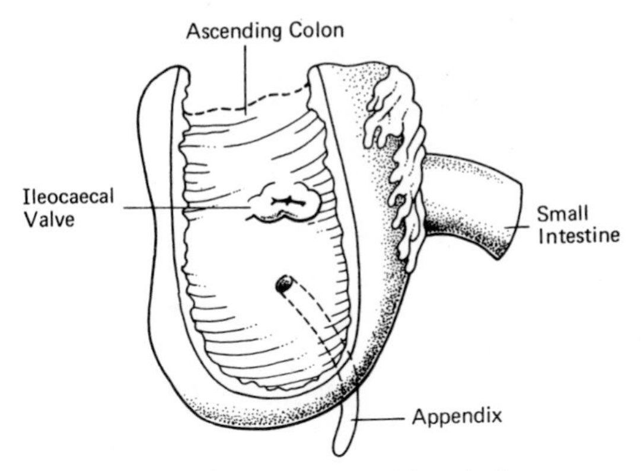

Fig. 150 The caecum opened from the front.

The appendix

The appendix (vermiform = worm-like) is attached to the blind end of the caecum and lies in the right iliac fossa. Sometimes it lies to the side of the caecum, sometimes tucked up behind it, and at others its tip hangs down into the pelvic cavity. It is about 9 cm ($3\frac{1}{2}$ in) long. It is lined by mucous membrane and contains lymphoid tissue in its walls.

Inflammation of the appendix (appendicitis) is a relatively common condition which may affect both children and adults. It is characterized by pain and tenderness in the right iliac fossa. The main danger is that the organ may perforate and liberate its infected contents into the peritoneal cavity causing peritonitis.

The ascending colon

This passes upwards from the caecum through the right lumbar region and is held in position on the posterior abdominal wall by

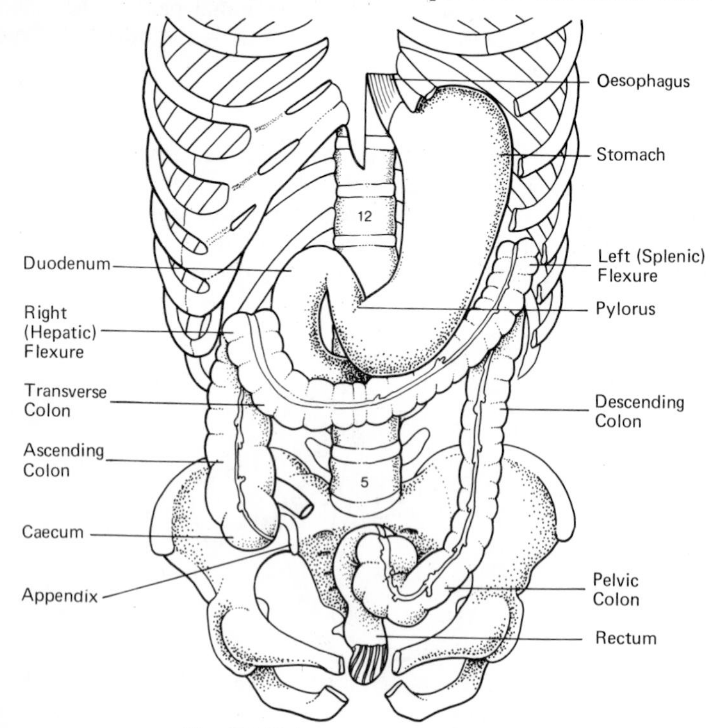

Fig. 151 The stomach and large intestine.

peritoneum. It reaches the under surface of the liver, where it turns sharply to the left at the **right** or **hepatic flexure** to become the transverse colon.

The transverse colon

As its name implies, this passes across the abdominal cavity as a loop which may fall well below the umbilicus. It rises as it reaches its left extremity and comes in contact with the spleen, where it turns sharply downwards at the **left** or **splenic flexure** to continue as the descending colon.

The transverse colon lies in the fold of peritoneum which extends downwards from the greater curvature of the stomach called the great omentum. It is, therefore, freely movable within the abdominal cavity.

The descending colon

This passes downwards in the left lumbar region. It is anchored to the posterior abdominal wall by peritoneum and, therefore, like the ascending colon, is not movable.

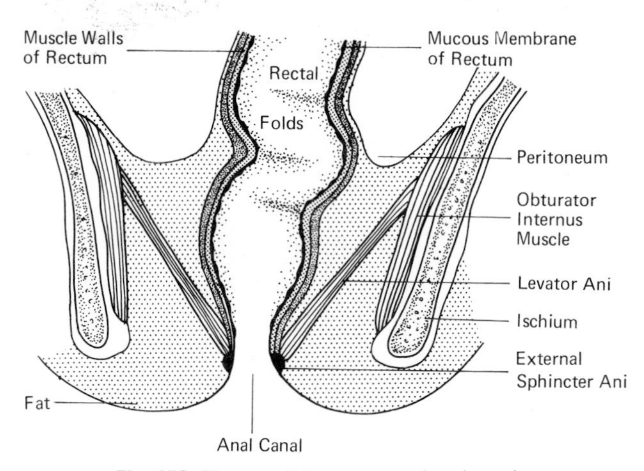

Fig. 152 Diagram of the rectum and anal canal.

The pelvic or sigmoid colon

This is the continuation of the descending colon and, having a mesentery, is movable. It lies mainly in the pelvic cavity and passes into the rectum.

The rectum

This lies in the pelvic cavity and is situated in the concave hollow formed by the anterior surface of the sacrum. It is about 13 cm (5 in) long and opens to the exterior at the anus.

The anus

This opening is guarded by a sphincter muscle, the anal sphincter, which is under the control of the will so that it can be relaxed when it is desired to empty the rectum of its contents.

Structure of large intestine. The large intestine, like the small intestine, has four coats; peritoneal or serous, muscular, submucous and mucous.

The characteristic feature, however, is that the longitudinal muscles do not form a continuous layer over the whole gut but are arranged in three separate bands (taenia coli). These bands are somewhat shorter than the length of the large intestine which accounts for its sacculated appearance.

When the longitudinal fibres reach the rectum they spread out over its whole surface.

The mucous membrane contains no villi and is not thrown into folds like the small intestine. It is lined by columnar epithelium.

11 The Physiology of the Digestive System

The changes in the food taking place in the mouth and its rapid conveyance to the stomach have already been discussed.

Functions of the stomach

The secretions of the stomach

The glands in the mucous membrane of the stomach secrete the **gastric juice**. This secretion is continuous so that even when the stomach is at rest and no meal has been taken for a period of many hours, e.g. before breakfast, there is always some gastric juice present. This is called the 'resting juice' and usually amounts to about 20 ml.

The intake of food into the stomach is followed by a considerable increase in the output of gastric juice. (This results partly by a reflex mechanism through the vagus nerve and partly by means of a chemical messenger or hormone called **gastrin**. This is formed by the action of certain foodstuffs on the gastric mucous membrane and is carried by the blood to the gastric glands causing them to increase their activity. The psychological stimulus of the sight or smell of food also causes a flow of gastric juice similar to the effect on the salivary glands (see page 198).

The gastric juice is a clear watery fluid and, in contrast to the saliva, is acid in reaction. The acidity is due to the presence of hydrochloric acid (HCl) secreted by the glands of the gastric mucosa. Under fasting conditions about 1 litre of juice is secreted per day, but with normal food intake it may reach 5 litres.

Two important enzymes are present:

1. *Pepsin.* This is formed mainly in the glands in the fundus and body of the stomach. In the presence of hydrochloric acid it commences the digestion of proteins, converting them into less complex

substances called peptones. This is the first stage of protein digestion; further stages take place in the small intestine. (Pepsin is actually formed by the action of hydrochloric acid on pepsinogen.)

2. *Rennin.* This is an enzyme which curdles milk. The process consists of the conversion of the soluble protein of milk, caseinogen, into the insoluble form called casein.

It will be recalled that food entering the stomach has been made alkaline by the saliva and that it also contains ptyalin. The ptyalin continues to act in the stomach for a short time, but as the food becomes acid the action of ptyalin in converting cooked starches into maltose is stopped.

Hydrochloric acid has a slight antiseptic action and is able to kill a number of bacteria, thereby preventing their entry into the intestine. Its functions may be summarized:

1. It neutralizes the alkaline saliva, acidifies the food and stops the action of ptyalin.
2. It aids the action of pepsin in the conversion of proteins into peptones.
3. It acts as an anti-microbial barrier to many bacteria.

The hydrochloric acid in the stomach is concentrated enough to kill bacteria ingested with food. The acid also softens fibrous foods and converts pepsinogen into pepsin, the active enzyme. Why does the acid not corrode the lining of the stomach itself? The epithelial cells are fused together and a barrier is formed by sheets of fat molecules in the cell membrane. Furthermore there is a continual shedding and replacement of cells so that the lining of the stomach is completely renewed every three days.

In some persons hydrochloric acid is absent from the gastric juice, a condition known as achlorhydria or achylia. Excess of hydrochloric acid is called hyperchlorhydria and a diminished amount hypochlorhydria. Achlorhydria always occurs in pernicious anaemia.

Movements of the stomach

The movements of the stomach can be observed by X-rays after the administration of a meal containing barium.

When food first enters the stomach, the pyloric sphincter muscle contracts and prevents its exit until the appropriate degree of digestion has occurred. The pylorus then relaxes at intervals and allows small quantities to enter the duodenum.

In due course the stomach is emptied, but it is clear that the time taken to empty will depend on the nature or digestibility and the amount of food taken; an ordinary meal takes about five hours to leave the stomach completely. During this period it is constantly churned and mixed with the gastric juice by waves of peristalsis passing from the upper part of the stomach towards the pylorus. The food which leaves the stomach to enter the duodenum is in a semi-liquid state and this partially digested material is called **chyme.**

Peristaltic movements taking place in an empty stomach give rise to the sensation of hunger.

The blood-forming functions of the stomach

The stomach has another very important function not directly concerned with digestion.

Gastric juice contains a substance known as the **intrinsic factor** (of Castle). This acts on certain foodstuffs containing vitamin B_{12} or cyanocobalamin, sometimes called the extrinsic factor. The interaction of these two factors enable the vitamin B_{12} to be absorbed and carried to the liver where it is stored. Vitamin B_{12} is essential for the proper development of the red cells in the bone marrow (see also page 166).

SUMMARY OF FUNCTIONS OF STOMACH

1. It acts as a reservoir for food.
2. It secretes the gastric juice containing:
   (*a*) Hydrochloric acid, which neutralizes the alkaline saliva, acidifies the food, aids the action of pepsin and is an antimicrobial barrier.
   (*b*) Pepsin, which converts proteins into peptones.
   (*c*) Rennin, which curdles milk by converting caseinogen into casein.
3. It secretes the intrinsic factor essential for the absorption of the erythrocyte-maturing factor vitamin B_{12} (Cyanocobalamin).
4. It is where water, glucose, alcohol and some drugs are absorbed.
5. It breaks down food into chyme.

Symptoms of gastric disease include epigastric pain, flatulence, vomiting and the vomiting of blood (haematemesis). Haematemesis may result from erosion of the gastric mucous membrane, which is sometimes due to aspirin, chronic ulceration or cancer of the stomach. Flatulence may occur as the result of the bad habit of swallowing air.

Digestion in the small intestine

(*a*) *In the duodenum.* The acid chyme from the stomach is mixed with the alkaline bile and the pancreatic juices in the duodenum (page 232). These juices continue to act on the food during its passage through the small intestine.

(*b*) *In the jejunum and ileum.* The small intestine has a number of glands in its mucous membrane which secrete the intestinal juice or **succus entericus.** This is an alkaline fluid containing the following ferments:

1. **Erepsin**, which converts peptides into amino-acids.
2. **Invertase**, converting cane-sugar into glucose.
3. **Maltase**, converting maltose into glucose.
4. **Enterokinase**, an enzyme which converts the trypsinogen in the pancreatic juice into trypsin.

The most important function of the small intestine, however, is the absorption of almost all of the products of digestion.

Functions of the large intestine

When the food is taken into the stomach the ileocaecal valve relaxes and the semi-liquid contents of the lower part of the small intestine are passed into the caecum by peristaltic action. This is a part of what is called the **gastrocolic reflex.**

Secretion. The large intestine has only one secretion, *mucin,* which lubricates the faeces and facilitates their passage through the rectum and anus.

Digestion. Many bacteria are present in the large intestine. These act on the various food residues which have not been digested or absorbed in the small intestine.

Absorption. The contents of the small intestine are in a liquid state, while the faeces when excreted are almost solid. It is clear, therefore, that during their passage through the large intestine the faeces lose water. This water is absorbed by the walls of the colon into the blood. Some glucose and salt may also be absorbed, and use is made of these facts in the administration of rectal salines and rectal glucose. A number of drugs administered in the form of suppositories, e.g. aminophylline, are absorbed from the rectum.

Excretion. Excess of calcium, iron and drugs of the heavy metal type, such as bismuth, are excreted from the walls of the large intestine and mix with the faeces. Iron given by mouth makes the faeces black in colour.

In order that active peristalsis may take place it is essential that there should be a sufficient food residue of undigested material or **'roughage'**. This is mainly supplied by the cellulose of vegetables and fruit (see page 249).

Defaecation. The rectum is a sac in which the faeces accumulate until there is a sufficient bulk to evacuate.

The gastrocolic reflex which has already been mentioned is very important and has the following results:

1. Relaxation of the ileocaecal valve and the passage of contents from the small intestine into the large intestine.
2. General peristaltic action throughout the whole of the colon.
3. The contents of the pelvic colon enter the rectum.

The process of emptying the rectum is called defaecation and is performed in the following way:

When a quantity of faeces reaches the rectum it produces a distension of the cavity and stretching of its walls. This causes impulses to pass along the nerves to the spinal cord and thence to the brain, where they arouse the conscious sensation of the desire to defaecate. If this call is neglected, the rectum accommodates itself to its contents by relaxation of the muscle in its walls, impulses cease to travel along the nerves and the desire passes off. The arrival of more faecal matter in the rectum causes further distension and another set of impulses are conveyed to the nervous system, again producing the conscious desire to evacuate the bowel.

Failure to respond to this natural process is a common form of constipation. It will be clear that the longer the faeces remain in the rectum the harder and more solid they will become since the rectum has the power of absorbing water. Hard faeces are more difficult to evacuate than soft ones.

The act of defaecation, while reflex in the infant, in the adult is under the control of the will and is carried out in response to the desire to empty the bowel produced by distension of the rectum with faeces.

The following actions occur:

1. The sphincter muscle of the anus relaxes.
2. The muscular walls of the rectum contract.
3. The muscles of the floor of the pelvis contract.
4. The pressure within the abdomen is raised:
   (*a*) by holding the breath and contracting the diaphragm;
   (*b*) by contracting the muscles of the abdominal wall.

It will be noted that a number of these actions are increased in force if the squatting posture, which is the primitive and natural one, is adopted.

The injection of an enema into the rectum has the effect of rapidly distending its walls and so initiating the mechanism whereby the desire to defaecate is produced. It also helps to soften and break up hard masses of faeces.

SUMMARY OF FUNCTIONS OF LARGE INTESTINE

Preparation of the faeces for evacuation from the rectum by:

1. Secretion of mucin.
2. Absorption of water, glucose and salts.
3. Excretion of calcium, iron, etc.
4. Bacterial decomposition of cellulose.

Diseases of the large intestine include dystentery, ulcerative colitis and cancer. The main symptoms of colonic disease include diarrhoea, or sometimes constipation and the passage of blood and mucus in the stool.

The faeces

The faeces are normally a semi-solid, paste-like mass coloured brown by stercobilin, a pigment derived from the bilirubin and biliverdin of the bile. Water, even after the absorption which takes place in the colon, still forms 65 to 70 per cent of the total bulk of the faeces. The remainder consists mainly of undecomposed cellulose, some fatty acids, protein residue (skatol, histidine, indole and tryptophane), dead bacteria and epithelial cells. Some salts of calcium and iron and also bismuth, if it has been administered, are present. (NOTE: Iron and bismuth turn the faeces black.) The surface of the faeces is lubricated by the mucin secreted by the large intestine.

If the faeces are passed too rapidly through the intestines insufficient time will be allowed for the absorption of water which accounts for the watery character of the stool in some cases of diarrhoea. In others, irritation of the bowel results in pouring out of fluid from the mucous membrane.

ABSORPTION OF FOODSTUFFS

Ingestion and digestion are the first two phases of the physiological processes occurring in the alimentary tract. The third phase is that of absorption. A number of details concerning absorption have already been mentioned.

Absorption from the mouth

There is normally no absorption from the mouth, but a few drugs may be absorbed into the blood through the mucous membrane if allowed to dissolve under the tongue (e.g. glyceryl trinitrate, isoprenaline).

Absorption from the stomach

The only substances normally absorbed from the stomach are water, glucose and alcohol. The rapid effects of alcohol on subjects unaccustomed to its use clearly indicate that it is absorbed before it has time to reach the other parts of the alimentary tract. Some drugs may also be absorbed by the stomach.

Absorption from the small intestine

The majority of the digested foods are absorbed in the small intestine.

(*a*) *Carbohydrates.* All sugars and starches are eventually converted by the action of various enzymes into glucose, and in this form are absorbed into the capillaries in the villi of the small intestine. Glucose is conveyed by the portal vein to the liver, where it is stored as glycogen.

(*b*) *Protein.* The digestive ferments, pepsin, trypsin and erepsin,

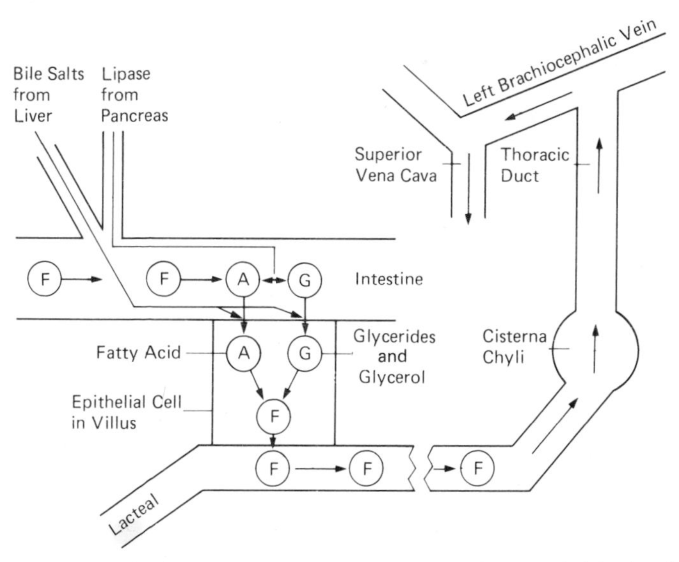

Fig. 153 Diagram illustrating the splitting of fats by lipase of pancreatic juice into fatty acids, glycerides and glycerol, which are absorbed with the aid of bile salts, reconstituted into fats in cells of villi, passed into lacteals and thence to thoracic duct.

break down proteins into peptones, polypeptides and, finally, amino-acids, in which form they are absorbed into the blood in the villi of the small intestines and, like glucose, conveyed to the liver.

(*c*) *Fats.* Lipase converts fats into fatty acids and glycerol. For the proper absorption of digested fats the presence of bile salts is necessary. These act by lowering surface tension. Fatty acids and glycerin are absorbed into the villi of the small intestine and are there recombined to form saturated fats. These, unlike glucose and amino-acids, enter the lacteals and not the blood stream directly. By the lacteals the fat is carried to the cisterna chyli and then by the thoracic duct to the left brachiocephalic vein, where it enters the blood (see also page 182). The small intestine also absorbs iron, vitamin B_{12} and calcium, the latter in the presence of vitamin D.

Absorption from the large intestine

The only substance normally absorbed from the large intestine is water. It has been seen, however, that its mucous membrane is also capable of absorbing glucose and salts.

12 Accessory Organs of Digestion

In addition to the tongue, the teeth and the salivary glands, which are structures accessory to the digestive system, two important abdominal organs remain to be described: (*a*) the liver and bile ducts, (*b*) the pancreas.

The liver

The liver is the largest organ in the body and weighs between 1275–1550 g (45–55 oz). It is situated in the upper part of the abdominal cavity immediately below the diaphragm. It occupies mainly the right hypochondrium and epigastrium, but extends partly into the left hypochondrium.

The shape of the liver is irregular. It may be conveniently divided into right and left lobes and described as having superior, inferior, anterior and posterior surfaces.

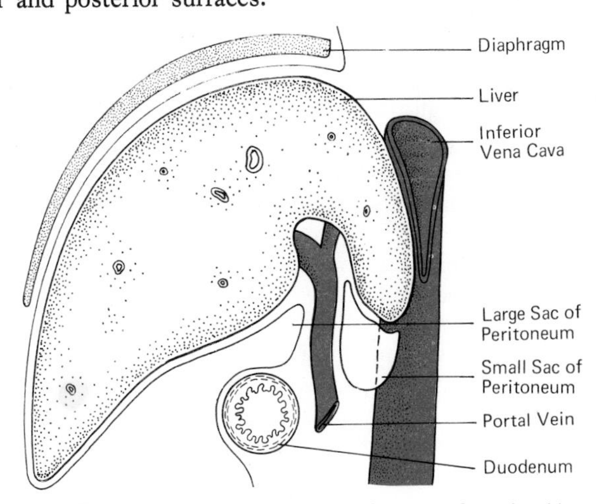

Fig. 154 The liver and some of its relations seen from the side.

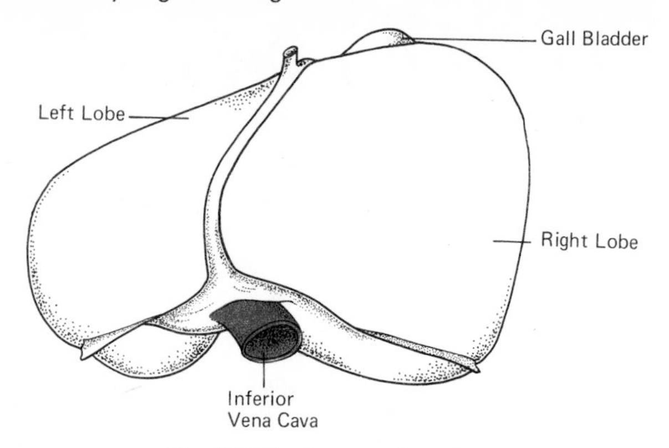

Fig. 155 The liver seen from above.

The **superior** surface is in contact with the under surface of the dome-shaped diaphragm which separates it from the bases of the lungs.

The **inferior** surface is related to other abdominal viscera, including the kidney and right (hepatic) flexure of the colon on the right and the fundus of the stomach on the left.

The **anterior** surface is separated from the right lower ribs and costal cartilages by the margin of the diaphragm and, in the midline, is related to the anterior abdominal wall.

The **posterior** surface crosses the vetebral column in the midline and is also related to the aorta, inferior vena cava and lower end of the oesophagus.

In the centre of the inferior surface is the **hilum** or gate of the liver. This is also called the **portal fissure** and through it the important blood vessels, the hepatic artery and portal vein, enter and the bile ducts leave the liver.

Attached to the under surface of the right lobe of the liver is the gall-bladder.

The liver has a fibrous capsule and the greater part of its surface is covered by peritoneum.

Structure of liver. The liver is built up of a large number of lobules consisting of columns of liver cells. Each lobule has a special blood supply and from it passes a small bile duct which carries the bile secreted by the liver cells and unites with similar ducts from other lobules until finally the larger bile ducts are formed.

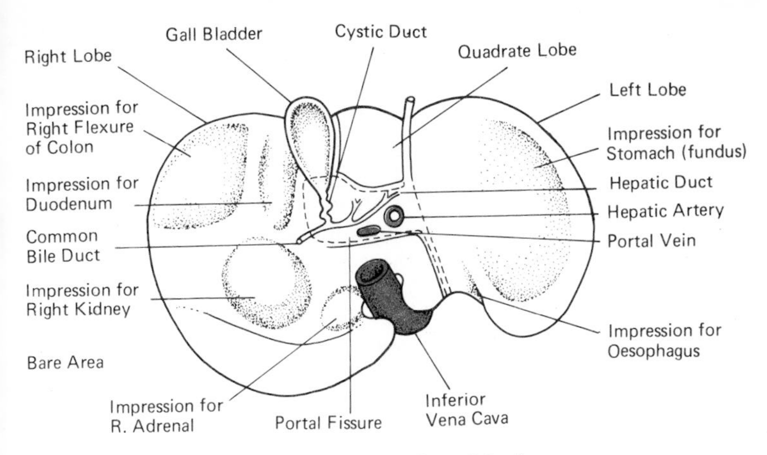

Fig. 156 The under-surface of the liver.

Blood supply

The liver has a double set of vessels bringing blood to it which enter at the portal fissure, and a single set removing blood from it to the inferior vena cava.

Blood is brought to the liver by: (*i*) the hepatic artery, (*ii*) the portal vein.

Hepatic artery. This is a branch of the coeliac axis coming from the abdominal aorta. This artery conveys oxygenated blood which nourishes the liver cells and supplies them with oxygen necessary for the execution of their other functions.

Fig. 157 Liver, showing liver lobules and radial arrangement of the cells from the centre of each lobule.

It divides into many branches in the liver substance, a small branch ultimately passing to each lobule. Here it forms a network of capillaries around the cells, the blood from which enters the vein situated in the centre of the lobule called the **central vein.**

Portal vein. This conveys venous blood, poor in oxygen but rich in nourishment, from the stomach, spleen and intestines, and enters the liver at the portal fissure. Its smallest branches lie between the lobules and are called the **interlobular veins**. These also divide into sinusoids situated around the cells of the lobule which finally pour their blood into the central vein where it mixes with the blood brought to the liver by the hepatic artery but which has also become venous, having given up its oxygen to the liver cells.

The central veins pass through the lobules to join the tributaries of the hepatic veins which unite to pour their blood into the inferior vana cava by which it is returned to the right side of the heart.

THE BLOOD SUPPLY OF THE LIVER

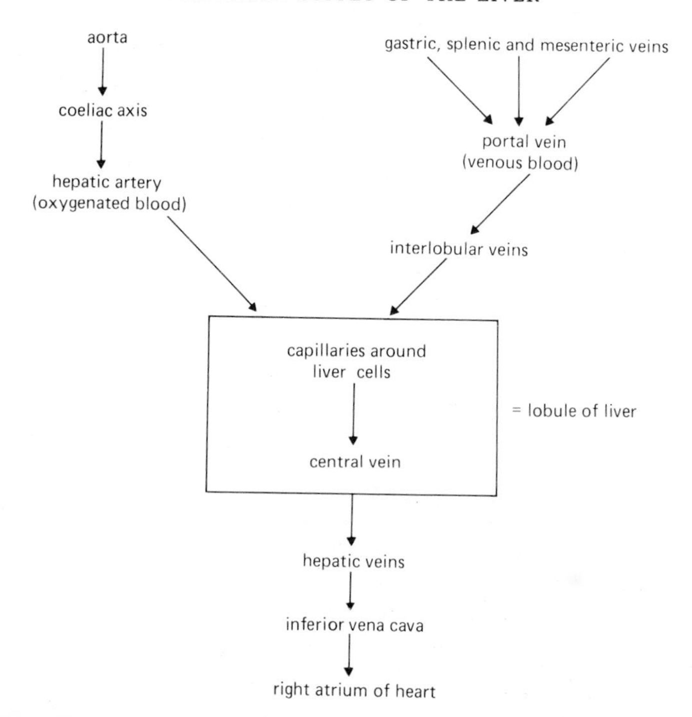

The bile ducts

The smallest bile channels are formed in the lobules between the liver cells. These gradually unite with those of larger size and the bile finally leaves the liver by two main ducts from the right and left lobes. These two ducts unite in the portal fissure to form the **hepatic duct**. After a course of two inches the hepatic duct is joined by the **cystic duct** from the gall-bladder and continues downwards closely related to the hepatic artery and the portal vein as the **common bile duct**. This passes behind the first part of the duodenum and is then buried in the head of the pancreas. It enters the second part of the duodenum at a small papilla called the **ampulla of Vater**, where it is joined by the pancreatic duct (of Wirsung).

The gall bladder

This is a pear-shaped sac, with a capacity of 30 ml (1 fl oz), situated on the under surface of the right lobe of the liver. The stalk of the pear is represented by the **cystic duct** which unites with the hepatic duct from the liver to form the common bile duct.

The gall-bladder acts as a reservoir for the storage of bile which, after leaving the liver by the hepatic duct, passes up the cystic duct to the interior of the gall-bladder where it remains until it is required for purposes of digestion. After a meal, especially if it contains fat, the gall-bladder contracts and discharges the bile via the

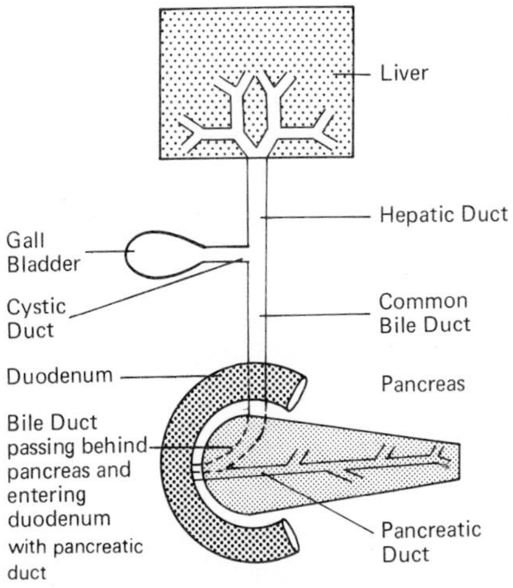

Fig. 158 Diagram illustrating the biliary apparatus.

cystic duct and common bile duct into the duodenum. During its stay in the gall-bladder the bile is concentrated by the extraction of water.

Structure of gall-bladder. The gall-bladder consists of three coats.

1. The outer peritoneal coat continuous with the peritoneum covering the liver (which binds the gall-bladder in position on the under surface of the liver).
2. A muscular coat, contraction of which enables the gall-bladder to empty itself into the common bile duct.
3. An inner coat of mucous membrane consisting of epithelial cells continuous with and similar to those lining the bile ducts.

Inflammation of the gall-bladder is known as cholecystitis and is usually associated with gallstones (cholelithiasis).

Main functions of the liver

These may be summarized:

1. The secretion of bile.
2. The storage of glycogen.
3. The breakdown of protein with the formation of urea.
4. The production of plasma proteins (albumin and globulin).
5. The desaturation of fats.
6. The storage of vitamin B_{12} and iron.
7. The production of heat.
8. Other general metabolic functions including the destruction of various toxic substances and drugs and the production of heparin, prothrombin and fibrinogen—substances which are concerned in the clotting of blood.

1. Secretion of bile. Bile is formed by the liver cells and therefore may be regarded as the secretion of the liver. It is alkaline in reaction and its important components are: (*a*) the bile pigments, bilirubin and biliverdin, (*b*) the bile salts. In addition, it contains water (over 90 per cent), mucin and cholesterol.

Bile pigments. The colour of bile, which is due to the pigments bilirubin and biliverdin, varies from yellow-brown to green. These substances are formed in the liver, spleen and bone marrow from the haemoglobin derived from worn-out red blood corpuscles which have been destroyed in the spleen. In the process, however, iron is separated and saved in the liver and spleen for further use in the preparation of fresh haemoglobin for new red cells. The bile pigments are converted in the bowel into stercobilin which colours the faeces but, apart from this, are waste products having no other function. Bacteria in the bowel convert some of the stercobilin into urobilinogen which is excreted in the urine.

Bile salts. Bile has a bitter taste largely due to these salts (called sodium glycocholate and sodium taurocholate). The bile salts have important functions (*i*) of assisting the digestive action of the pancreatic ferments, lipase, amylase and trypsin, and (*ii*) of aiding the absorption of fat from the intestine.

These salts, by lowering surface tension, cause fats to break up or emulsify into small droplets allowing fat-digesting enzymes to work more efficiently and convert them into fatty acids and glycerol, in which form they are absorbed.

In the condition known as jaundice the bile pigments accumulate in the blood, giving the skin and mucous membranes a yellow colour. At the same time, they appear in the urine, which is turned a dark brown. Absence from the faeces results in a pale clay-coloured stool. The absence of bile salts leads to an excess of fat in the faeces.

2. Storage of glycogen. Carbohydrate taken in the food is broken down by the digestive juices into the simple sugar, glucose, and, in this form, is absorbed into the tributaries of the portal vein by which it is conveyed to the liver. Here it is converted into a similar substance called glycogen (a polysaccharide) and stored. Glycogen is re-converted into glucose according to the requirements of the body and discharged into the blood stream. Both these processes are carried out by the action of enzymes. During physical activity, glycogen is liberated from the liver by the action of adrenalin.

3. Formation of urea. The end-products of protein digestion in the alimentary canal are the amino-acids which are absorbed by the small intestine and taken to the liver by the portal vein. In the liver the nitrogen-containing portion of the amino-acids is removed and is converted into urea which, in turn, is carried to the kidneys and excreted in the urine. This process is sometimes called de-amination or de-nitrification of the amino-acids.

4. Production of plasma proteins. The important plasma proteins albumen and globulin are synthesized by the liver from the amino-acids derived from the protein in the diet.

5. Desaturation of fats. Fats are split up by the lipase of the pancreatic juice into fatty acids and glycerin and are absorbed in this form. In the villi of the small intestine, however, these substances are recombined into fats and carried by the thoracic duct to the blood which conveys them to the fat stores of the body, especially the subcutaneous tissues and the omentum.

Fats are stored in the form called saturated fats. (This means that they cannot take any more hydrogen into their composition.) Before the saturated fats can be used by the tissues of the body they are conveyed to the liver which performs a chemical change (the removal of hydrogen) and they are then called unsaturated fats (see also page 222).

*6. **Storage of Vitamin B_{12} and iron.*** These functions of the liver are concerned with blood formation (see page 166). The absorption from the small intestine of erythrocyte-maturing factor in the diet is aided by the intrinsic factor secreted by the stomach. Iron in the liver is derived from the haemoglobin of worn out red corpuscles which have been destroyed in the spleen. The haemoglobin residue after extraction of iron is excreted by the liver in the form of the bile pigments, bilirubin and biliverdin.

*7. **Production of heat.*** It will be clear that a large number of chemical processes are carried out by the liver. This work involves the production of a certain amount of heat which, combined with the work of the muscles and other organs, helps to maintain the heat of the body.

*8. **Other metabolic functions.*** A number of drugs are altered in composition and toxic substances are destroyed by the action of the liver cells. Heparin, fibrinogen and prothrombin are also produced in the liver.

The pancreas

The pancreas is a gland somewhat similar in structure to the salivary glands. It lies transversely across the posterior abdominal wall at the level of the first and second lumbar vertebrae and is situated behind the stomach (see Fig. 159).

The pancreas has a head, a body and a tail. The **head** is situated to the right and fits into the C-shaped curve of the duodenum. Buried in the substance of the gland is the termination of the common bile duct as it joins the pancreatic duct to form the ampulla of Vater, which, at its opening into the duodenum, is guarded by the sphincter of Oddi.

The **body** is that part of the gland lying in front of the bodies of the lumbar vertebrae. The **tail** of the gland is situated to the left and is in contact with the hilum of the spleen.

The splenic artery, arising from the coeliac axis, runs along the upper border of the pancreas and enters the spleen at the hilum.

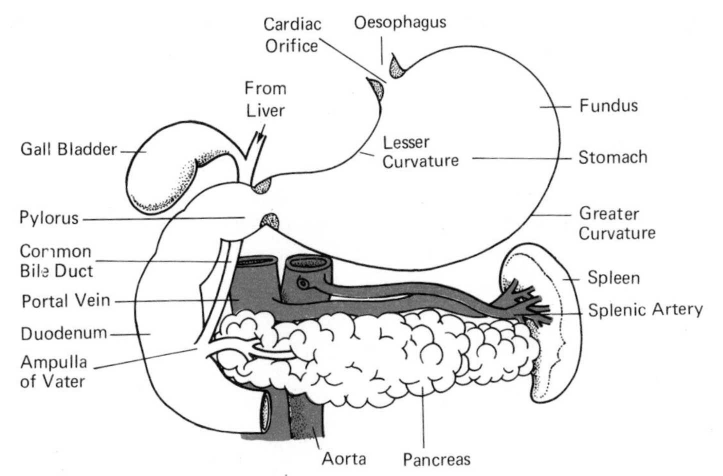

Fig. 159 The stomach, pancreas and bile ducts.

The anterior surface of the pancreas is covered by peritoneum.

Structure of pancreas. The pancreas consists of a number of lobules formed by secretory tubules (acini) lined with columnar epithelium. From the lobules small ducts emerge which unite with each other and eventually reach the main **pancreatic duct**, or duct of Wirsung. This passes from left to right in the centre of the organ and, in the head of the pancreas, joins the termination of the common bile duct to enter the second part of the duodenum at the ampulla of Vater.

Between the acini of the gland are clumps of cells differing in character and appearance from those of the secretory epithelium. These are called the **islets of Langerhans** which mainly secrete **insulin**, the internal secretion of the pancreas, directly into the blood (see also page 303).

There are, in fact, two types of cell, viz. the alpha (α) cells which secrete glucagon and the beta (β) cells which secrete insulin.

Functions of the pancreas

The pancreas is a gland having two different types of secretion:

1. An **internal secretion**—insulin—from the islets of Langerhans which is absorbed directly into the blood. The function of insulin is to enable the tissues to use sugar (page 238).
2. An **external secretion** which leaves the gland by the pancreatic duct.

The external secretion is an important digestive juice and contains the following enzymes:

(*a*) Amylase or amylopsin, which converts all starches into maltose.

(*b*) Lipase or steapsin, which splits fats into fatty acids and glycerin.

(*c*) Trypsin, which converts peptones into amino-acids.

These enzymes enter the duodenum by the pancreatic duct and mix with the food which has left the stomach. They continue to act on the chyme during its passage through the small intestine.

QUESTIONS

1. What is an enzyme? Describe the part enzymes play in the digestion of food.
2. Describe the large intestine and its functions.
3. What do you mean by a sphincter muscle? Give any two examples and describe how they work.
4. Give an account of the digestive juice found in the alimentary tract.
5. What do you know of the functions of the pancreas?
6. Give an account of the teeth in the child and the adult.
7. Describe the gall-bladder and the bile ducts. Give some account of the appearance and functions of bile.
8. Give an account of the structure of the liver and describe its functions.
9. Describe the anatomy of the stomach. What are its functions?
10. Describe the tongue and its functions.
11. Describe the functions of (*a*) gastric juice, (*b*) bile, (*c*) pancreatic secretion.
12. Give an account of the structure and functions of the duodenum.
13. What do you understand by the term excretion? Give examples and describe one organ concerned in the process of excretion.
14. Name the different parts of the alimentary canal. Explain the functions of the small intestine.
15. Give a description of the mouth. What do you know of its functions?
16. State what you know of the functions of the mouth. Describe what you would see when it is open.
17. Give an account of the salivary glands. What is the value of their secretions?
18. Describe the structure of the small intestine and comment on its functions.
19. Give an account of the digestion, absorption and use of carbohydrates.
20. Give an account of the absorption of fat. How is it utilized in the body? State the important fat-soluble vitamins.

13 Metabolism, Diet and Nutrition

All the various activities of the body are a form of work. In order that work may be carried out energy is required. This energy is supplied by the process of combustion or oxidation of appropriate foodstuffs in the tissues which also results in the production of heat.

Further, the continuous work of the tissues means that wear and tear will take place and will require repair. Materials are, therefore, needed for this purpose and also to allow for growth.

Foodstuffs, therefore, may be subdivided into three main groups:

(*a*) Those providing energy and heat (carbohydrates and fats).

(*b*) Those required for body-building and repair (proteins).

(*c*) Those regulating vital processes (vitamins and salts).

The series of changes involving the building up and breaking down of substances for use in the body is called **metabolism**. The chemical changes involving the breaking down of worn-out tissues and their removal is called **catabolism**. The building up of fresh tissues from the nutritive materials supplied in the food is called **anabolism**.

In health there is a balance between anabolism and catabolism. Among other results it is clear that if anabolism is in excess of catabolism there will be a tendency to increase in weight; if catabolic processes are excessive loss of weight will ensue.

These changes involve the consumption of a large amount of oxygen which must be taken into the lungs, absorbed into the blood stream and conveyed to the tissues where the chemical changes take place.

Likewise over 90 per cent of the protein, carbohydrate and fat consumed in an ordinary diet is utilized by the body, the remainder being lost in the faeces.

Normally all the carbohydrate and fat, after passing through a series of chemical changes, are completely converted into carbon dioxide and water. Protein is partially converted into carbon dioxide and water and partly into urea and other substances.

Excretion of waste products

The end products of metabolism which are not required by the body are removed by the various excretory organs, viz.:

lungs:	carbon dioxide
	water (a small amount)
kidneys:	water (the greater part)
	urea
	salts and other substances e.g. drugs.
bowel:	food residue
	water (a small amount)
skin:	water (a moderate amount)
	salts (a small amount)

Energy requirements of the body. A convenient method of estimating the energy requirement of the body is to measure it in terms of heat. Likewise the energy value of the various foodstuffs can be expressed in the same terms.

The unit of heat used for this purpose is the Calorie (abbreviated to C). A Calorie may be defined as the amount of heat required to raise 1 litre (1000 ml or 1 kilogram) of water 1°C at 15°C.

[The calorie (with small 'c')—the amount of heat required to raise 1 ml of water to 1°C—is no longer used in dietetics and will not be referred to again.]

The average Calorie value of carbohydrate, fat and protein is:

1 gram of carbohydrate	=	4 C or 116 Calories per ounce.
1 gram of protein	=	4 C or 116 Calories per ounce.
1 gram of fat	=	9 C or 206 Calories per ounce.

That is, if 1 gram of carbohydrate is fully burnt or oxidized it will raise the temperature of 1 litre of water 4°C, and so on.

The energy requirement of the body in terms of Calories is expressed either as (*a*) the total for the period of twenty-four hours, or (*b*) for each pound or kilogram of body weight.

Energy may also be measured in joules (J); 1 calorie is approximately 4·2 J and 4200 J equal one kilocalorie or C.

In man, basically about 167,000 J per square metre of body surface per hour are lost at complete rest and this must be made up by the energy supplied by food for any muscular activity or work.

One megajoule (MJ) equals 1 million joules. The daily resting requirements of the average man is 7·5 megajoules. Therefore, the average daily requirements of a manual worker would be 18.8 megajoules (12.6 for a woman). Sedentary workers require less e.g. 10·5 for a man and 8·8 for a woman.

Calculations for these figures can be made from tables showing the composition of various foods and their energy values.

Basal Metabolism

Even when the body is at complete rest, such as when lying down or during sleep, a certain amount of energy is needed for the continuation of the vital processes, for example the beating of the heart, the activity of the brain and the maintenance of body temperature. During rest, however, the metabolic needs of the body are at their lowest. The respiratory rate and the pulse-rate are relatively slow. The oxygen intake and the carbon dioxide output are also at the minimum level.

The amount of energy required by the body in this state of rest is called the **basal metabolism**, which may therefore be defined as 'the minimum metabolic requirement of the body during the state of complete rest'. (The basal metabolism can be calculated by estimating the intake of oxygen and the output of carbon dioxide.)

The average adult requires 25 C per kilogram (2·2 lb) of body weight in 24 hours to supply the basal metabolic needs when at rest—roughly 1 C per kilogram, per hour.

Factors influencing metabolism

1. It will be clear that more energy will be required when active muscular work is being carried out.
2. The basal metabolism of individuals varies according to their size and is actually related to the surface area of their body. The larger the surface area, the greater the area from which heat can be lost and, therefore, in order to maintain a constant body temperature there must be more heat production than in a person of less surface area.
3. The metabolic rate of children is relatively greater than in adults. This is accounted for by growth.
4. An increase in the body temperature, i.e. in feverish states, results in an increase in metabolism.
5. A decrease in the external temperature (i.e. during cold weather) raises the metabolic rate, since the body must then produce more heat to maintain a constant temperature.
6. The thyroid gland plays an important part in metabolism and

in disease of the gland when its secretion is excessive (thyrotoxicosis) the basal metabolism is increased; in cases of under-secretion (myxoedema and cretinism) the basal metabolism is low (see page 291).

Each individual, therefore, requires food of such Calorie value that it will supply energy to maintain the basal metabolism and also the energy requirements of muscular exercise, growth, etc.

It has been seen that the average adult requires in 24 hours about 25 Calories per kilogram (2·2 lb) of body weight for basal metabolism, i.e.

A man weighing 10 stone needs 1500 C per day.
(10 stone = 140 lb = 60 kg × 25 C = 1500 Calories.)

An additional supply of energy will be needed for his daily activities, the amount of which will depend on whether he is a sedentary worker or a heavy labourer. Thus the total requirements per day are about:

	Calories
woman	2500
average man	3000
man doing very heavy work	4000–5000

A well-balanced diet for the average man would be made up in the following way to give the necessary energy value:

	grams		*Calories*
protein	100	=	400
fat	100	=	900
carbohydrate	500	=	2000
			3300

Carbohydrate metabolism

Carbohydrates consist of carbon, hydrogen and oxygen, but in a totally different chemical combination to that of fat. Carbohydrates are widely distributed throughout the animal and vegetable kingdoms but are especially found in the latter. They include starches, sugars and cellulose and are present in such substances as bread, potatoes, jam, fruits and vegetables.

The common carbohydrates are:

glucose, dextrose or grape sugar, a monosaccharide.
sucrose or cane sugar }
maltose or malt sugar } disaccharides.
lactose or milk sugar }

The disaccharides are more complicated in chemical composition than the monosaccharide—glucose—which is the simplest form of carbohydrate found in the body.

The polysaccharides are more complex in composition than either of the former. Included in this group are starch, cellulose and glycogen.

Carbohydrates are easily digested and broken down by enzyme action into glucose, in which form they are absorbed.

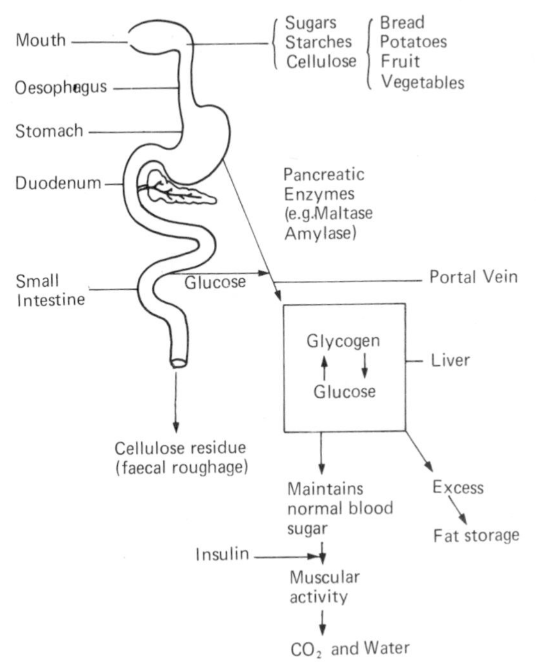

Fig. 160 Carbohydrate metabolism.

SUMMARY

1. Starches and sugars are all converted into glucose:
   (*a*) cooked starch into maltose by ptyalin of saliva.
   (*b*) all starch into maltose by amylase of pancreatic juice.
   (*c*) maltose into glucose by maltase of intestinal juice.
   (*d*) cane sugar into glucose and fructose by invertase of intestinal juice.
2. Glucose is absorbed by the stomach and small intestine and

conveyed by the portal vein to the liver where it is stored as glycogen.

3. When required, glycogen is reconverted into glucose which is conveyed to the tissues by the blood. In the presence of insulin, secreted by the islets of Langerhans of the pancreas, glucose is burnt into carbon dioxide and water. (NOTE: The normal fasting blood sugar is 80–120 mg per 100 ml.)

Fat metabolism

The term fats includes all fatty and oily foods. Like proteins, fats may be (*a*) of animal origin (fat meat, butter, cream and lard) or (*b*) of vegetable origin (olive oil, some kinds of margarine, and nut oils). They contain the elements carbon, hydrogen and oxygen (see also page 229).

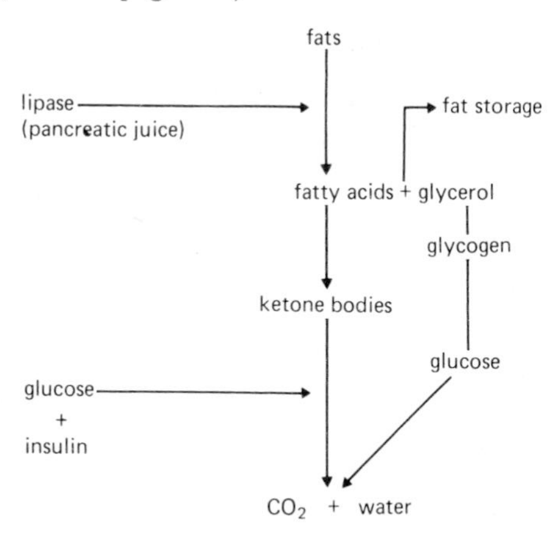

Many animal fats contain the vitamins A and D. These may also be introduced artifically into vegetable fats used in diet.

The fats of land mammals (as opposed to aquatic, e.g. whales) are highly **saturated**. Some vegetable oils (e.g. olive oil) are also highly saturated but others (e.g. corn oil) are rich in **unsaturated** fatty acids. A dietary excess of saturated fats may contribute to the formation of atheroma in the coronary arteries. A special margarine is recommended to certain patients with ischaemic heart disease since in ordinary margarine the unsaturated oils have become saturated during the manufacturing process (hydrogenation).

SUMMARY

1. Fats are split into fatty acids and glycerol by the enzyme lipase of the pancreatic juice.
2. In the presence of bile salts these are absorbed by the villi of the small intestine and recombined into fats.
3. From the lacteals of the villi they pass via the cisterna chyli and thoracic duct to the blood stream which takes them to the fat stores of the body.
4. When they are needed they are conveyed to the liver, altered in composition and taken to the tissues where they are converted into carbon dioxide and water—with the production of heat.
5. When the tissues are deprived of glucose, for example in starvation and some cases of diabetes, the oxidation of fat is incomplete and the intermediate products of metabolism, acetone and diacetic acid (ketones) pass into the blood stream and urine.

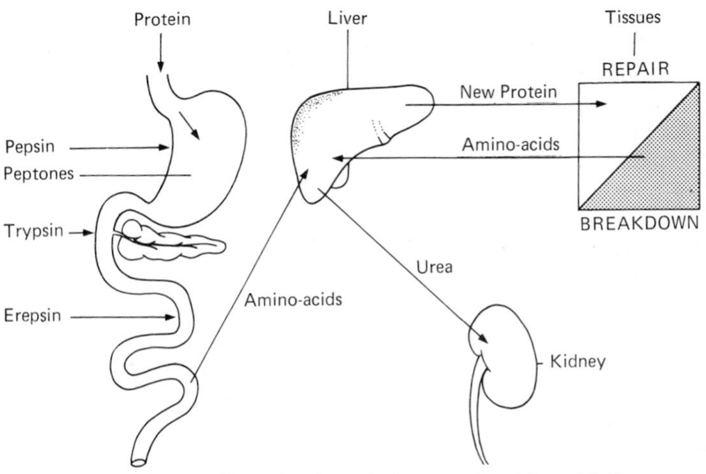

Fig. 161 Diagram illustrating the main features of protein metabolism.

Protein metabolism

It has already been stated that proteins are complex organic compounds containing nitrogen, sulphur and phosphorus in addition to carbon, hydrogen and oxygen (C, H, O, N, S and P), and that they enter into the composition of protoplasm itself. There are two types: (*a*) animal, and (*b*) vegetable.

Animal proteins. Meat, fish, eggs, milk and cheese are all rich in protein. Meat contains the protein of muscle called myosin. Eggs contain albumen and globulin. Milk contains caseinogen; and cheese, casein.

Vegetable proteins. Wheat, with peas, beans and lentils (the leguminous plants), contain a relatively good supply of protein but less than animal foods. The protein of what is called gluten; that of the leguminous foods, legumin.

SUMMARY

1. Proteins are first acted on by the enzyme pepsin (formed by the action of hydrochloric acid on pepsinogen) in the stomach and converted into proteoses and peptones.
2. Trypsin from the pancreatic juice and erepsin in the intestinal juice (succus entericus) converts peptones into amino-acids.
3. Amino-acids are absorbed by the villi of the small intestine and conveyed to the liver by the blood stream.
4. The nitrogen-containing portion of the amino-acids which are not required by the tissues is removed by the liver (deamination) conveyed to the kidneys and excreted in the urine as urea (page 283).
5. The remainder is used by the tissues for repair and body-building. (NOTE: The adult human body contains about 10 Kg of protein, most of which is in muscle.)

The metabolism of salts etc.

The salts of a number of chemical elements play an important part in the vital processes of the body. They are necessary for body-building and are an important constituent of bone. They help to maintain the appropriate balance of the fluids in the body and the reaction of the blood (page 284). Iron is necessary in the formation of haemoglobin.

The salts most commonly found in the body are the chlorides, phosphates and sulphates. Others include carbonates and oxalates.

Calcium. The main sources of calcium or lime salts are milk, cheese, green vegetables and animal foods. It is most plentiful in the body in the form of calcium phosphate. Its most important function is its presence in bone and teeth. It also plays a part in the clotting of blood and has an effect on the excitability of the nervous system. The metabolism of calcium is controlled by the parathyroid glands. Vitamin D is necessary for its absorption.

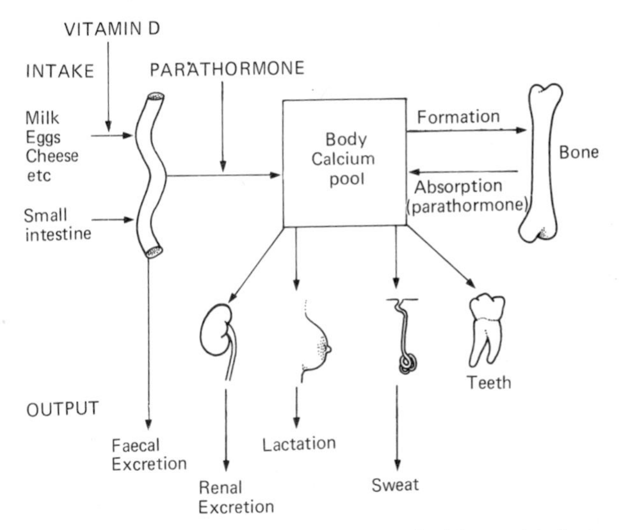

Fig. 162 Diagram illustrating main features of calcium metabolism.

Sodium and potassium. These are important and their metabolism is largely controlled by the cortex of the adrenal glands. Sodium chloride (common salt) is a constituent of most of the body fluids. Chlorides may be lost from the body in cases of excessive vomiting, hence the importance of their replacement by intravenous salines.

Iron. This is normally present in sufficient quantity in the diet. (egg-yolk, spinach, green vegetables and carrots) to supply the needs of the body. Its function is in the formation of harmoglobin and it is mainly absorbed in the duodenum and upper intestine and stored in the spleen and liver. Absorption is aided by the presence of ascorbic acid and hydrochloric acid.

In health about 10 per cent of food iron is absorbed, but when the body has become deficient in iron this may be increased to 50 per cent. The iron stores of the body normally amount to about 1 G, while the total amount circulating in haemoglobin is about 4 G. The quantity of iron absorbed daily from the diet is very small, consequently the body is very economical in its use, and the daily excretion is equally small. In females, however, there is a loss of about 30 mg during a menstrual period. Likewise, during pregnancy the mother contributes iron to the foetus to the extent of about 1 mg daily, so that additional iron is usually necessary.

Iodine. This is normally present in the minute quantities necessary in drinking-water. It is stored in and its metabolism controlled by the thyroid gland.

Role of water in the body

Water is an essential constituent of every tissue and cell in the body and plays an important part in all the vital processes. Its main functions are:

1. The tissues consist of about 60 per cent water and it forms the basis of the body fluids—blood plasma, lymph and the tissue fluids.
2. It acts as a solvent for the important salts of sodium, potassium and other minerals necessary in metabolism. There is a very important balance between salt and water intake and output which, in health, is maintained at a steady level but which may be seriously upset in disease. In health, any excess of the intake of water and salt above the minimum which is essential is rapidly excreted by the kidneys in the urine.
3. Water is, therefore, required for the elimination of excess of salts and waste products which can be dissolved in it. Urine and sweat both consist of over 90 per cent water.

Intake of water. Many foods consist largely of water but, in addition, 1500 ml (3 pints) a day are necessary to make a total of 2 to 2·5 litres (4 to 5 pints). As has already been pointed out, it is unlikely that in health too much will be taken as any excess will be rapidly excreted by the kidneys.

Output of water. Water is excreted from the body by:

(*a*) by the kidneys as urine,
(*b*) by the skin as sweat,
(*c*) by the lungs as water vapour,
(*d*) by the bowels in the faeces.

Normal intake daily			*Normal output daily*		
	ml	(fl. oz)		ml	(fl. oz)
fluid	1500	50	urine	1500	50
food	1100	40	skin	600	21
	——	—	lungs	400	15
	2600	90	faeces	100	4
				——	—
				2600	90

Distribution of water. It has been seen that water is distributed in:

(*a*) the blood plasma (extracellular fluid)
(*b*) the tissue fluid
(*c*) the cells (intracellular fluid)

In health there is a constant balance between the amount of intracellular fluid and the extracellular fluid. Likewise there is a balance between the salt content of these fluids. This is maintained largely by the process of osmotic pressure.

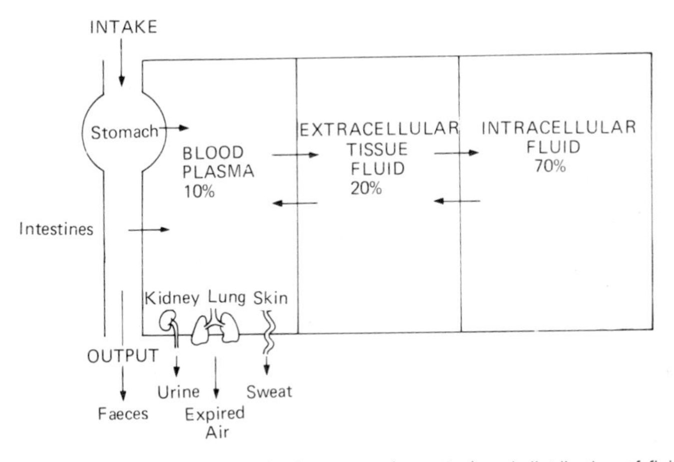

Fig. 163 Diagram illustrating the intake, output (excretion) and distribution of fluid in the body.

Osmotic pressure. When a salt such a sodium chloride is dissolved in water its minute particules or molecules exert a pressure on the walls of the structure containing the solution. The walls of the cells and the capillaries are membranes through which both water and molecules of salt can pass, although water passes more easily than salt molecules. The pressure exerted by the molecules on such a membrane is called osmotic pressure and the degree of pressure is dependent on the number of molecules present. In other words, the greater the concentration of the solution the more molecules it contains and the greater its osmotic pressure.

Solutions which have the same osmotic pressure are said to be **isotonic**. In human beings the tissues are isotonic with a solution containing 0·9 per cent sodium chloride in water which is known as normal saline. A stronger solution is called *hypertonic* and a weaker one *hypotonic*.

Water, salts and waste products are constantly passing backwards and forwards between the plasma, tissue fluids and cells. At the same time the osmotic pressures of all three are kept constant. This balance is maintained in the following way. If the osmotic pressure of the intracellular fluid rises and exceeds that of the tissue fluids surrounding it (i.e. the intracellular fluid becomes hypertonic on account of containing an increased number of salt molecules) water will pass from the weaker tissue fluid into the cells until their

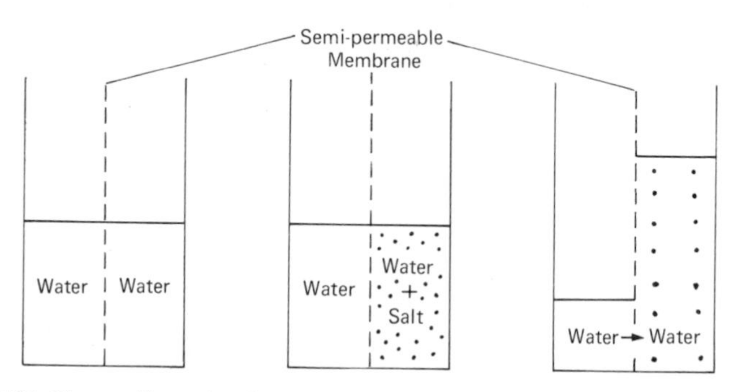

Fig. 164 Diagram illustrating the passage of water through a membrane into a strong (hypertonic) salt solution.

osmotic pressures are equal and become isotonic again. The opposite will happen if the tissue fluids or the plasma become hypertonic in respect of the cells, and water will pass from the cells until the osmotic pressures are equal.

In other words water passes through an animal membrane from the weaker to the stronger solution to dilute it until the fluid on both sides of the membrane becomes isotonic.

If, therefore, there is water deficiency in the body the plasma and tissue fluids will tend to become more concentrated and hypertonic, and water will be extracted from the cells which become dehydrated. On the other hand, if there is excess of salt in the body water is retained and not excreted by the kidneys. This accumulation of fluid in the tissues leads to oedema or dropsy. In this condition, therefore, it is customary to restrict the intake of further water and salt.

It will be recalled (page 171) that the plasma proteins help to maintain the osmotic pressure of the blood.

There may be serious loss of fluid from the body in cases of severe vomiting, diarrhoea and excessive sweating, leading to the condition known

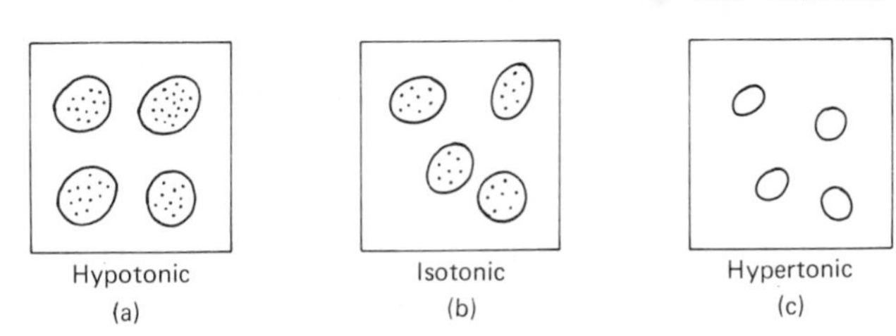

Fig. 165 Diagram illustrating the effects of various strengths of salt solution on red blood cells. (*a*) Cells swell because water enters them. (*b*) No change; electrolytes equal. (*c*) Cells shrink; water passes out into strong solution.

as dehydration. This is relieved by giving fluids (e.g. saline) by mouth per rectum or intravenously.

Requirements of a normal diet

1. It must have sufficient Calorie value (3000 C for the average man doing some physical work).
2. It must contain protein (one-fifth), fat (one-fifth) and carbohydrate (three-fifths) in roughly these proportions.
3. It must contain a certain proportion of fresh foods, green vegetables, etc., in order to provide:
   (*a*) vitamins,
   (*b*) roughage or waste products to stimulate the action of the bowel.
4. It must contain a proportion of salts, especially those of sodium, calcium, potassium and iron.
5. It must contain an adequate amount of water.
6. It must be palatable, attractive and easily digested.
7. Children require an adequate supply of milk.

Malnutrition. Malnutrition may consist of either under- or over-nutrition. Protein and calorie deficiency is common in under-developed countries whilst over-nutrition, causing obesity, is common in developed countries.

The vitamins

Vitamins or accessory food factors are substances the presence of which in the food is essential for normal health and growth. Our knowledge of their composition and nature has increased rapidly in recent years and it is now possible to manufacture a number of them in the laboratory. They are found in many natural foodstuffs but, in order to maintain health, only minute quantities, compared with the bulk of other articles of diet, are required.

The nomenclature for describing them has now become rather complicated. Originally they were described alphabetically. Subsequently, those whose composition was known were given chemical names or referred to according to the medical condition which resulted from their absence, and finally, now that they are manufactured, a number of proprietary names exist. Perhaps the simplest method in the first place is to learn and think of them by the alphabetical terminology. Increasing knowledge, however, has shown that more than one vitamin may be included in each alphabetical group.

Vitamin A (axerophthol)

Sources. This vitamin is found especially in milk, butter, egg-yolk, halibut- and cod-liver oil. It is also formed in the liver from carotene, a substance found in certain vegetables such as spinach, tomato, raw carrots and other green vegetables. Vitamin A is also necessary for normal health and growth.

Effects. Its deficiency produces:

1. A disorder of the eyes called xerophthalmia which is rare except in under-developed countries.
2. Night-blindness, the difficulty experienced by some people in seeing in the dark.

Vitamin B complex

A number of separate substances each having a different action have been obtained from this group of water-soluble vitamins. The best known are:

Vitamin B_1 or aneurine, thiamine.
Vitamin B_2 or riboflavine, lactoflavine.
Vitamin B_3 or nicotinic acid, nicotinamide.
Vitamin B_6 or pyridoxine.
Vitamin B_{12} or extrinsic factor (cyanocobalamin).

Also associated with the Vitamin B complex are folic acid, pantothenic acid and para-amino-benzoic acid. Deficiency in folic acid results in an uncommon type of anaemia which, however, responds to treatment when it is administered. Pyridoxine may be used in the treatment of radiation sickness.

Thiamine, aneurine (B_1). The anti-neuritic vitamin.

Sources. This is found in many natural foods including peas, beans and lentils, wholemeal bread, yeast, raw carrots and cabbage.

Most of the vitamin is lost when wheat is milled and the husk is removed to produce white flour.

Effects. Thiamine assists cellular enzymes in the breakdown of carbohydrates and also has an important role in the normal function of nervous tissue, especially in the peripheral nerves. Deficiency of the vitamin may lead to wet **beri-beri**, in which there is oedema due to cardiac failure, or dry beri-beri in which there is a polyneuritis.

Riboflavine (B_2). Deficiency of this vitamin is associated with soreness of the lips and redness of the eyes. It is present in liver, kidney and milk.

Nicotinic acid—niacin (B_3). *Sources.* This is closely associated with vitamin B_1 in its natural distribution, but can be manufactured chemically.

Effects. If deficient, a disease called pellagra, characterized by gastrointestinal disturbances, a skin eruption and often mental changes, develops.

Cyanocobalamin (B_{12}). In addition to its importance in the formation of red blood cells, a deficiency may also result in subacute combined degeneration of the spinal cord, a complication of pernicious anaemia. See pages 168, 217.

Vitamin C (ascorbic acid)

The anti-scorbutic vitamin.

Sources. This is especially found in fresh vegetables, tomato, and the acid fruits, e.g. orange, lemon, grape-fruit. It is present in milk, but is largely destroyed by pasteurization, hence the importance of tomato or fruit juice for children fed on cow's milk. It can also be manufactured and supplied in tablet form.

Effects. Deficiency results in scurvy, a disease which may affect either infants or adults. It is characterized by haemorrhages into the tissues, under the skin and from the gums. It aids the absorption of iron in the duodenum.

Vitamin D

The anti-rachitic vitamin—(**calciferol**)

Sources. This very important vitamin is more complicated than some of the others because it has two distinct sources:

(*a*) Its natural distribution especially in cod-liver oil and, to some extent, in butter and eggs.

(*b*) The body is able to manufacture it for itself by the action of sunlight on a substance in the skin called dehydrocholesterol.

Vitamins

Vitamin	Names	Sources	Effects	Notes
A	*Axerophthol*	Milk, butter, cream, egg-yolk, cod-liver oil, spinach, tomato, carrots	Prevents night-blindness Prevents eye disease (xerophthalmia)	Found in vegetables as carotene which is probably converted into vitamin A by liver.
B_1	*Thiamine* *Aneurine* Anti-neuritic	Peas, beans, lentils, wholemeal bread, husks of cereals (rice, wheat, oats, barley), yeast, cabbage, raw carrot,	Prevents beri-beri	Riboflavine (vitamin B_2) is also associated with this vitamin.
B_3	*Nicotinic acid* (*Niacin*)		Prevents pellagra	
B_{12}	*Cyanocobalamin*	meat, liver	Prevents pernicious anaemia	The erythrocyte-maturing factor.
	Folic acid	Liver, kidneys, yeast, green vegetables	Erythrocyte-maturing factor	Absence causes megaloblastic anaemia.
C	*Ascorbic acid* Anti-scorbutic	Acid fruits: orange, lemon, grape-fruit, tomato, swede, cabbage	Prevents scurvy	Easily destroyed by heat and alkalis.
D	*Calciferol* Anti-rachitic	Cod-liver oil, eggs, butter, milk, cream	Prevents rickets and osteomalacia Necessary for calcium absorption	Also manufactured in the body by the action of sunlight (ultra-violet rays) on the skin.
E	*Tocopherol*	Wheat germ, olive oil	? Prevents abortion in some animals	—
K	*Menaphthone*	Liver, fats, spinach, other vegetables	Necessary for production of prothrombin	Absence leads to excessive bleeding.
P	*Hesperidin* *Rutin*	—	Affects permeability of capillaries?	—

It is the ultra-violet rays of sunlight which have this action and, 'artificial sunlight' has the same

Effects. Lack of vitamin D causes rickets, a disease of young children characterized by deformities of the bones which are deficient in calcium. It plays an important part in the calcium metabolism of the body and the proper absorption of calcium from the intestine.

Vitamin E (tocopheryl acetate)

This is present in the germ of wheat and its deficiency is said to result in a tendency to abortion in some animals. It may possibly have some effect on the nervous system.

Vitamin K (menaphthone)

This is present in liver, spinach and other green vegetables. It is necessary for the production of prothrombin (page 173) in the body and, therefore, its deficiency results in a tendency to increased bleeding, such as may occur in some cases of obstructive jaundice, steatorrhoea and haemorrhagic disease of the newborn.

Vitamin P (hesperidin)

Little is known about this vitamin, but the functions of the capillaries appear to be influenced by it.

Roughage

Adequate peristalsis in the bowel only occurs when there is a sufficient residue for the muscle in its wall to act upon. This is because the normal stimulus to peristaltic action is stretching of the walls by the bowel contents. The undigested faecal residue which performs this function is sometimes referred to as roughage. In the normal diet it is mainly provided by the cellulose found in green vegetables, salads, fruit and wholemeal bread.

Insufficient roughage leads to sluggish peristalsis and a tendency to constipation. In some conditions, however, it is necessary to provide a 'low residue' diet in which, therefore, these articles should be avoided.

The service of food

It is a matter of common sense that food should be palatable, attractive and easily digested. The importance of the psychological factor in stimulating the flow of the saliva and gastric juice has been referred to (pages 198, 215). This is especially desirable when the appetite is otherwise poor or the digestion disturbed by disease.

So often good food is spoiled by lack of variety, bad cooking or indifferent service.

Approximate Times for Retention of Foods in Stomach

	Hours		*Hours*
Pork, roasted	5	Mutton, roasted	3–3½
Veal	4½	boiled	3
Wheaten bread	3–4	Poultry, boiled or roasted	2½–4
Apples	3–4	Potatoes	2½–3½
Beef, boiled	3	Ham, boiled	2–3
roasted	3–4	Oysters, raw	2
Cabbage, carrots or turnips	3–4	Milk	2
Cheese	3–4	Fish, boiled	1½–2½
Eggs, fried or boiled hard	3–3½	Rice, sago, tapioca	1–2
Eggs, raw	2	Tripe	1

Milk

1. *Human milk* is the natural food for all young infants.
2. *Cow's milk* is a valuable and necessary food for all children. It is a useful article of diet for adults and especially for sick persons. It can be modified so that it is suitable for young infants but can only be a 'second best' to breast milk.

The following table shows the composition of human and cow's milk in percentages:

	Protein	Fat	Sugar (*lactose*)	Salts	Water
Human milk	1·5	3·5	6·5	0·2	88·3
Cow's milk	3·5	4·0	4·5	0·7	87·3

It will be noted first of all that the protein of cow's milk is at least double that in human milk. The first step, therefore, in approximating cow's milk to human milk in composition is to dilute it with an equal quantity of water. This will equalize the amount of protein but will have reduced the relative amounts of both fat and sugar, so that cream and lactose must be added to the milk-water mixture. The protein of cow's milk, however, is less digestible.

Although milk is the natural food for infants, and can supply most of the energy requirements of older children and adults if taken in sufficient quantity (1200–1800 ml, 2–3 pints daily according to age), its contents of salts, iron and vitamins is insufficient to maintain health in the latter group unless supplemented by other foods. Further, a long-continued milk diet becomes very monotonous.

Dangers of milk. Although an excellent food which should be included in all diets, milk may be a source of danger to health. It forms a good breeding-ground for bacteria and may become infected (*a*) by tuberculosis from a tuberculous cow, and brucellosis from an infected cow, (*b*) by germs such as those of typhoid and scarlet fever in the handling and distribution.

Composition of milk. The foodstuffs provided by milk are:

1. **Protein**. The main protein of milk is caseinogen which is converted by the rennin of the gastric juice into the insoluble form—casein.
2. **Fat** is in the form of minute globules and can be removed as cream.
3. **Sugar**. The sugar of milk is the di-saccharide, lactose.
   Vitamins are present in varying quantities, but vitamin C is easily destroyed by heat.

Bread. This is made from flour and contains approximately 6 per cent protein and 55 per cent carbohydrate, with a small amount of fats and salts. In white bread the husk is removed from the wheat and, unless specially treated, lacks the vitamins present in the husk (particularly vitamin B), a defect absent from wholemeal bread. The protein of bread is called gluten.

Eggs. The white of egg consists mainly of egg-albumen and egg-globulin, proteins which have already been mentioned. The yolk, in addition to protein, contains fats, salts and some vitamins.

Cheese. This is made from milk, consists mainly of protein (casein 33 per cent) and fat (27 per cent) and is, therefore, a very valuable food.

Beverages

Tea and **coffee** contain aromatic substances which give them their flavour. They have a stimulating effect due to the presence of caffeine and are useful for providing fluid and acting as vehicles for milk, but have no food value in themselves. Tea, in addition, contains tannin or tannic acid, a substance which hardens protein and makes it less digestible. The amount is increased in strong or over-brewed tea which, though never desirable, should always be avoided with a meal containing meat.

Alcohol. The use or value of alcoholic beverages is a matter of considerable controversy, most of which is finally settled by personal opinion. Certain physiological facts are clear:

1. Alcohol is not a true stimulant of the nervous system. It acts by depressing the highest centres of the brain, thereby lessening the individual's normal control, self-consciousness and anxiety.
2. The immediate effect is a dilatation of the blood vessels of the skin, producing a sensation of warmth without any actual increase in temperature.
3. A similar effect is produced on the mucous membrane of the stomach, which may at the time improve the appetite, but ultimately tends to injure it (gastritis).
4. It has some food value since it is oxidized in the tissues into carbon dioxide and water with the production of heat and, therefore, of energy. Further, it is quickly absorbed from the stomach and small intestine.
5. Taken in large quantities over a long period it can undoubtedly act as a poison and may have serious effects on the mental and physical health of the individual. In some people (chronic alcoholics) alcohol must be regarded as a drug of addiction.

Condiments. Mustard, pepper, sauces, salt, etc., are all flavouring materials which help to stimulate the appetite partly by their flavour and partly by their slightly stimulating effect on the stomach, but when taken in excess may be harmful.

QUESTIONS

1. How and in what form are the waste products of metabolism got rid of from the body?
2. What is the value of fat as a food and what becomes of it after it is eaten?
3. How and where are proteins (*a*) digested and (*b*) absorbed in the body? Give a short account of their fate after absorption.
4. What is a carbohydrate? Give two examples. Describe the means by which carbohydrate is (*a*) digested and (*b*) utilized in the body.
5. How are the various foodstuffs absorbed after digestion? How are they utilized in the body?
6. What are vitamins? Mention some of the sources and the value to health of any two of them.
7. How is water used by the human body? What organs are concerned with fluid excretion?

8. What are the essential constituents of a normal diet? Discuss the importance of the substances you mention.
9. Name good sources of the following and explain their importance in normal nutrition (*a*) first-class protein, (*b*) vitamin C, (*c*) iron, (*d*) fat.

14 The Skin. Regulation of Body Temperature

The skin

Structure

The skin or integument is the outer covering of the body and has the following parts:

(*a*) The outer layer or epidermis.
(*b*) The inner layer or dermis (corium or cutis vera) containing:
sweat glands,
sebaceous glands,
nerve endings,
blood vessels, lymphatics and capillaries.
(*c*) The appendages: nails and hair.

The epidermis. This consists of stratified squamous epithelium which is clearly divided into superficial and deep layers. The superficial or horny layer is formed by flattened cells which have lost their nuclei and whose protoplasm has been converted into a hard substance called **keratin**. These cells are actually dead and the most superficial layers are constantly being shed and replaced by cells from the deeper layers.

The cells of the deepest layers are rounded and contain nuclei. The flattening of the cells gradually becomes apparent as the surface is approached. The epidermis contains no blood vessels, but lymph circulates between the cells of the deepest layers and nourishes them. Dark pigment (melanin) may be present in the deepest layers of the epidermis.

The dermis. The dermis or true skin is the inner layer and consists of fibrous and connective tissue. It contains sweat glands, sebaceous glands, nerve fibres and their endings, blood vessels and capillaries.

The **sweat glands** are found in the skin all over the body, but are most abundant in the axillae, palms of the hands, soles of the feet and on the forehead. Each gland consists of a tube coiled up into a ball lying in the dermis from which the free end passes as a duct through the epidermis and opens on its surface by a small opening or pore. The coiled end is surrounded by a network of capillaries.

The **sebaceous glands** are especially numerous in the skin of the face, scalp and all hairy parts of the body. Each consists of a number of small alveoli lined by epithelium from which a duct leads through the epidermis to the surface, usually opening into a

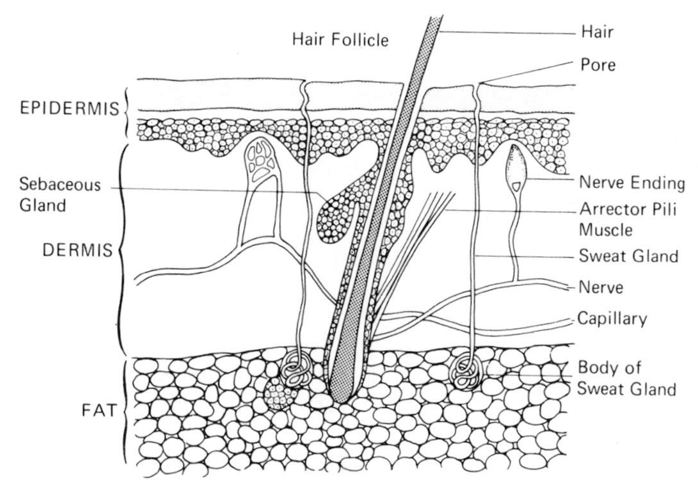

Fig. 166 Section through the skin (diagrammatic).

hair follicle. They form a greasy secretion called **sebum**, which helps to keep the skin supple and the hairs from becoming dry and brittle. The ceruminous or wax glands in the meatus of the ear are a modified form of sebaceous gland (see page 344).

The **nerve endings** in the skin are mainly sensory, although some motor nerves are distributed to the blood vessels (vasodilator and vasoconstrictor). The sensory nerve fibres end in small bulbous enlargements, the tactile corpuscles, which are therefore the true end-organs of the cutaneous nerves.

The nails. These are modifications of the epidermis and are horny plates corresponding to the claws and hoofs of other animals. Each consists of a closely packed mass of keratin which grows from

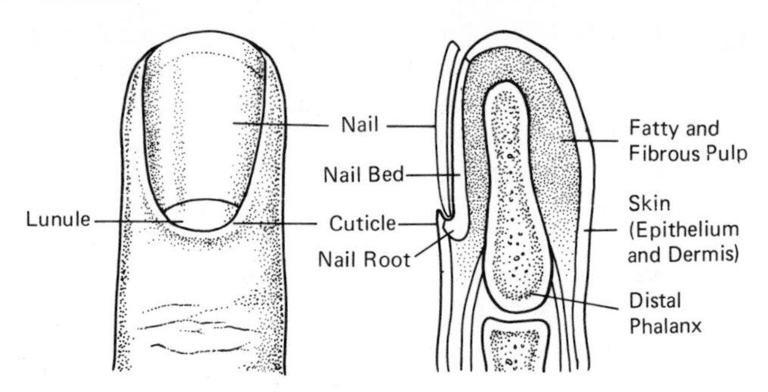

Fig. 167 Diagram of a nail and nail bed.

special epithelial cells situated at its root and lies on its bed formed by the dermis.

The hairs. These structures develop in the hair follicles of the skin. The hair follicle is a depression in the skin having at its deepest part a small papilla from which the hair grows. The portion of the hair contained within the follicle is called the root, and to it is attached a minute muscle (the arrector pili), contraction of which causes the hair to become raised or erect—an effect more marked in animals than in man.

The sweat

Sweat is a clear, colourless fluid containing 99 per cent water, with salts, especially sodium chloride, and minute quantities of other waste products.

The secretion of sweat is controlled by the action of the sympathetic nerves. There is a constant secretion of sweat from the glands, but, at rest, it evaporates so quickly that it is unobserved and is then called **insensible** perspiration. When the amount formed is increased or its evaporation is delayed the sweat becomes visible and it is described as **sensible** perspiration.

NOTE: There are two types of sweat gland, viz (*a*) *ecerine,* found over most of the body, which secrete a clean watery fluid; and (*b*) *apocrine,* mostly in the axillae and genital regions, producing a viscid milky secretion.

Functions of the skin

1. It forms a protective covering for the body.
2. It contains the end-organs of the sensory nerves (pain, touch and temperature).

3. It secretes sebum.
4. It is capable of absorbing small amounts of oily substances.
5. It gives origin to the hair and nails.
6. It contains dehydrocholesterol, a substance converted into vitamin D by the action of sunlight (page 247).
7. By secreting sweat it:
   (*a*) acts as an excretory organ, removing waste products, viz. water and salts:
   (*b*) plays an important part in the regulation of body temperature.
8. Some drugs are absorbed through the skin.

Regulation of body temperature

In health, the temperature of the body is maintained at an average level of 98·4° F (37°C), but shows a slight variation during the day, being a little lower in the early morning.

The constant level of body temperature is maintained by a balance between the heat produced and the heat lost by the body. It is controlled by a **heat-regulating centre** situated in the brain.

1. Heat production

Heat is produced in the body by the chemical processes of metabolism. This takes place chiefly in the muscles, but also to a lesser extent in the liver and other glandular structures. The amount formed is greatly increased by muscular exercise.

2. Heat loss

Heat is lost principally through the skin, and to a lesser degree by the expired air, the urine and the faeces. Loss of heat through the skin takes place in the following ways:

(*a*) by **radiation** to the surrounding air;
(*b*) by **conduction** to the clothing;
(*c*) by **convection** of the air circulating round the body;
(*d*) by **the evaporation** of sweat.

The loss of heat depends on two main factors both of which are under the control of the autonomic nervous system: (*i*) The amount of blood circulating in the vessels in the skin. (*ii*) The amount of sweat formed and the rate of its evaporation.

The rapid evaporation of any liquid is a physical process which results in a fall in temperature. This is clearly illustrated by placing a few drops of ether or methylated spirit on the skin, which immediately begins to feel abnormally cold as the liquid evaporates.

The body makes use of this fact in the regulation of its temperature by the evaporation of sweat. Increased heat production by muscular exercise or a rise in external temperature results in increased sweating. There is therefore more sweat available for evaporation and consequently greater heat loss.

It has been stated that the amount of blood circulating in the skin is also important. In cold weather the blood vessels of the skin are contracted as a result of the activity of the vasoconstrictor nerves. In hot weather or after muscular exercise they are dilated by the action of the vasodilator nerves so that there is more blood near the surface of the body which can lose heat by the processes of radiation, conduction and convection. Further, this increase in blood supply corresponds with the increase in the activity of the sweat glands.

Sudden chilling of the body surface results in shivering and the production of 'goose-flesh', which are reflex efforts on the part of the body to increase heat by muscular action. The latter is due to the contraction of the arrectores pilorum muscles attached to the hairs of the skin.

Loss of heat from the body is also prevented by clothing, which hinders radiation, conduction and convection. Heat is also retained by the subcutaneous layer of fat.

QUESTIONS

1. Describe the structure and functions of the skin.
2. Give an account of the mechanisms concerned with the regulation of body temperature.

15 The Respiratory System

The essential features of respiration are the transference (*i*) of oxygen from the atmosphere to the tissues, and (*ii*) of carbon dioxide (CO_2) from the tissues to the outer air. In addition to this interchange of gases, some water vapour is also excreted from the body.

It must be clearly understood, however, that there are two phases in the interchange of gases:

1. **External respiration**, or the absorption of oxygen from the air into the blood and the excretion of carbon dioxide from the blood into the air. This takes place in the lungs.

2. **Internal or tissue respiration**, in which the oxygen is transferred from the blood to the tissues of the body which at the same time give up carbon dioxide. This change takes place through the walls of the capillaries. It follows that external respiration is mainly a function of the respiratory system and the act of breathing; while internal respiration is entirely dependent on the blood flowing in the capillaries and the efficiency of the circulation. The importance of the circulatory system in maintaining the respiratory process is shown by the following summary:

LUNGS:

1. Absorption of oxygen from the air
2. Excretion of CO_2 from the blood

CIRCULATORY SYSTEM:

1. Arteries: carry oxygen in the haemoglobin of the red cells to the tissues.
2. Capillaries: tissues absorb oxygen and give up CO_2 to the blood.
3. Veins: convey venous blood containing excess of CO_2 and a deficiency of oxygen from the tissues to the lungs.

The air passages

The respiratory system consists of the lungs and air passages. In its course to the lungs the air passes through the following structures:

nasal cavities
pharynx
larynx
trachea
bronchi
bronchioles
alveoli or air cells

(bronchi, bronchioles, alveoli or air cells: the lungs)

Nasal cavities

1. The anterior nares or nostrils form the entrances to the nasal cavities. Small hairs inside the anterior nares act as a filter for dust in the inspired air.
2. The nasal cavities are separated into right and left portions by the nasal septum (formed above by the perpendicular plate of the ethmoid, behind by the vomer and in front by the carti-

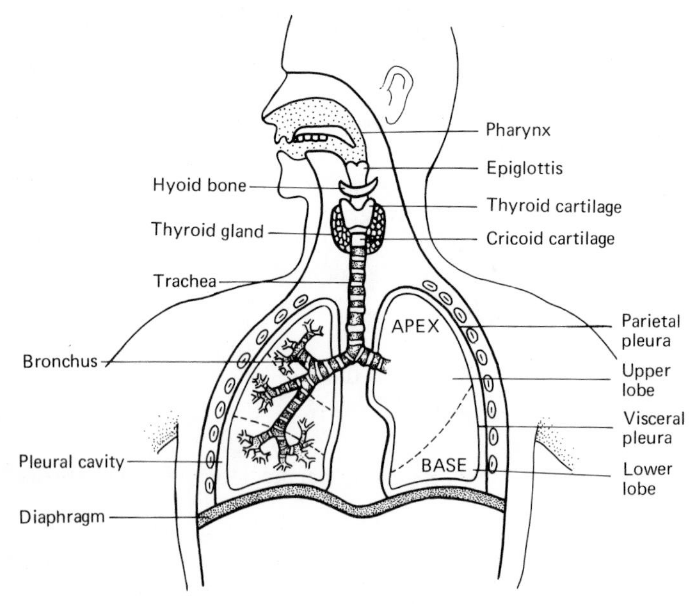

Fig. 168 The respiratory system.

lage of the septum). Each cavity is lined by mucous membrane covered with ciliated columnar epithelium and is plentifully supplied with blood. The surface area of the nasal mucous membrane is increased by the presence of three (upper, middle and lower) turbinate bones (conchae) which project medially from the lateral wall of each cavity. The importance of this increased surface area is that the air entering the respiratory tract may be warmed and moistened before reaching the lungs. The floor of the nasal cavities is formed by the upper surface of the hard palate, and the roof by portions of the frontal, ethmoid and sphenoid bones. The maxilla forms the main part of the lateral wall.

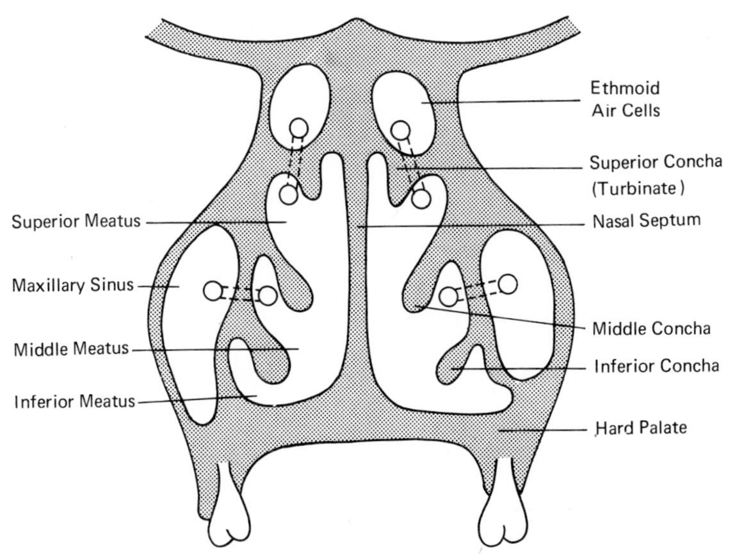

Fig. 169 Diagram of the nasal cavities, showing the conchae (turbinate) bones, and the ethmoid and maxillary sinuses. (Note that the maxillary sinus opens into the middle meatus.)

3. The posterior nares are situated at the back of the nasal cavities and constitute the entrances to the nasopharynx.

In addition to the anterior and posterior nares, each nasal cavity has the following openings:

1. The maxillary antrum (of Highmore) (Fig. 46).
2. The frontal, ethmoidal and sphenoidal air sinuses (Fig. 139).
3. The nasolacrimal duct which conveys the tears from the conjunctival sac to the nose (see also page 352).

Pharynx

The **nasopharynx** lies at the base of the skull, immediately behind the nasal cavities, which open into it via the posterior nares. On its lateral walls are the openings of the Eustachian tubes, which connect it to the middle ears. In its posterior wall is the lymphoid tissue known as the nasopharyngeal tonsil or, when enlarged, as 'adenoids'. The nasopharynx is continuous below with the oropharynx.

The **oropharynx** is continuous in front with the buccal cavity ('mouth') and below with the laryngeal part of the pharynx. The tonsils lie in its lateral walls.

The **laryngeal part** of the pharynx is continuous with the oesophagus. Near its upper end, on each side, is a small recess, the pyriform fossa, in which foreign bodies may lodge.

Larynx

Besides acting as part of the air passages the larynx is modifed in structure to enable it to perform the special function of voice production (page 272). It is situated in the mid-line of the neck between the pharynx above and the trachea below. It is placed in front of the oesophagus and corresponds with the levels of the fourth, fifth and sixth cervical vertebrae.

Structure of Larynx. The larynx consists of a framework of the following hyaline cartilages:

(*a*) The **thyroid** cartilage, which is the largest and consists of

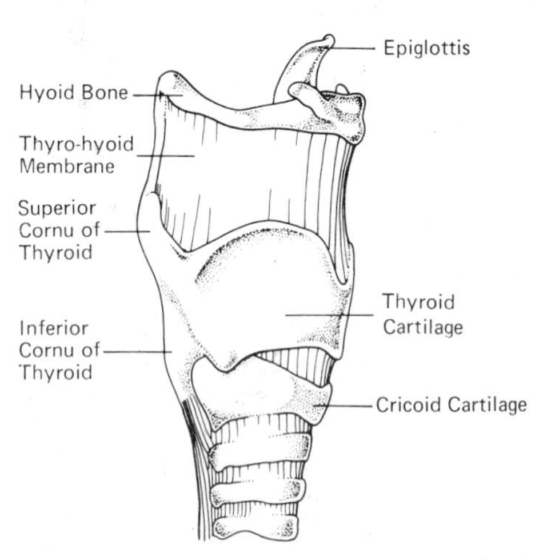

Fig. 170 The cartilages of the larynx (right lateral view).

two side wings united in the mid-line in front to form an angular projection, sometimes called the Adam's apple.

(*b*) The **cricoid** cartilage, situated below the thyroid cartilage, forms the lowest part of the larynx which it connects to the trachea. It is circular in shape with an expansion at the back, giving it a resemblance to a signet ring.

The lobes and upper poles of the thyroid gland lie on each side of the cricoid and thyroid cartilages.

(*c*) The **arytenoid** cartilages are two small structures situated on the upper surface of the expanded signet portion of the cricoid cartilage. They are shaped like pyramids and give attachment to the posterior ends of the true vocal cords. In front the vocal cords are attached to the posterior surface of the thyroid cartilage. The tension of the cords is varied by muscles which rotate the arytenoid cartilages and, in this way, the pitch of the voice is altered.

(*d*) The **epiglottis**. This is a leaf-shaped plate of yellow elastic fibrocartilage situated in an upright position between the base of the tongue and the upper opening of the larynx. Its lower stalk-like portion is attached by a ligament to the thyroid cartilage. The main function of the epiglottis is to prevent food from entering the larynx during the act of swallowing.

(*e*) The **hyoid** is a horse-shoe-shaped bone lying between the mandible above and the larynx below. It is situated at the base of the tongue and gives attachment to this and various other muscles. It does not actually take any part in the formation of the true larynx.

The larynx is lined by mucous membrane which, except over the vocal cords, is covered with ciliated columnar epithelium.

The vocal cords. The **true vocal cords** are fibro-elastic bands extending from the posterior aspect of the thyroid cartilage in front to the arytenoid cartilages behind. The **false vocal cords** are two loose folds of mucous membrane situated above the true cords which do not appear to play any special part in voice production (see Fig. 134, page 193).

The trachea

The trachea or windpipe is 12 cm ($4\frac{1}{2}$ in) long and about 2·5 cm (1 in) in diameter. Its upper half is situated in the mid-line of the neck, its lower half in the superior mediastinum of the thorax. It lies in front of the oesophagus and ends opposite the fourth thoracic (dorsal) vertebra, where it divides into the two main bronchi (see Fig. 168).

The trachea consists of a number of C-shaped rings of cartilage connected by fibrous tissue and having the opening of the C posteriorly. It is lined by mucous membrane and therefore the posterior wall, which lies in front of the oesophagus, consists of mucous membrane unsupported by cartilage. The function of the rings of cartilage is to keep the windpipe permanently open so that its walls do not collapse like those of the oesophagus. It is also lined with ciliated columnar epithelium and cells which secrete mucus. Its upper four rings are crossed by the isthmus of the thyroid gland.

Tracheostomy. This is an operation which consists of making an opening in the upper rings of the trachea below the cricoid cartilage and inserting a tube in cases of laryngeal obstruction and other conditions in which there is severe respiratory embarrassment.

The bronchi

The trachea ends by dividing or bifurcating into the two main bronchi opposite the level of the fourth thoracic vertebra. Each bronchus (right and left) passes to the corresponding lung. From each main bronchus, numerous smaller bronchi are given off like the branches of a tree, and the smallest bronchial tubes are called **bronchioles**.

The structure of the bronchi is similar to that of the trachea, consisting of incomplete hoops of hyaline cartilage lined with mucous membrane covered with ciliated columnar epithelium. In addition, they have some plain involuntary muscle in their walls. The structure of the bronchioles is similar to that of the bronchi, but they contain no cartilaginous hoops. Instead, the muscular tissue is more developed.

It is spasm of the involuntary muscle fibres found in the walls of the bronchi and bronchioles which occurs in asthma. The contraction of these muscles causes narrowing of the bronchial tubes and consequently difficulty in the passage of air through them. This spasm is relaxed by adrenaline (page 297), aminophylline and similar drugs.

Alveoli or air sacs. Each bronchiole terminates in an irregular sac made up of a number of air pockets. These pockets are lined with a delicate layer of flattened epithelial cells and are surrounded by numerous capillaries, through the walls of which the interchange of gases takes place. The blood in the capillaries is conveyed to the lungs by the pulmonary artery.

The lungs

The lungs are a pair of conical-shaped organs, each enveloped in a serous membrane, the pleura. They occupy the greater part of the thoracic cavity. The lungs are separated from each other by the mediastinum which contains the heart and great vessels, the oesophagus and, in its upper part, the trachea. Each lung is divided by deep fissures into lobes. The right lung has three lobes, upper, middle and lower; and the left two, upper and lower. The lung is described as having a mediastinal and a costal surface, an apex and a base.

The outer or costal surface is in contact with the wall of the pleural cavity which consists of the ribs and intercostal muscles and is lined by pleura.

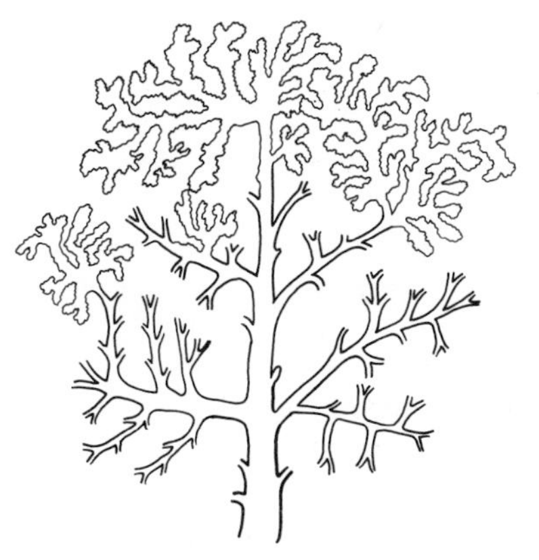

Fig. 171 Terminal air cells of a bronchiole.

The medial surface of the lung is applied to the mediastinum. Its chief feature is the presence of the **hilum** where the main bronchus and pulmonary artery enter and the pulmonary veins leave the lung. Also at the hilum are lymph nodes which may be enlarged by disease, e.g. tuberculosis or cancer.

The **apex** rises into the root of the neck for about one inch above the clavicle. The **base** is concave and is related to the upper surface of the diaphragm.

Structure of lungs. On examination the lung feels spongy and, if a portion is dropped into water, it will float because of the air which it contains.

Each lobe of the lung is made up of a larger number of small lobules consisting of the alveoli and their bronchioles, which join with each other to form the larger bronchi. In addition, there is a framework of fibrous or interstitial tissue in which run the blood vessels and lymphatics. The pulmonary artery which supplies the lung with blood arises from the right ventricle of the heart, while the pulmonary veins pour their blood into the left atrium.

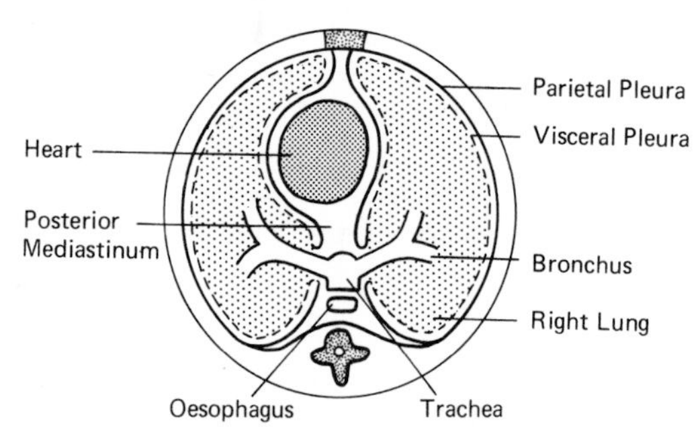

Fig. 172 Horizontal section of the chest, showing normal positions of the heart and lungs.

Groups of lymph glands which drain the lung are situated at the hilum of the lung and at the bifurcation of the trachea.

The pleura

The pleura is a serous membrane which, like the pericardium, consists of two layers, the visceral and the parietal. The **visceral layer** forms the outer covering of the lung which it encloses completely except at the hilum, where it is reflected over the structures entering the lung and becomes continuous with the parietal layer. The **parietal layer** lines the interior of the chest wall and upper surface of the diaphragm. The two layers are smooth and shiny and are moistened by a small amount of serous fluid resembling lymph which acts as a lubricant so that the two surfaces can glide smoothly over each other during the act of breathing. In disease states, the two layers of the pleura may become separated by fluid (pleural effusion) or by air (pneumothorax).

Respiratory movements

The renewal of the air in the lungs is secured by the respiratory movements of **inspiration** (breathing in) and **expiration** (breathing out). The thorax may be regarded as a completely closed box which alters its size and shape with each respiration. With inspiration,

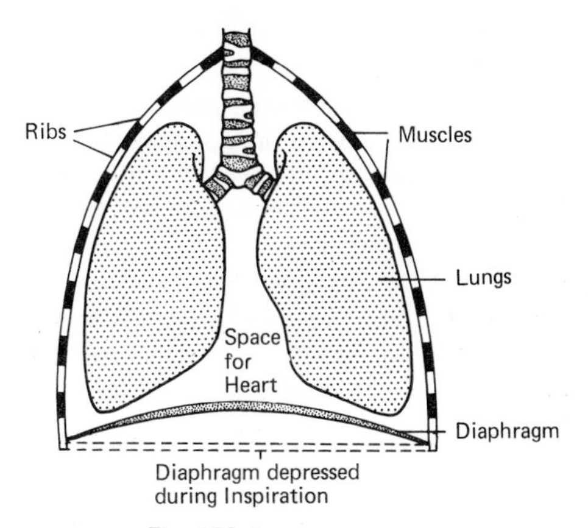

Fig. 173 The chest in section.

the cavity of the thorax is enlarged and the lungs, being elastic, expand to fill up the increased space. This expansion of the lungs causes air to be sucked in through the upper air passages and trachea.

With expiration, the capacity of the thorax returns to its former size and air is expelled from the lungs.

The increase in the size of the thoracic cavity during inspiration is brought about by two factors:

1. Upward movement of the ribs.
2. Downward movement of the diaphragm.

Upward movement of the ribs results mainly from contraction

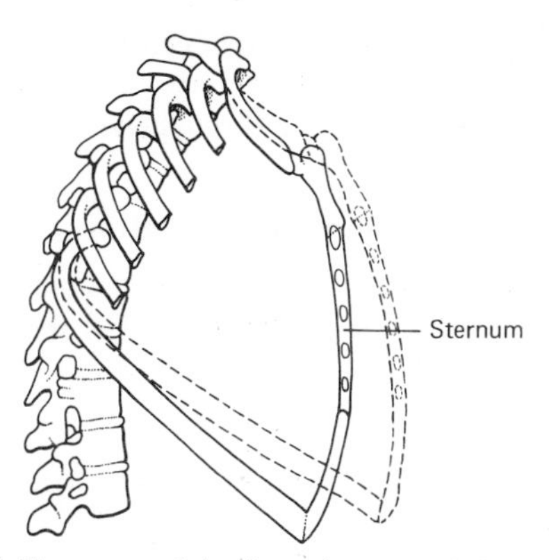

Fig. 174 Movements of the ribs and sternum during respiration.

of the intercostal muscles. In forced, deep inspirations the muscles of the neck and shoulder girdle may be brought into operation, viz. trapezius, sternomastoid and pectoralis major muscles.

When at rest the diaphragm is dome-shaped, having its concavity towards the abdomen. When the muscle of the diaphragm contracts during inspiration it becomes flattened and, therefore, depressed towards the abdominal cavity.

During quiet expiration the chest returns to its resting size mainly on account of the elasticity of the lungs and the chest wall and by the upward pressure of the abdominal contents on the diaphragm as it relaxes. In forced expiration, such as occurs in coughing and after very deep breaths, other 'accessory' muscles are employed.

Normal respiration, therefore, is a combination of two sets of movements:

thoracic = thoracic breathing,
diaphragmatic = abdominal breathing.

In men quiet respiration is mainly carried out by movements of the diaphragm, while in women the thoracic type of respiration usually predominates.

Another important function of the respiratory movements is to aid in the return of venous blood to the heart (page 160).

SPECIAL RESPIRATORY MOVEMENTS

1. Sighs and yawns are types of prolonged inspiration.
2. Cough is a forcible expiration usually preceded by a prolonged inspiration. The sound of a cough is produced by forcing air through the narrow opening between the vocal cords.
3. Hiccough is a noisy inspiration caused by muscular spasm of the diaphragm at irregular intervals. The noise is produced by the sudden sucking of air through the vocal cords.

Physiology of respiration

The stages of respiration are:

1. Ventilation of the lungs so that air moves freely in and out.
2. Interchange of gases between the blood and the air in the alveoli.

In addition there is:

1. Excretion of water vapour (and certain drugs such as general anaesthetics).
2. Supply of air to the larynx for the purpose of voice production.

	Inspired or atmospheric air (%)	*Expired air* (%)
oxygen	20	16
carbon dioxide	0·04	4
nitrogen	79	79

Atmospheric air is a mixture of gases and, as a result of oxygen being absorbed and carbon dioxide being excreted by the lungs, it follows that the amount of oxygen in expired air is diminished, while the amount of carbon dioxide is increased. With quiet breathing, the oxygen uptake and carbon dioxide excretion is about 250–300 ml per minute. The percentage of nitrogen remains constant. Actually the amount of carbon dioxide in expired air is 100 times greater than in atmospheric air.

The following is a summary of the differences between expired air and inspired air:

1. Expired air contains less oxygen and more carbon dioxide.
2. Expired air is warmed to body temperature.
3. Expired air is saturated with water vapour. These minute droplets of water may pick up bacteria during their passage through the respiratory tract and become a source of infection (droplet infection).

Movement of air in respiratory tract

It has been seen that the expansion of the chest by the movements of the thorax and diaphragm causes air to enter the lungs with each inspiration. Further, by an added effort, a forced inspiration will result in still greater expansion and and an additional amount of air will enter the lungs.

In the same way, a normal expiration can be supplemented by a forced expiration. Even after a forced expiration, however, some air still remains in the alveoli of the lungs.

The amount of air passing in and out of the lungs with ordinary quiet breathing is called **tidal air** and measures about 500 ml. The additional volume taken in by forced inspiration is called the **inspiratory reserve**. That expelled by forced expiration after an ordinary breath is referred to as the **expiratory reserve**, while that remaining in the alveoli is the **residual air**.

The term **vital capacity** may be defined as the volume of air that can be expelled by the deepest possible expiration after the deepest possible inspiration.

	ml	
Tidal volume	500	quiet breathing
Inspiratory reserve	2500	forced inspiration
Expiratory reserve	1000	forced expiration
Vital capacity	4000	
Residual volume	1000	always left in lung
Total lung capacity	5000	

It will be seen from the table that the vital capacity (VC) is the sum of the tidal volume, the inspiratory reserve and the expiratory reserve—also that the total lung capacity (TLC) is the sum of the vital capacity and the residual volume (RV), i.e.

$$TLC = VC + RV$$

The term hyperpnoea is sometimes used to express an increased depth of respiratory movement. Apnoea means a temporary cessation of breathing. Difficult or laboured breathing is called dyspnoea.

The regulation of respiration

The normal rate of respiration in adults is 14 to 18 breaths per minute. In children the rate is more rapid and in infants approaches 40 per minute. The rate is increased in certain diseases, e.g. pneumonia, and may also be abnormally slowed, especially in some cases of poisoning, e.g. morphine.

Respiration is controlled by nervous impulses and by the chemical composition of the blood.

Nervous control. Although for a short time the rate and depth of respiration can be controlled by the will, ordinarily it is an automatic act under the unconscious control of the nervous system. Situated in the medulla oblongata of the brain is a collection of nerve cells called the *respiratory centre*. From this centre, nerve connections pass to the diaphragm and the respiratory muscles.

Chemical control. It has been seen that carbon dioxide passes from the tissues into the blood and thence to the lungs where it is excreted. The respiratory centre is specially sensitive to the amount of carbon dioxide (carbonic acid) in the blood. If the amount rises as a result of more being formed in the tissues, such as during muscular exercise, the respiratory centre is stimulated. It therefore sends out impulses to the respiratory muscles to produce deeper and quicker breathing in order that carbon dioxide can be excreted

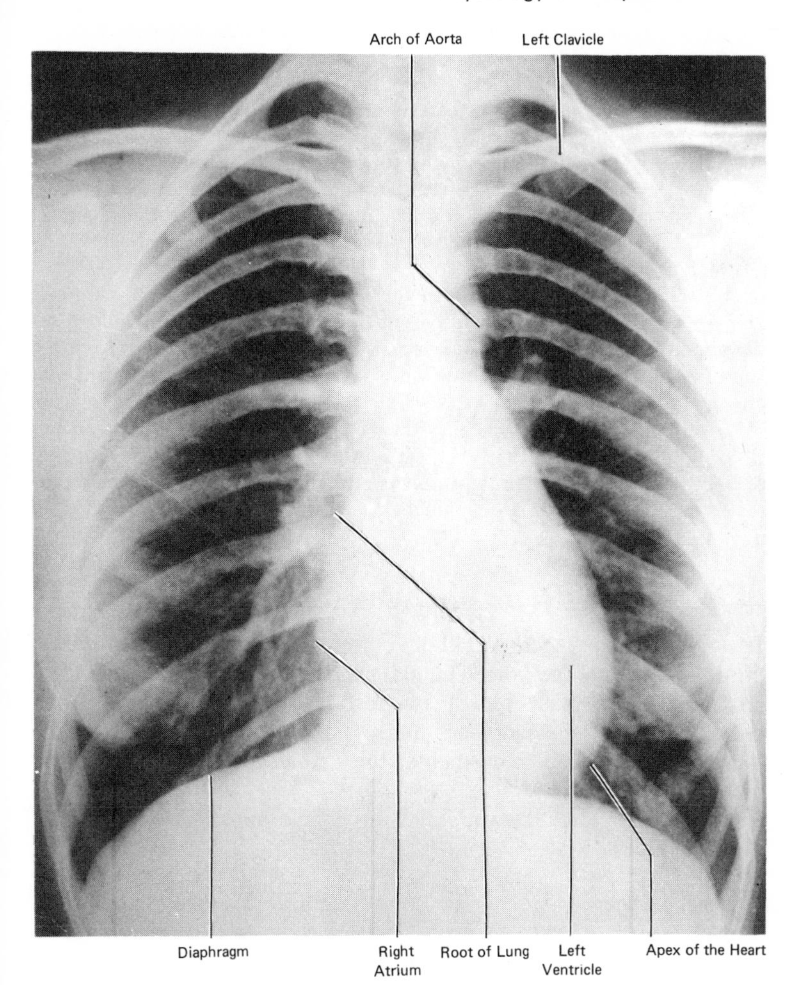

Fig. 175 X-ray of chest.

more rapidly by the lungs and thus lower the amount in the blood to its normal level. In other words, the function of the respiratory centre is to send out impulses to the respiratory muscles which maintain the rate and depth of breathing so that the level or concentration of carbon dioxide in the blood remains constant. (See also page 7.)

There are several very important practical applications of this fact.

(1) Carbon dioxide may be administered mixed with oxygen during and after a general anaesthetic. It will be clear from the facts just mentioned that the effect of inhaling carbon dioxide will be to increase the amount in the alveolar air and, thus, in the blood. The raised level of carbon dioxide in the blood acts as a stimulus to the respiratory centre which sends out impulses causing increased ventilation of the lungs, hence a more rapid excretion of the anaesthetic and at the same time a greater intake of oxygen. Carbon dioxide is sometimes used to produce increased ventilation of the lungs for other purposes.

(2) When administering oxygen the nurse must be careful to ascertain the percentage of oxygen required and the appropriate type of apparatus for administering it. In some respiratory disorders (e.g. chronic bronchitis) the amount of carbon dioxide in the blood may be persistently raised and the respiratory centre no longer sensitive to it. Deficiency of oxygen (hypoxia) may then be supplying the stimulus to the respiratory centre. If too high a concentration of oxygen is given, this hypoxic stimulus is removed. Consequently breathing becomes shallower and there is further retention of carbon dioxide, which can result in coma and ultimately death.

Voice production

It is convenient to consider this subject in connection with the respiratory system. The voice sounds are produced in the larynx. They are modified in character by the resonance afforded by the nasal cavities and the accessory air sinuses, and finally, by means of the tongue, lips and jaw movements, the actual sounds of speech are produced.

The human voice has the following characteristics:

1. Loudness
2. Pitch
3. Quality

1. **Loudness**. The fact that the loudness of the voice can be varied is evident, and it is dependent upon the force of the air currents expelled from the lungs through the vocal cords which produce sound by the vibrations caused by this force. The vibration of the vocal cords is called phonation.

2. **Pitch**. By pitch is meant the variation in note, i.e. a high note or a low note. This is dependent on two factors, the length and tension of the vocal cords. In children the vocal cords are relatively short and, therefore, the pitch of the voice is high. Alterations in pitch can be produced by voluntary action by using certain of the muscles of the larynx which increase or decrease the tension of the

vocal cords. Variation in pitch by the alteration of tension can easily be demonstrated by twanging a stretched string, wire or elastic, viz. the greater the tension of the cord or string, the higher the note produced.

3. **Quality**. The quality of a note is due to the resonance produced in the mouth, nose and accessory nasal sinuses in the skull. The difference in quality of sound is easily demonstrated by 'speaking through the mouth' and 'speaking through the nose'. The soft palate plays an important part in this act, and if imperfectly formed (cleft palate) or paralysed (e.g. in diphtheria) a typical nasal voice develops. Full use of the possible variations in quality is made in singing.

Speech

The sounds of the spoken word or articulation are modifications of the primary laryngeal sound which are brought about by movements of the lips, tongue and jaw working either independently or together.

There are two types of broken sound:

1. **Vowels**: sounds produced with the mouth open and with the vocal cords vibrating continuously without interruption.

2. **Consonants**: there is a sharp interruption or curtailment of the vocal cord vibration. Some consonant sounds are produced mainly by the movement of the tongue against the teeth, e.g. *t* and *d*, are called dentals. The sounds *p* and *b* are dependent upon closure of the lips and are called labials. Throaty sounds, *g* and *k*, are gutterals.

Whispering. In the act of whispering the sound is produced entirely by movement of the air in the mouth. The vocal cords are relaxed (or open) and do not vibrate, i.e. there is no phonation. The formation of word sounds in whispering is carried out by the movements of the mouth and tongue.

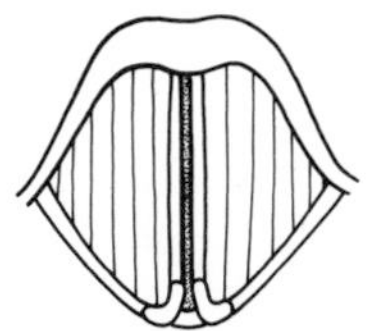
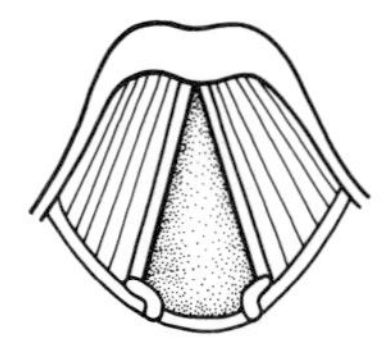

Fig. 176 The glottis, showing the vocal cords closed (*left*) on phonation and open (*right*) during inspiration.

PRACTICAL CONSIDERATIONS. Inflammation of the vocal cords intereferes with their contraction and vibration, and in consequence the voice becomes hoarse or is entirely lost, so that whispering only is possible. If it is necessary to rest the vocal cords the patient must be instructed to speak only in whispers.

QUESTIONS

1. What is the purpose of respiration? What conditions determine its frequency?
2. Describe the nose and its chief functions.
3. Describe briefly the organs concerned in the production of speech.
4. Describe the lungs and the processes of respiration
5. Describe the chief muscles concerned in respiration and their action.
6. Give an account of the physiology of respiration.
7. Give an account of the anatomy of the lungs. What is the mechanism by which air enters the lungs?

16 The Urinary System

The urinary system is one of the four excretory systems of the body, the other three being the bowels, the skin and the lungs. The following structures take part in its formation:

The kidneys—the excretory organs.
The ureters—the ducts draining the kidneys.
The bladder—the urinary reservoir.
The urethra—the channel to the exterior.

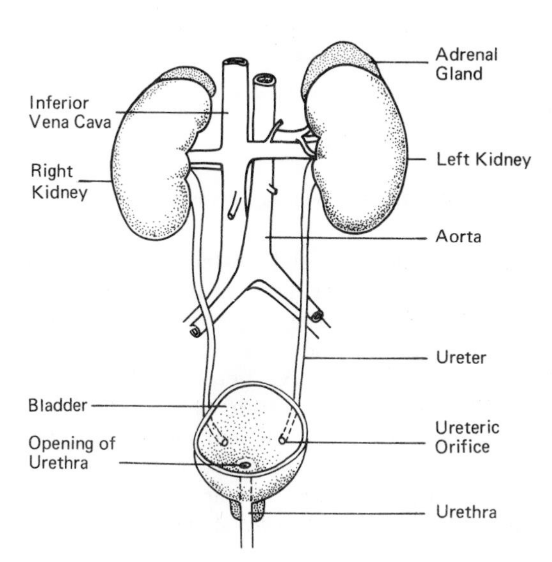

Fig. 177 Urinary tract—front view. The top of the bladder is cut off to show the openings of the ureters and urethra.

The kidneys

The kidneys are a pair of organs 11 cm ($4\frac{1}{2}$ in) in length, 5 cm (2 in) in width and about 3 cm ($1\frac{1}{4}$ in) in maximum thickness, lying almost vertically on the posterior abdominal wall level with the last thoracic (dorsal) and upper three lumbar vertebrae. Each kidney weighs about 150 grams (5 ounces) and is described as bean-shaped. The shape is so characteristic, however, that the term kidney-shaped is frequently used to describe other objects. The kidneys are dark red in colour.

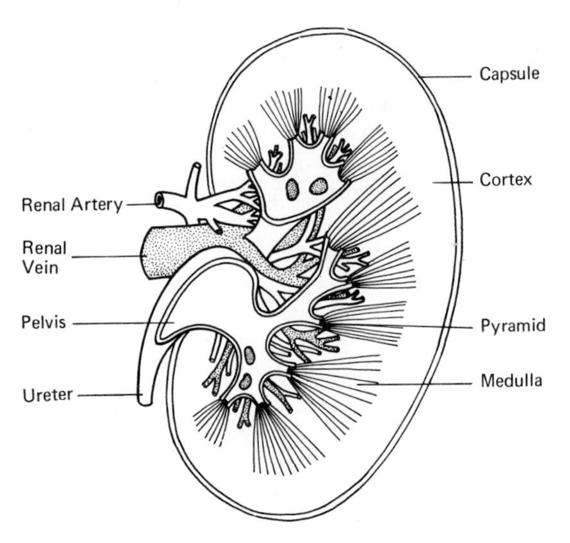

Fig. 178 Section through the kidney.

Each kidney has anterior and posterior surfaces and a lateral convex and a medial concave border. In addition, the uppermost part is called the superior pole and the lowest part the inferior pole. The adrenal (suprarenal) glands are situated on the upper pole of each kidney. The medial concave border is important because it contains a notch known as the **hilum** of the kidney through which the renal artery enters and the renal vein and ureter leave the kidney.

Relations. The kidneys lie embedded in fat (the perinephric fat) on the posterior abdominal wall but are situated behind the peritoneum. The right kidney is actually slightly lower in position than the left.

The following important structures are related to the kidneys.

Right kidney:
posterior surface: part of the diaphragm, psoas muscle
upper pole: right (suprarenal) adrenal gland
anterior surface: liver, duodenum
lower pole: right or hepatic flexure of colon
medial border: inferior vena cava, right ureter

Left kidney:
posterior surface: part of the diaphragm, psoas muscle
upper pole: left (suprarenal) adrenal gland
anterior surface: stomach, spleen, tail of pancreas
lower pole: left or splenic flexure of colon
medial border: aorta, left ureter

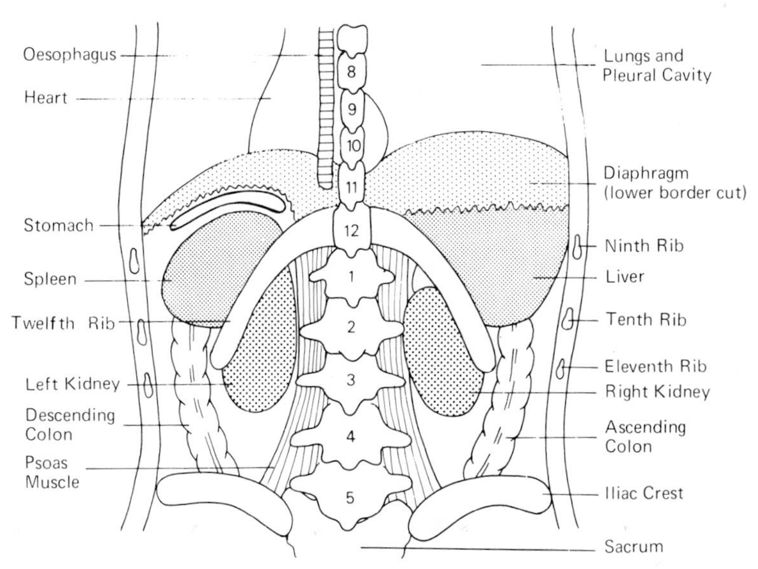

Fig. 179 Diagram showing position of kidneys, liver and spleen from behind.

(It is not necessarily intended that the student should attempt to learn these relations mechanically, but they are included in order that, with the aid of diagrams or an anatomical model, a mental picture of these important structures may be obtained.)

Structure. Each kidney is surrounded by a smooth fibrous capsule. When the kidney is cut in half it is seen to consist of two layers, and an open collecting portion called the *pelvis*, which is the upper expanded end of the ureter.

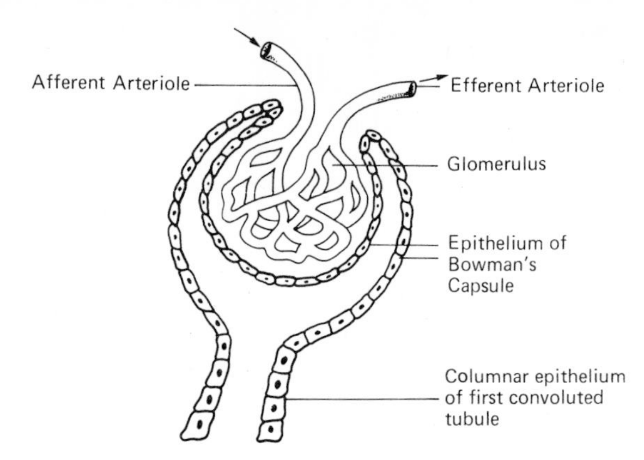

Fig. 180 Malpighian body, highly magnified.

The *cortex* or outer part is reddish-brown in colour. The inner part or *medulla* is darker and shows a series of projections into the pelvis called the *pyramids*. These number about a dozen in each kidney and their apices, directed into the pelvis, end in nipple-shaped papillae.

Microscopically the kidney substance is seen to be composed of:

1. Malpighian bodies.
2. Tubules, (*a*) first convoluted tubule
   (*b*) the loop of Henle
   (*c*) second convoluted tubule
   (*d*) collecting tubule
   } the parenchyma
3. Blood vessels and supporting fibrous tissue, the interstitial tissue.

Malpighian bodies. Each tubule begins in the cortex in a blind expanded end (*Bowman's capsule*) which is indented by a bunch of capillaries called a *glomerulus*. The two parts together constitute a Malpighian body (Fig. 180).

The **first convoluted tubule** is situated in the cortex and from it the long **loop of Henle** passes to the medulla and then returns to the cortex to become the **second convoluted tubule**. From this, straight **collecting tubules** pass to the medulla and open at the apices of the pyramids into the pelvis of the kidney.

Each unit of a glomerulus and its associated tubules is called a **nephron**. Each kidney contains about a million nephrons. All the tubules of the kidney are lined by a single layer of cubical epithelium and are surrounded by capillaries.

Blood supply. The kidneys are plentifully supplied with blood from the renal arteries which are branches of the abdominal aorta. Blood leaves the kidneys by the renal veins which enter the inferior vena cava.

NOTE. Narrowing (stenosis) of the renal artery is a cause of hypertension.

The ureter

The duct of the kidney, which collects the urine secreted by the organ and conveys it to the bladder, consists of two parts, the pelvis of the kidney and the ureter proper.

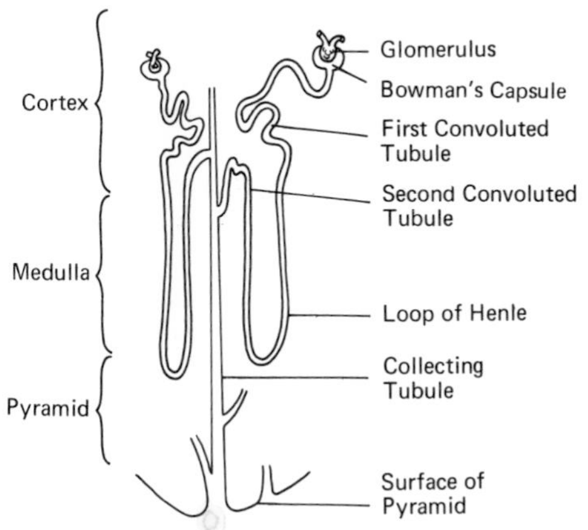

Fig. 181 Diagram of a kidney tubule.

The **pelvis** of the kidney is the upper expanded portion of the ureter which lies mainly in the interior of the kidney. It commences as a number of funnel-shaped channels, which surround the papillae of the pyramids, called **calyces** (singular = calyx).

The ureter is a tube about 26 cm (10 in) long, consisting of an outer fibrous coat, a middle muscular layer and an inner lining of transitional epithelium. The upper or abdominal part passes downwards on the posterior abdominal wall on the surface of the psoas muscle. Crossing the common iliac artery, it passes forward to enter the bladder. The lower portion is called the pelvic part and, in the female, is closely related to the side of the cervix of the uterus. The muscle of the wall of the ureter undergoes peristaltic contraction.

PRACTICAL CONSIDERATIONS. The ureter is of importance in the condition known as renal calculus, in which stones are formed in the kidney. Such stones are frequently passed down the ureter and this movement results in attacks of severe pain in the loin, often accompanied by the passage of blood in the urine from injury to the wall of the ureter (renal colic).

Inflammation of the pelvis of the kidney is called pyelitis.

The bladder

The bladder, or reservoir in which the urine is received from the kidneys via their ducts, is a pear-shaped muscular sac lined by mucous membrane. It is situated in the pelvic cavity immediately behind the symphysis pubis. It lies in front of the rectum in the male, but in the female is separated from the rectum by the uterus and vagina.

In addition to anterior, lateral and posterior walls it has a superior and an inferior surface, the lowest part of which is called the **base** and surrounds the opening of the urethra.

Examination of the interior of the bladder reveals three openings or orifices, viz. those of the two ureters and the urethra. These are arranged in the form of a triangle, called the **trigone** of the bladder, with its apex below at the urethral opening (Fig. 177).

Structure. The bladder is a muscular sac made up of four layers:

1. The outer serous coat of peritoneum.
2. The muscular coat (unstriped or involuntary muscle).
3. The submucous coat.
4. The mucous coat, consisting of transitional epithelium.

It is important to remember, however, that the serous peritoneal coat only covers the superior surface.

The bladder must be considered in its two natural states, viz. (1) empty, and (2) full. It has been seen that the empty bladder lies behind the symphysis pubis. When the bladder fills, however, the base remains fixed while the superior surface rises above the symphysis pubis behind the lower part of the anterior abdominal wall. In doing so it raises with it the serous coat of peritoneum. This is important because it is possible to pass a trocar and cannula into the full bladder immediately above the symphysis pubis without risk of injuring the peritoneum or infecting the peritoneal cavity.

When the bladder becomes over-distended it may reach as high as the umbilicus and, of course, can easily be felt in this state as a rounded tumour arising out of the pelvis.

The bladder can contain up to 600 ml (20 fl. oz) or urine without being over-distended, but in the majority of instances the organ is emptied when it is holding 180–300 ml (6–10 fl. oz). An over-distended bladder may contain as much as 1200–1800 ml (2–3 pints) of urine.

The urethra

The urethra is the canal conveying the urine from the bladder to the exterior. It differs in the two sexes since, in the male, it also plays a part in the reproductive functions.

The **female urethra** is a short tube 4 cm ($1\frac{1}{2}$ in) in length, which leaves the base of the bladder at the trigone and reaches the exterior between the labia minora immediately in front of the opening of the vagina. As it leaves the bladder it is surrounded by a sphincter muscle.

The **male urethra** is a channel about 20 cm (8 in) long leading from the bladder to its external orifice or the meatus of the urethra at the extremity of the penis. It has three parts:

1. The pelvic or prostatic portion.
2. The perineal or membranous portion.
3. The penile or spongy portion.

The first part, as it leaves the bladder, is guarded by a sphincter muscle and is surrounded by the prostate gland.

The portion situated in the perineum is especially liable to injury or rupture by falling astride a structure such as a bar, gate or chair.

PHYSIOLOGY OF THE URINARY SYSTEM

Composition of the urine

Normal urine is a clear, yellow or amber-coloured fluid having a specific gravity varying between 1015 and 1025. It is usually slightly acid in its reaction to litmus (turns blue litmus red) and has a faintly aromatic odour.

The daily amount secreted is about 900 to 1800 ml (30 to 60 fl. oz). This very considerable variation which may occur in health is dependent on two main factors:

1. Fluid intake.
2. The fluid loss from the skin by evaporation of sweat.

In hot weather, the quantity of urine is decreased, it becomes more concentrated, of higher specific gravity and has greater intensity of colour. In cold weather, when the secretions of sweat is scanty, the converse is true.

Urine consists of water and dissolved solids:

water = 96%
solids = 4%
urea (2%)
urates
uric acid
chlorides, phosphates, sulphates, oxalates } of { sodium, potassium, calcium

The urea, uric acid and urates are end-products of protein metabolism. The salts are mainly derived from vegetable foods and common salt (sodium chloride) taken in the diet.

It has been stated that normal urine is clear. This is true when applied to freshly passed urine which has body temperature, but if allowed to stand and cool, it is quite common for a deposit of phosphates or urates to form, especially if the urine is concentrated. Urates dissolve when the urine is warmed and phosphates can be dissolved by the addition of dilute acetic acid.

CHARACTERISTICS OF URINE

Colour:	*yellow*, due to urochrome pigment	
Odour:	*aromatic* when fresh: *ammoniacal* after bacterial decomposition of urea	
Volume:	1–1½ litres per 24 hours	
Specific gravity:	1.001–1.040	
pH	about 6 (slightly acid)	
Main constituents:	*water*	
	inorganic salts	chloride
	sodium	phosphates
	calcium	bicarbonate
	potassium	sulphate
	organic compounds	
	urea	ammonia (secreted by renal tubules)
	creatinine	
	uric acid	

Abnormal constituents

1. Protein (albumin) } in many conditions, especially disease or
2. Blood } injury of the renal tract.
3. Sugar (glucose) in diabetes.
4. Acetone and diabetic acid in diabetes, starvation and acidosis.
5. Bile in jaundice.

Functions of the kidneys

The primary function of the kidneys is to help in keeping the composition of the blood constant by the excretion either of abnormal constituents or the excess of substances normally present. In performing this function they carry out the following important work:

1. The excretion of water.
2. The excretion of the end-products of protein metabolism.
3. The excretion of salts (electrolytes).
4. The excretion of drugs, toxins and chemical substances which may be harmful.
5. Assist in regulating the reaction (pH) of the blood.

1. Excretion of water. A certain quantity of water is necessary to maintain the normal composition of the blood and tissue fluids. In addition to that taken by mouth it has been seen that water is also an end-product of the metabolism of proteins, carbohydrates and fats. Water is also excreted by the skin, bowels and lungs.

2. Excretion of end-products of protein metabolism. Urea (2 per cent of urine) is the most important of these end-products and forms half of the total of the solid constituents of the urine. It is formed in the liver by the removal of the nitrogen-containing fraction of the amino-acids by the process of de-amination or de-nitrification (page 229).

Uric acid and urates are also derived from the protein found in the nuclei of cells and come partly from those in the food (exogenous—from without) and partly from tissues of the individual which are removed in the general process of wear and tear (endogenous—from within).

3. Excretion of salts. The most important salts excreted in the urine have already been mentioned, viz. the chlorides, phosphates, sulphates and oxalates of sodium, potassium and calcium. These are derived mainly from the food and especially vegetables.

Salts may be acid, neutral or alkaline in reaction to litmus and the correct and accurate balance of those circulating in the blood plays an important part in the regulation of its reaction. The blood

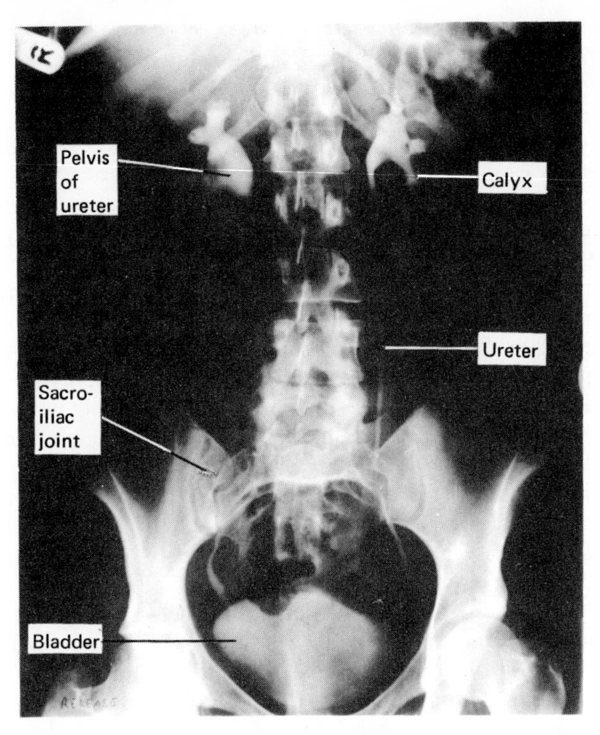

Fig. 182 Pyelogram showing pelvis and ureters outlined after injection of radio-opaque substance.

is always very slightly alkaline and it is one of the very important functions of the kidneys to keep this constant.

One of the most important salts circulating in the blood is sodium bicarbonate, an alkali, which in the presence of acids sets free carbonic acid and becomes a neutral salt. In this way any excess of acid is neutralized and the reaction of the blood remains constant. Sodium bicarbonate is, therefore, called 'the alkali reserve' of the blood. If this alkali reserve is reduced we use the term 'acidosis', but it must be remembered that this is really a misnomer as the blood can never become acid in circumstances compatible with life.

By excreting acid substances and acid salts the kidneys aid in maintaining the reaction of the blood by sparing the use of sodium bicarbonate (alkali reserve) which would otherwise be used to neutralize them.

4. Excretion of drugs, toxins and other chemical substances. Many drugs and toxins are excreted by the kidneys. This fact enables a number of drugs to be given over long periods in suitable doses without the risk of their accumulating in the blood and the consequent production of poisonous symptoms. In other words, the kidneys help to maintain such substances, deliberately administered, at a constant level in the blood during the period over which they are given and, when finally discontinued, to excrete the remains of the drug from the circulation.

It is very important for the body to maintain the amount of sugar (glucose) in the blood at a constant level. Abnormal symptoms are produced if the blood-sugar becomes unduly low (hypoglycaemia) or excessively high (hyperglycaemia). Normally no sugar is excreted by the kidneys, for unless very excessive quantities are taken the

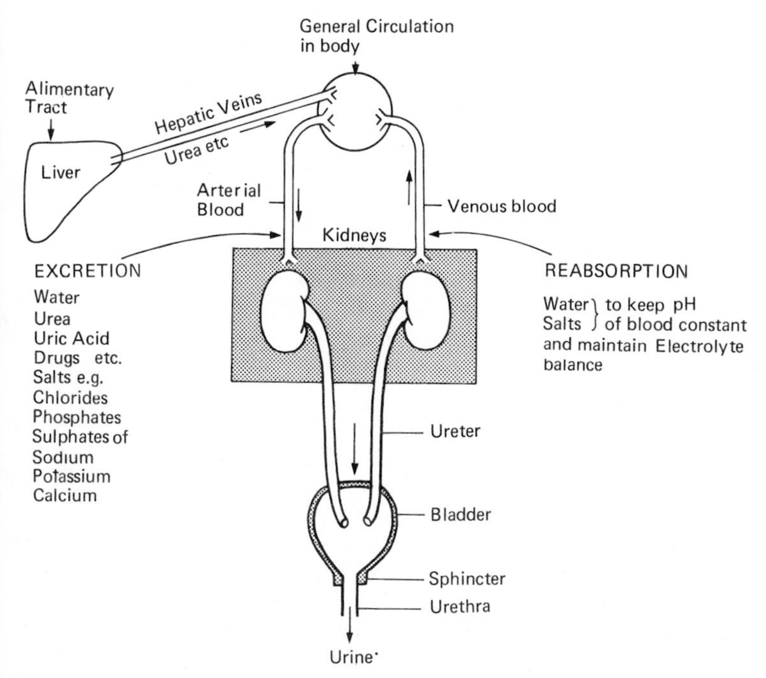

Fig. 183 Diagram illustrating renal function.

body is able to use it and convert it into carbon dioxide and water and to store the excess as glycogen in the liver. In diabetes, however, the deficiency of insulin makes it impossible for the tissues to utilize sugar, hence it tends to accumulate in the blood and the blood-sugar rises. When the blood-sugar, and consequently the sugar in the glomerular filtrate, rises above a certain level the kidneys commence to excrete sugar because the tubules cannot reabsorb any more.

It would be very wasteful if the kidneys always excreted sugar and therefore the renal cells have a standard level of concentration or 'threshold value', and it is only when the blood-sugar rises above this that the normal kidneys will excrete sugar.

The kidneys have a threshold value for many other substances and it is this factor which is so important in keeping the composition of the blood constant.

Formation of the urine

Three processes are employed by the kidneys in the production of urine:

1. Filtration
2. Secretion
3. Absorption

Filtration. This is a simple physical process which takes place through the glomeruli. Water, salts and other substances are filtered from the blood in the glomeruli and pass into the first convoluted tubules.

Secretion. This takes place mainly in the convoluted tubules and is an active, vital process performed by the cells of the cubical epithelium lining the tubules. These cells select either abnormal substances or normal substances from the blood when their concentration exceeds their threshold value, and pass them into the lumen of the tubules.

Absorption. If all the fluid and salts passing from the glomeruli into the tubules by the process of filtration were allowed to pass to the exterior in the urine there would be a serious loss of valuable material from the body and the composition of the blood would be affected. In order to prevent this, some of the water and salts are reabsorbed into the circulation by the kidney cells, especially those in the loop of Henle.

In other words, filtration is a purely mechanical process which does not take into account the requirements of the body, whereas the cells of the renal tubules look after the needs of the body (*a*) by excreting those substance which the body does not require but which will not pass through the glomeruli by the process of simple filtration, (*b*) by retaining those substances which are useful to the body but which have passed from the blood through the glomeruli, i.e. the renal tubules reverse to a large extent the effects of glomerular filtration and return much of the filtered material back into the blood.

The product resulting from these processes passes via the collecting tubules into the pelvis of the kidney and thence down the ureter by peristaltic action of its walls. It enters the bladder from the uretic openings in a regular series of squirts and remains in the bladder until the organ is emptied.

Inflammation of the kidneys (nephritis), the failure of renal function (uraemia) and other types of kidney disease are of great importance and should be studied from medical textbooks.

Micturition

Micturition is the act of emptying the bladder or passing urine. Three important facts must be recalled:

1. The bladder is a muscular sac.
2. The urethral orifice is surrounded by a sphincter muscle.
3. The bladder is supplied with nerves which pass to and from the lumbo-sacral region of the spinal cord. There is a further set of connections between this area in the spinal cord and the brain.

Micturition is primarily a reflex act which, after infancy, can be consciously controlled by impulses from the higher centres of the brain.

As urine accumulates in the bladder the muscle fibres of its walls become gradually stretched. In other words the pressure within the bladder (intravesical pressure) is raised. When the pressure is raised to a certain level and the fibres are stretched a definite amount, the sensory or afferent nerves of the bladder are stimulated and an impulse passes to the spinal cord and thence to the higher centres where it is interpreted as the desire to micturate (the sensory side of the act). At this point the will can control the act and, if necessary, postpone emptying the bladder.

In due course, however, the motor cells of the brain send down, via the spinal cord and efferent nerves to the bladder, impulses which cause the bladder muscles to contract and, at the same time, the sphincter muscle of the urethra to relax. The bladder then empties itself. This mav be assisted by holding the breath and contracting the diaphragm and muscles of the abdominal wall, thereby increasing the pressure within the abdomen. The increased pressure of the abdominal contents on the bladder obviously helps in the process of emptying.

PRACTICAL CONSIDERATIONS. In infants, before the higher centres are functioning, the act is a reflex one; the sensory stimulus passing to the spinal cord being followed immediately by the motor response of micturition. A similar state of affairs may also occur in disease of the brain and spinal cord, and in cases of unconsciousness.

In states of semiconsciousness the patient may react to fulness of the bladder by becoming restless and fidgety. The nurse who observes this and supplies a urinal will do much to prevent incontinence and keep such a patient dry.

A defect in the nervous mechanism may result in the bladder failing to empty. In such instances retention of urine occurs and the bladder becomes over-distended and can only be relieved by the passage of a catheter. Retention of urine may also be caused by mechanical obstruction to the outlet from the bladder by enlargement of the prostate gland or by a narrowing (stricture) of the urethra which impedes the passage of urine to the exterior.

QUESTIONS

1. How is water excreted from the body?
2. Describe the functions of the kidneys. What may influence the quantity of urine by normal persons?
3. How is urine formed and how is it excreted from the body? What is the composition of normal urine?
4. Describe the urinary tract. Give an account of the functions of the kidney.
5. By which routes are waste products voided from the body? Describe the structure of the kidney.

17 The Ductless Glands or Endocrine System

It has been seen that many of the activities of the body are controlled by the nervous system. One of the characteristics of this system is the rapidity of the response to various kinds of stimuli.

There exists in the body a series of organs which also exercise the most important control over its activities, especially those of a slower character which are only manifest over long periods, such as metabolism, growth and the formation of character and temperament. These are called the endocrine organs or ductless glands. They have the capacity of elaborating special chemical substances called hormones which pass directly into the blood stream. They are, therefore, sometimes referred to as the organs of internal secretion.

A **hormone** may be defined as a chemical messenger secreted by a ductless gland which reaches its destination by the blood stream and which has the power of influencing the activity of other distant organs.

In some instances an organ may have both an internal secretion which enters the blood directly and an external secretion which leaves it by a duct. The pancreas is an example. The internal secretion of the pancreas, insulin, passes into the blood, while the pancreatic juice reaches the duodenum via the pancreatic duct.

A number of hormones have been isolated and their chemical composition is known. A few can actually be prepared artificially in the laboratory. As more are discovered the subject is becoming very complicated, and although they are mentioned, the student can hardly be expected to have a detailed knowledge of all of them.

Some hormones, such as adrenaline, have an immediate action; others like the pituitary growth hormone exercise their influence over many years.

The function of the ductless glands has been studied in a number of ways. Before any hormones were separated, some knowledge of their action was obtained by observing the effects of disease. It was found that two sets of symptoms existed, namely, those produced by over-secretion (hypersecretion) of the affected gland and those resulting from under-secretion (hyposecretion). Clearly, therefore, the action of the gland in health is to maintain a balance between these two effects.

Another method of study is to obtain the hormone itself either by recovering it from the gland or manufacturing it in the laboratory and to consider the effects produced by its administration both in health and in disease.

A third method is to study the results of the removal of the gland in animals.

The most important glands in the endocrine system are:

(*a*) The anterior pituitary.
(*b*) The adrenal cortex.
(*c*) The thyroid gland.
(*d*) The sex glands or gonads.

All of these are controlled by the trophic hormones of the anterior pituitary.

A second group consists of:

(*a*) The posterior pituitary.
(*b*) The adrenal medulla.
(*c*) The parathyroid glands.
(*d*) The insulin secreting cells of the Islets of Langerhans in the pancreas.

These are controlled by other stimuli including nervous impulses.

The liver and the stomach also produce certain internal secretions but any necessary mention of them is made in connection with these organs.

Disorders of the thyroid gland are not uncommon and easier to understand; this gland is therefore considered first.

The thyroid gland

The thyroid gland is situated in the lower part of the neck. It consists of *two lobes* lying one on either side of the trachea and joined together by *an isthmus* which passes in front of the trachea just below the cricoid cartilage. The lobes have upper and lower poles, the upper pole extending to the side or wing of the thyroid cartilage. It receives its plentiful blood supply from branches of the carotid and subclavian arteries. The recurrent laryngeal nerves lie close to the lower poles and may be accidentally injured during the operation of thyroidectomy.

When examined under a microscope the thyroid gland is seen to consist of a number of closed vesicles or small chambers lined by epithelium which contain a jelly-like substance called colloid.

Functions

The active hormones of the gland, which can be obtained from colloid, are called **thyroxine** and **triodothyronine**. They contain a high proportion of iodine (65 per cent) and one of the functions of the thyroid gland is to restore iodine in the body. This it obtains from the blood in minute amounts absorbed from the food. Traces of iodine are found in sea fish and in most drinking water, but in some inland districts, e.g. the Swiss Alps, the Andes and the Himalayas, it may be deficient—which accounts for 'endemic' goitre being prevalent in these areas. The secretion of thyroxine is regulated by the thyroid-stimulating hormone (TSH) secreted by the anterior lobe of the pituitary gland.

The main functions of the thyroid gland are:

1. It exercises control over the general metabolism of the body.
2. It affects the irritability of the nervous system.
3. It plays a part in keeping the skin and hair in good condition.
4. It cooperates with other ductless glands in the general endocrine balance of the body.
5. It has an effect on body growth and mental development in infancy.
6. Storage of iodine.

Illustration of some of these functions is obtained by consideration of the effects of disease.

*1. **Over-secretion*** (hypersecretion). This is generally associated with enlargement of the gland which is called *goitre*, and the disease produced is known as thyrotoxicosis (exophthalmic goitre or Graves' disease). It is characterized by protrusion of the eyeballs (exophthalmos), a rapid pulse, increased sweating and general nervousness. The appetite is often good but the patient is usually thin, indicating that the rate of general metabolism is increased.

The interaction with other ductless glands is shown by the frequency with which disturbances of menstruation occur.

*2. **Under-secretion*** (hyposecretion). The effects of hyposecretion differ in adults and infants. The condition produced in adults is called **myxoedema**. In contrast with the appearance of extreme

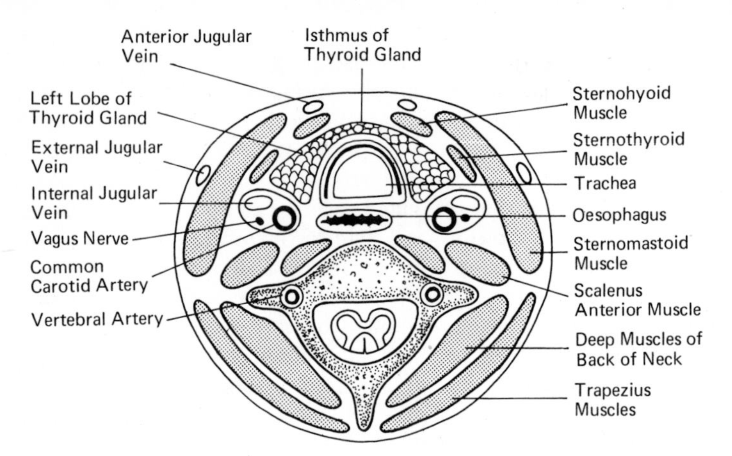

Fig. 184 Transverse section of the neck at the level of the seventh cervical vertebra, showing the position of important structures.

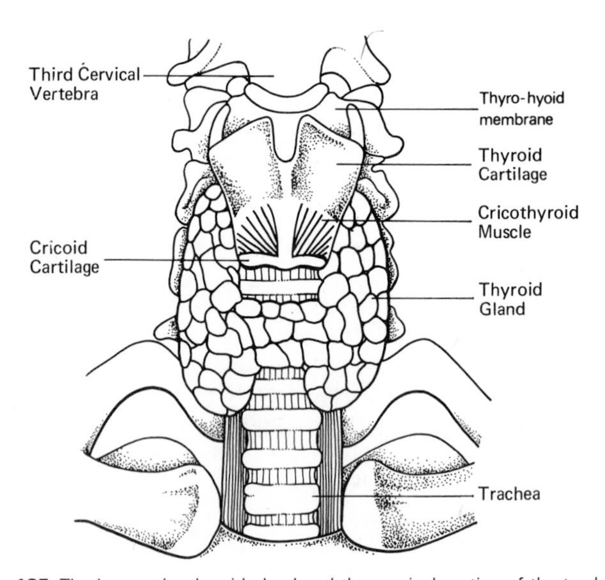

Fig. 185 The larynx, the thyroid gland and the cervical portion of the trachea.

nervousness seen in hyperthyroidism, the patient becomes mentally dull, while speech and movement are slow. Instead of being thin, the patient shows a tendency to put on weight. The skin is dry and thick while the hair and eyebrows are scanty. The temperature of the body tends to be subnormal. In fact the whole rate of metabolism and body activity appears to be slowed down. The excretion of cholesterol in the bile is diminished and, therefore, the blood cholesterol may be raised in this condition, but lower than normal in hyperthyroidism.

In infancy the condition known as **cretinism** results from thyroid deficiency. A cretin shows dwarfism or stunting of growth and failure of mental development.

Over-secretion (thyrotoxicosis) may be treated by removing a portion of the gland by operation or by giving 'anti-thyroid drugs'. Myxoedema and cretinism are both treated by administering thyroxine, thereby supplying the missing factor.

Another hormone secreted by the thyroid gland is calcitonin which lowers the amount of calcium in the blood plasma by reducing the extraction of calcium from bone into the blood.

The parathyroid glands

These are four small glands, about the size of a pea, situated in the neck, one behind each of the four poles of the thyroid gland. Although anatomically so closely related to the thyroid gland their function is entirely different. They secrete a hormone called **parathormone**, the function of which is to control the calcium metabolism of the body.

Decreased activity (hypoparathyroidism) results in a condition known as **tetany**, in which the body is unable to mobilize and use calcium, and the calcium content of the blood is low (Normal: 9–11 mg per 100 ml.). It is characterized by muscular spasms and increased irritability of the nervous system. Complete removal of the parathyroids will result in death.

Increased activity (hyperparathyroidism) produces an increase of calcium in the blood and urine and renal calculi or a disease of the bones (osteitis fibrosa) which is not altogether surprising when it is recalled that bones contain a large amount of calcium.

The adrenal or suprarenal glands

The adrenal glands are two small yellowish bodies about 1 inch in length, situated on the upper pole of each kidney. They are plenti-

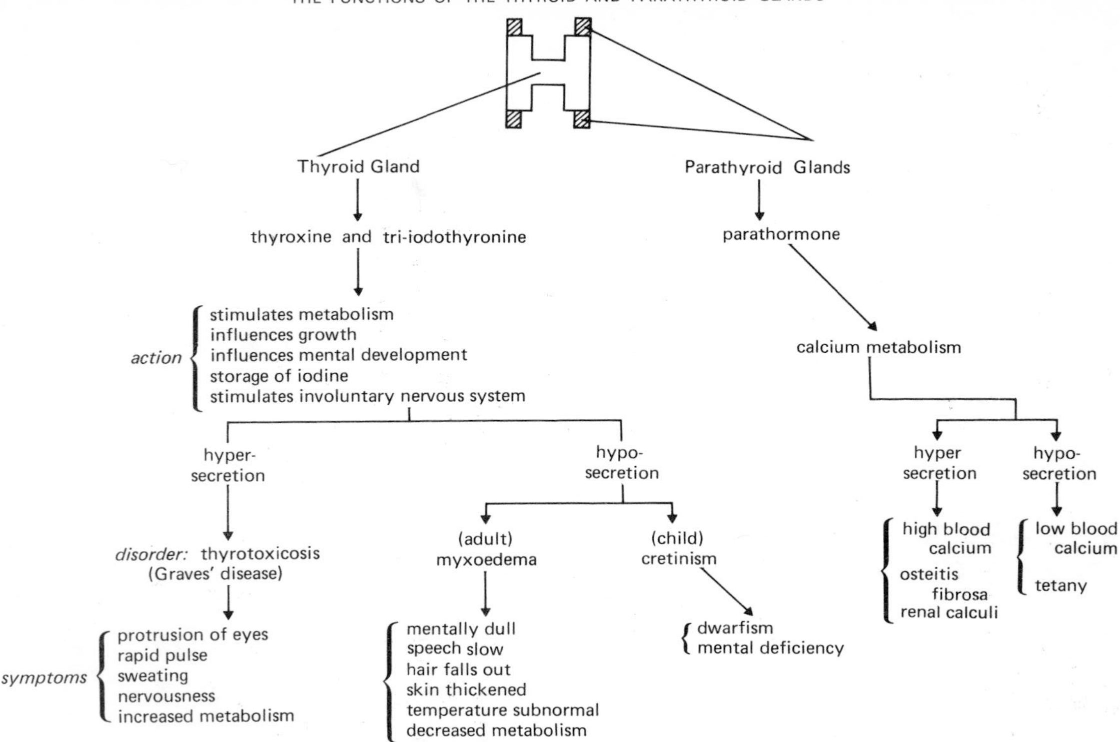
THE FUNCTIONS OF THE THYROID AND PARATHYROID GLANDS
Thyroid Gland
thyroxine and tri-iodothyronine
action
stimulates metabolism
influences growth
influences mental development
storage of iodine
stimulates involuntary nervous system
hyper-secretion
disorder: thyrotoxicosis (Graves' disease)
symptoms
protrusion of eyes
rapid pulse
sweating
nervousness
increased metabolism
hypo-secretion
(adult) myxoedema
mentally dull
speech slow
hair falls out
skin thickened
temperature subnormal
decreased metabolism
(child) cretinism
dwarfism
mental deficiency
Parathyroid Glands
parathormone
calcium metabolism
hyper secretion
high blood calcium
osteitis fibrosa
renal calculi
hypo-secretion
low blood calcium
tetany

fully supplied with blood from the aorta and renal arteries, and also with sympathetic nerve fibres from the coeliac (solar) plexus (page 314). Each gland consists of a yellowish outer portion of cortex and a darker interior or medulla, the functions of which must be considered separately.

Functions of the cortex

The adrenal cortex is essential to life and secretes a number of hormones, each having different functions. Some of these have been extracted from the gland and some can be made in the laboratory.

Chemically they belong to a class of fatty or wax-like substances called steroids. The most important ones fall into three main groups:

1. The **mineral corticoids**, of which aldosterone, fludrocortisone and the synthetic deoxycortone (DOCA) are examples. These act on the tubules of the kidney in such a way that:
   (*a*) sodium and chloride are retained in the body.
   (*b*) excess of potassium is excreted.

   They therefore help to maintain the water and electrolyte balance of the body.
2. The **gluco-corticoids**. These have a number of actions, one of the main ones being to influence carbohydrate metabolism.
   (*a*) They assist in the conversion of carbohydrate into glycogen.
   (*b*) They increase the blood sugar.
   (*c*) They help in the utilization of fat.
   (*d*) They decrease the number of lymphocytes and eosinophils in the blood.
   (*e*) They reduce the rate at which certain connective tissue cells multiply and so tend to suppress the natural reaction to inflammation and to delay healing.

   Cortisone and hydrocortisone are hormones of this group.
3. **Sex or gonad-like hormones** similar to those produced by the ovary and testis (oestrogens and androgens). These influence growth and sex development.

The output of hormones from the cortex of the adrenal is controlled by another hormone secreted by the pituitary gland known as the adreno-cortico-trophic hormone (ACTH) or corticotrophin. Unlike the medulla of the adrenal the secretions of the cortex are not regulated by nervous impulses.

Cortisone, hydrocortisone and many similar synthetic substances (e.g. prednisone and prednisolne) are all used in clinical medicine.

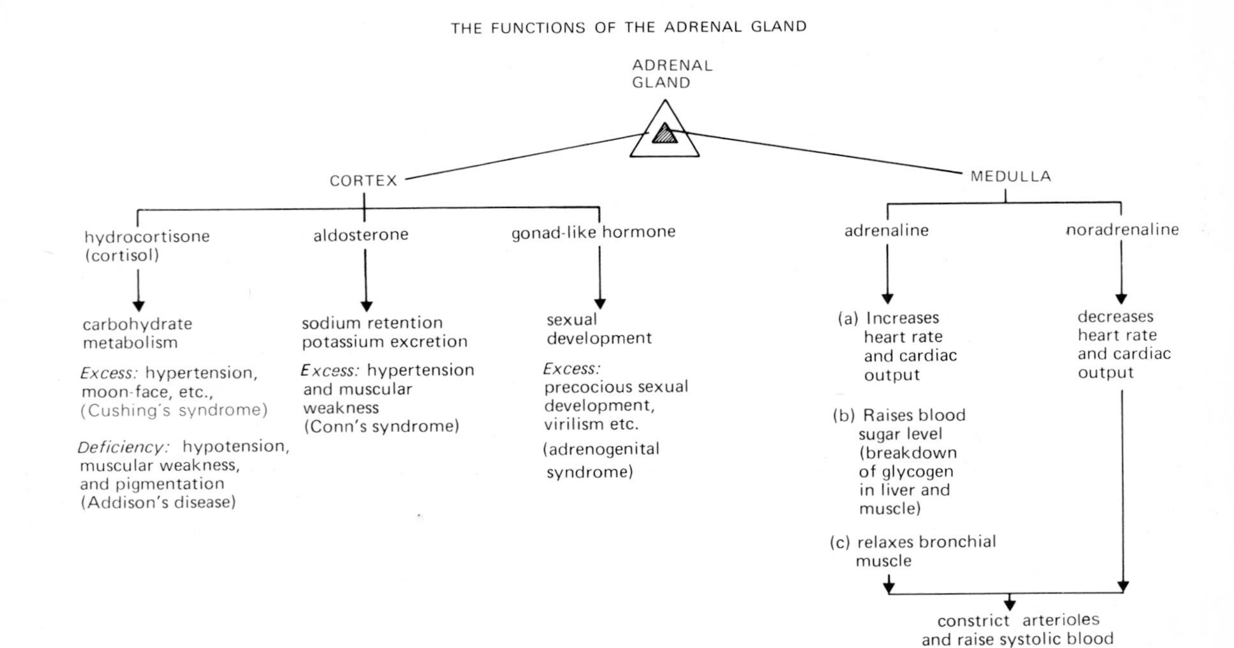
THE FUNCTIONS OF THE ADRENAL GLAND
ADRENAL GLAND
CORTEX
MEDULLA
hydrocortisone (cortisol)
carbohydrate metabolism
Excess: hypertension, moon-face, etc., (Cushing's syndrome)
Deficiency: hypotension, muscular weakness, and pigmentation (Addison's disease)
aldosterone
sodium retention potassium excretion
Excess: hypertension and muscular weakness (Conn's syndrome)
gonad-like hormone
sexual development
Excess: precocious sexual development, virilism etc.
(adrenogenital syndrome)
adrenaline
(a) Increases heart rate and cardiac output
(b) Raises blood sugar level (breakdown of glycogen in liver and muscle)
(c) relaxes bronchial muscle
noradrenaline
decreases heart rate and cardiac output
constrict arterioles and raise systolic blood pressure
Excess: hypertension (phaeochromocytoma)

When these are administered in large or excessive doses, among the complications observed is a disturbance of salt and water balance, which is, as mentioned above, a function of the mineral corticoids and not of the gluco-corticoids.

Large doses of cortisone, in addition to the normal actions mentioned, also depress the secretion of corticotrophin (ACTH) from the pituitary.

It must also be remembered that although the individual hormones mentioned each have separate effects on metabolism, in health, normal adrenal cortical secretion represents the sum of all these individual actions. In disease, if the adrenal cortex is destroyed or if the secretion of corticotrophin (ACTH) fails there will be no adrenal cortical hormones produced.

Functions of the adrenal medulla

The medulla secretes a substance known as **adrenaline** or epinephrine which can also be synthesized in the laboratory. It is very closely connected in function with the involuntary nervous system. In fact it may be broadly stated that injection of adrenaline produces the effect of general stimulation of the sympathetic system. For example, it causes a constriction of the arteries of the body resulting in a rise of blood pressure and also an increase in the rate of the heart; it stimulates the liver to convert glycogen into glucose which is liberated into the blood stream; it relaxes the involuntary muscles of the bronchi.

Adrenaline in the body really constitutes a reserve mechanism that comes into action at times of stress. It is poured into the blood at the onset of danger, anger or excitement and is responsible for many of the changes which accompany these emotions. Blanching of the skin due to constriction of the arterioles means diversion of the blood to the muscles where it is most needed. The increased blood pressure and force of the heart-beat result in better circulation both in the muscles and the brain, while the liberation of glucose supplies the muscles with the necessary fuel for increased activity.

The adrenal medulla also secretes an allied substance called noradrenaline. This also constricts blood vessels and raises blood pressure.

Destructive disease of the adrenal glands in adult life results in a condition known as Addison's disease. This is characterized by low blood pressure, digestive disturbances and a brown pigmentation of the skin and mucous membranes. There is an excessive loss of sodium from the body and dehydration is common. Both cortex and medulla are affected in this disease, which may be due to atrophy or tuberculosis of the glands.

Oversecretion of hydrocortisone results in Cushing's syndrome, i.e. rounding of the face ('moon face'), obesity, hypertension, diabetes and osteoporosis.

The pituitary gland (hypophysis)

This is a single gland about 1 centimetre in diameter, situated at the base of the brain in the saddle-shaped depression in the sphenoid bone known as the sella turcica. Its attachment to the brain is by a short stalk placed just behind the optic chiasma where the optic nerves from each eye meet.

The pituitary gland consists of two parts, the anterior and posterior lobes, which have different modes of development and entirely different functions.

Functions of the anterior lobe

A number of different hormones are produced by this part of the gland.

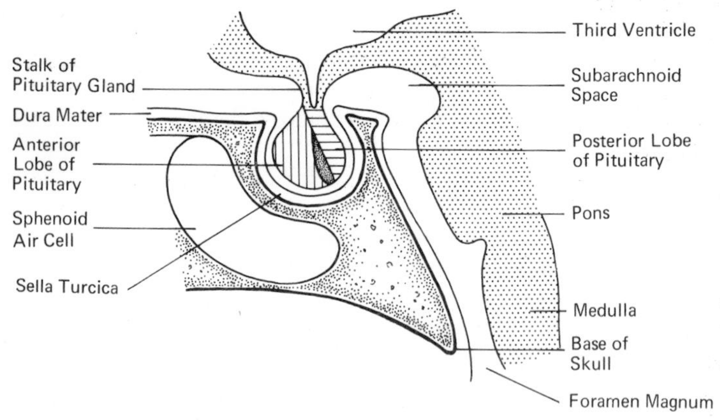

Fig. 186 Diagram of pituitary gland.

1. There is, for example, a hormone which plays an important part in the control of growth. ***Over-secretion*** (hyperpituitarism) in childhood leads to excessive growth in the length of bones and the condition known as gigantism is produced. Almost all the so-called giants are instances of anterior pituitary disorder. Disturbance of anterior lobe secretion in adult

life, when the normal growth of bone has ceased, still results in some bone deformity. The bones of the face, skull, hands and feet tend to enlarge while the lips become thick and the skin coarse. This condition is known as acromegaly. **Under-secretion** (hypopituitarism) results in obesity and disorders of carbohydrate metabolism; sexual development and activity may also be defective.

2. The anterior lobe of the pituitary secretes substances which act on the ovary and are referred to as gonad-stimulating hormones.
3. Another hormone—**prolactin**—is responsible for the flow of milk after childbirth.

In some respects the anterior lobe of the pituitary may be regarded as the **master gland** of the endocrine system since it appears to control the activities of so many of the others including the thyroid, adrenals, pancreas and sex glands.

SUMMARY OF THE HORMONES PRODUCED BY THE ANTERIOR LOBE OF THE PITUITARY GLAND

1. The growth hormone (GH) (somatrotrophin).
2. The thyroid-stimulating hormone (TSH). This appears to stimulate the growth and activity of the thyroid gland.
3. The adreno-cortico-trophic hormone (ACTH). This hormone is a protein substance which stimulates the cortex of the adrenal gland to secrete its own hormones, e.g.:
   (*a*) Mineral corticoids, e.g. aldosterone.
   (*b*) Gluco-corticoids, e.g. hydrocortisone (see page 295).
4. The gonado-trophic hormones (GTH). These are essential for the normal development of the sex organs and stimulate the production of the various sex gland hormones. They are:
   (*a*) The follicle-stimulating hormone (FSH). In the female this stimulates the ovarian follicles to produce oestrogen. In the male it stimulates the production of spermatozoa.
   (*b*) The luteinizing hormone (LH). In the female this stimulates the corpus luteum of the ovary to produce progesterone. In the male it stimulates the testes to produce testosterone.
5. The lactogenic hormone (prolactin) which helps to control the secretion of milk from the breast.

Although the actions of the various hormones which have been separated have been individually described, in health all work

THE FUNCTIONS OF THE PITUITARY GLAND

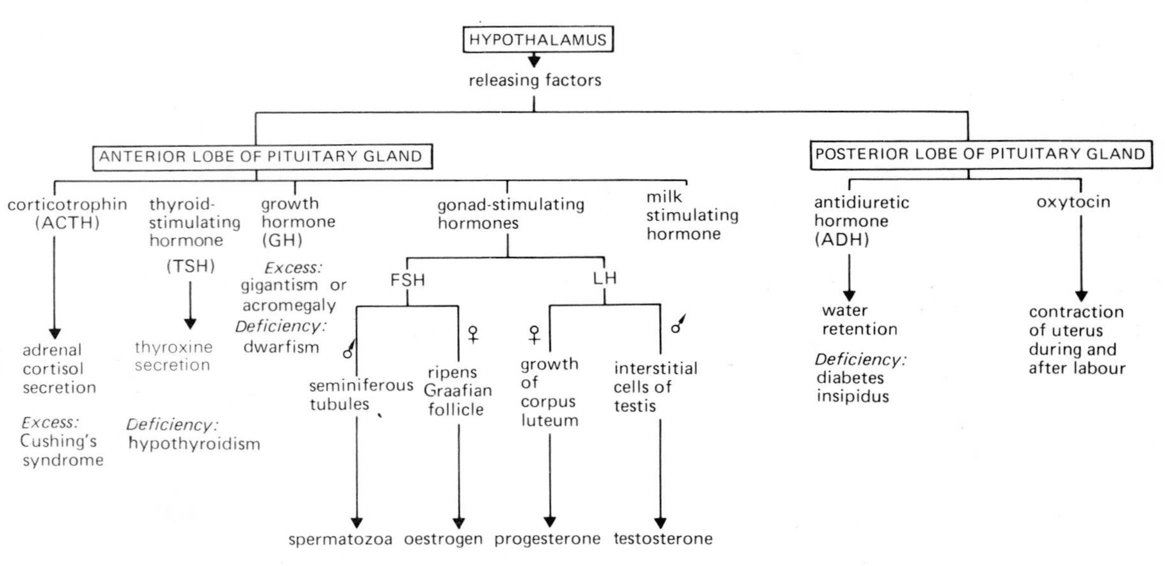

together. When diseased, all the functions of the whole anterior lobe may be affected, e.g.

over-secretion (hyperpituitarism) as in acromegaly,
under-secretion (hypopituitarism) as in Simmond's disease.

In view of the importance of the growth hormone, as already mentioned, the results will be modified according to whether a child or adult is affected.

The anterior pituitary gland is under the influence of the hypothalamus, which secretes *releasing factors* for each of the trophic hormones. They are named after the hormones which they release from the pituitary gland; GH-releasing hormone releases growth hormone. FSH-releasing hormone releases FSH, and so on. The releasing factors travel from the hypothalamus to the anterior pituitary gland in the blood of a portal system which runs in the pituitary stalk.

Functions of the posterior lobe

The posterior lobe secretes two hormones:

1. **Vasopressin** (antidiuretic hormone; ADH). The main effects of this are:
   (*a*) to concentrate the urine by increasing the absorption of water in the distal renal tubules and collecting ducts,
   (*b*) to raise the blood pressure, and
   (*c*) to cause contraction of involuntary plain muscle, especially of the intestines and bladder.
2. **Oxytocin**. This stimulates the plain muscles of the uterus during and immediately after labour. It also stimulates the lactating breast to eject milk.

Oxytocin (pitocin) is used only in obstetrics. Vasopressin is used in the treatment of diabetes insipidus, in which large volumes of dilute urine are passed.

The sex glands or gonads

The sex glands consist of the **ovaries** in the female and the **testes** in the male. In addition to their ordinary reproductive functions, viz. the production of ova by the ovary and spermatozoa by the testis, both produce hormones which have more general effects.

Internal secretions of the ovary

These are mainly concerned with the reproductive functions and include:

(*a*) Oestrogen. This plays an important part in the regulation of menstruation and also in the development of the secondary sexual characteristics of the female. Its secretion is considerably influenced by the follicle-stimulating hormone (FSH) of the anterior lobe of the pituitary. (It is formed by the Graafian follicle.)

(*b*) Progesterone, which appears to sensitize the mucous membrane of the uterus and prepare it for the reception of the fertilized ovum. (It is formed by the cells of the corpus luteum.)

Oestrogens help to diminish some of the unpleasant symptoms which may occur at the menopause, and also inhibit the secretion of milk after childbirth. They are also useful in the control of cancer of the prostate gland in the male. Synthetic oestrogens are available and include stilboestrol and ethinyloestradiol.

Progestogens have a number of uses in gynaecology including some cases of recurrent abortion and amenorrhoea. They are also used as oral contraceptives when combined with oestrogens.

THE FUNCTION OF THE OVARY

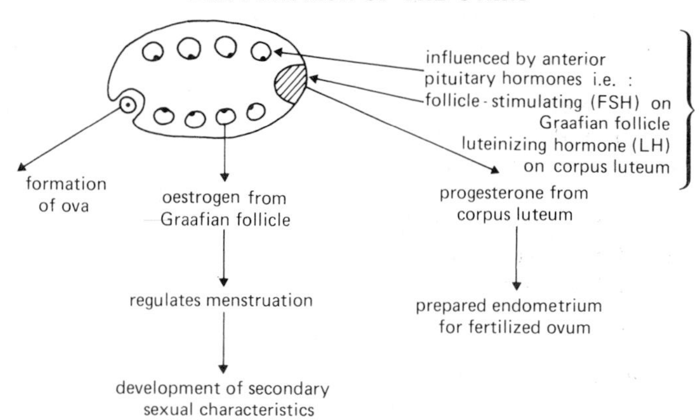

Internal secretions of the testis

These appear to control the secondary sexual characteristics of the male, such as the distribution of hair, the development of a deep voice, etc. They are called androgens, the most important being **testosterone.**

Testosterone is the main male hormone which is sometimes described as an androgen. Its secretion is responsible for the changes which take place at puberty.

In addition, it has various metabolic effects including protein building and an increase in the blood calcium when given therapeutically by injection. Androgens are of value in the treatment of cancer of the breast.

The thymus gland

This is situated in the anterior mediastinum of the thorax and is placed behind the sternum in front of the heart and pericardium and the arch of the aorta. It consists largely of lymphoid tissue and takes a part in the formation of lymphocytes.

The gland is relatively large at birth and continues to increase in size until puberty, when it gradually diminishes. Its function is obscure, but it plays a part in the general development of immunity and in the production of lymphocytes. It is enlarged in the disease known as myasthenia gravis.

The pineal gland

The pineal gland of body is a small reddish-grey structure about the size of a pea situated between the under-surface of the cerebrum and the mid-brain, just in front of the cerebellum. It is a rudimentary organ having no known function in man.

The pancreas

The pancreas lies on the posterior abdominal wall. It has a head which is situated in the C of the duodenum, a body passing transversely across the abdominal aorta and bodies of the lumbar vertebrae, and a tail which reaches the hilum of the spleen.

It has:

1. An external secretion—the **pancreatic juice**—containing the enzymes amylase, lipase and trypsin, which is poured into the duodenum at the ampulla of Vater by the pancreatic duct, at its point of junction with the common bile duct.

2. An internal secretion—**insulin**—formed by the islets of Langerhans, which passes directly into the blood and is concerned in the metabolism of sugar.

When examined under the microscope the functionally different parts are clearly distinguishable. The glandular (digestive) portion consists of secretory tubules lined by columnar epithelium. The endocrine portion consists of clumps of cells scattered between the secretory tubules. These clumps of cells are called the islets of Langerhans.

Insulin is essential for the normal metabolism of sugar in the body. Without insulin the muscles are unable to use the sugar circulating in the blood which supplies them with one of their sources of energy. It has been seen that:

1. Sugar in the form of glucose is normally used by the tissues and in order to produce energy is broken down into carbon dioxide and water.
2. Normally the blood-sugar level remains constant. Any excess of sugar is stored in the liver as glycogen (page 229).

THE FUNCTIONS OF THE PANCREAS

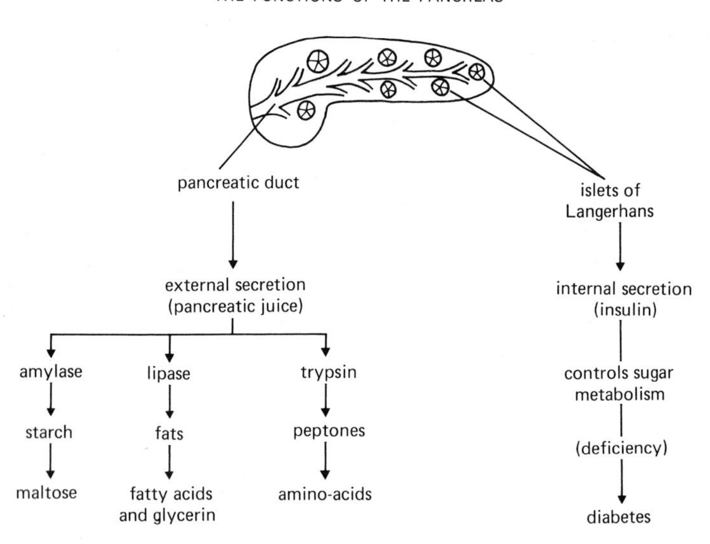

3. If the blood-sugar rises above its normal or 'threshold' level the excess of sugar is excreted by the kidneys and sugar appears in the urine (glycosuria) (page 286).
4. The full oxidation of fat also involves the normal consumption of glucose by the tissues. If sugar is not properly utilized there is a defect in the burning up of fat and ketone bodies are formed.

Deficiency of insulin due to disease of the islets of Langerhans results in diabetes, a condition in which the blood-sugar is high and sugar is passed in the urine. In severe cases the disturbed metabolism of fat results in ketosis and the presence of ketone bodies (acetone and diacetic acid) in the urine.

QUESTIONS

1. Give an account of the pancreas and of its functions.
2. Give an account of the thyroid gland and its functions.
3. Describe the functions of the following glands: (*a*) the thyroid gland (*b*) the suprarenal gland, (*c*) the ovaries.
4. What is an endocrine gland? Give an account of the functions of any three of them.

18 The Nervous System

The nervous system is the most highly developed and perhaps the most important of all the systems of the body. Not only does it correlate the activities of the other systems but also in the brain are situated the sites of consciousness, thought, memory, speech and the will to carry out purposeful actions. These factors all contribute to the formation of the personality of the individual.

The nervous system may be subdivided into three main portions:

1. The brain and spinal cord, or **central nervous system.**
2. The nerves or **peripheral nervous system** which form the connections between the central nervous system and the various organs and muscles of the body.
3. The **autonomic** (sympathetic and parasympathetic) **nervous system.**

The brain itself is the most important part. It is the centre which receives impulses or sensations which are stored and interpreted in the mind. The accumulation of these stored impulses forms the basis of memory. Not only does the brain receive impulses but it also transmits them via the nerves to the various parts of the body.

Sensory impulses travelling towards the brain are called **afferent** impulses. Impulses resulting in movement or action of some sort, travelling away from the brain, are called motor or **efferent** impulses.

The tissue of which the nervous system is constructed has already been described (page 30).

Coverings of the brain and spinal cord

1. **The meninges.**
2. **The cerebrospinal fluid.**

The meninges

The central nervous system lies within the skull and neural canal of the vertebral column (vertebral canal). In view of its great importance in maintaining the life of the individual and of the delicacy of its structure it is most carefully protected. In addition to the hard bony protection afforded by the skull and vertebral column, the brain and spinal cord have the following coverings called meninges:

1. The dura mater or outer layer.
2. The arachnoid mater or middle layer.
3. The pia mater or inner layer.

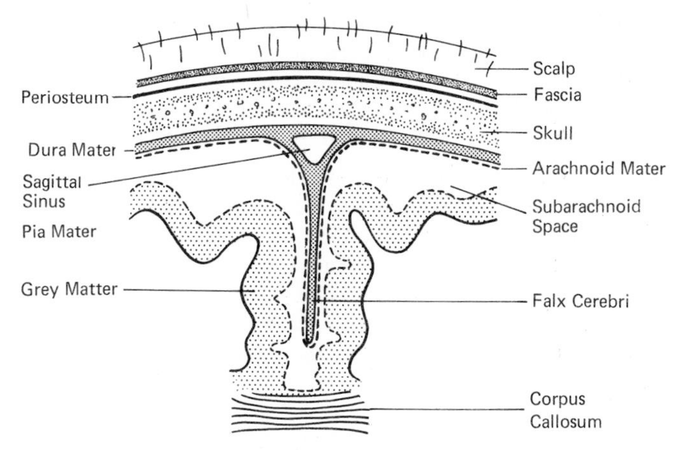

Fig. 187 Diagram of the scalp and meninges.

Between the arachnoid mater and pia mater is a fluid called the cerebrospinal fluid which also has important protective functions.

The **dura mater** is a strong fibrous membrane which lines the skull and also encloses the spinal cord in a membranous tube within the neural canal. In the skull it actually forms the periosteum lining the inner aspect of the cranial bones

In addition to lining the interior of the skull it has two important folds projecting into the cranial cavity which help to support the brain and to maintain it in position.

1. The **falx cerebri**, a sickle-shaped fold which lies vertically in the mid-line and separates the right from the left cerebral hemisphere.
2. The **tentorium cerebelli**, which lies horizontally and forms

a tent-like roof for the posterior cranial fossa of the skull, thereby separating the cerebrum above from the cerebellum below.

The important venous sinuses—sagittal, cavernous and transverse (lateral) sinuses—are contained between layers of the dura mater (page 152).

The **arachnoid mater** is a delicate transparent membrane, enveloping the brain and spinal cord, and is situated between the dura mater and the pia mater (arachnoid = spider-like).

The arachnoid mater is closely applied to the dura mater, but is separated from the pia mater by a narrow space, the **subarachnoid space,** which contains the cerebrospinal fluid.

Between the under surface of the cerebellum and the medulla oblongata this space is enlarged to form the **cisterna magna,** of importance because a needle is sometimes passed between the occiput and atlas vertebra into this great cistern in order to withdraw cerebrospinal fluid (cistern puncture). Great care must be taken not to insert the needle too far as it will damage the medulla oblongata, possibly with fatal results.

The **pia mater** is the innermost layer of the meninges and is closely applied to the exterior of the brain and spinal cord. It is a very delicate structure and carries numerous small blood vessels which supply the surface of the brain and spinal cord.

It follows closely the surface of the brain and dips into all the fissures between the convolutions, whereas the arachnoid mater forms only a bag-like covering for the whole of the central nervous system.

Cerebrospinal fluid

This is a clear, colourless fluid (specific gravity 1005) which occupies:

1. The subarachnoid space between the arachnoid mater and the pia mater.
2. The ventricles of the brain.

It is secreted by the **choroid plexus,** a structure consisting of a network of fine blood vessels situated in the lateral ventricles of the brain. This fluid, therefore, not only covers the surface of the brain and spinal cord but is also contained in the hollow interior of the brain and in the central canal of the spinal cord.

In many respects cerebrospinal fluid resembles a dilute form of lymph and contains, among other substances, protein, glucose and salts.

It has the following **functions**:

1. It acts as a 'water cushion' protecting the brain and spinal cord from jars and shocks due to body movement.
2. It conveys nutritive material to the brain and spinal cord.
3. It removes waste products from these structures.

There is a communication between the ventricular system of the brain and the subarachnoid space through which the cerebrospinal fluid can flow, called the median aperture of the fourth ventricle (**foramen of Magendie**) situated in the roof of the fourth ventricle.

Since the fluid is secreted by the choroid plexus from the blood it follows that some drugs, e.g. sulphonamides, barbiturates, absorbed into the blood stream may also reach the cerebrospinal fluid.

Lumbar puncture. Owing to the intimate association of the cerebrospinal fluid with the nervous tissues and the meninges, disease of these structures is often associated with changes in the fluid, which may be removed for examination by the procedure known as lumbar puncture. This consists of passing a special needle through the skin and between the spines of the second and third or third and fourth lumbar vertebrae into the subarachnoid space. It will be seen later that the spinal cord itself does not extend below the level of the first lumbar vertebra, so that there is no risk of injuring it by this procedure.

Bleeding may occur into the subarachnoid space (e.g. fractured skull; rupture of an aneurysm of the circle of Willis). At first this will be red in colour but after a day or two the haemoglobin is converted into a bilirubin-like pigment which stains the CSF yellow (xanthochromia).

ANATOMY OF THE BRAIN

The brain is that portion of the central nervous system which lies within the cavity of the skull. It consists of the following parts:

1. The two cerebral hemispheres or cerebrum
2. The cerebellum
3. The mid-brain }
4. The pons } the brain stem
5. The medulla oblongata }

The cerebrum

This consists of the right and left **cerebral hemispheres** and is the largest part of the brain. It occupies the anterior and middle fossae of the skull. The two hemispheres are separated in the mid-line by a fold of dura mater, the **falx cerebri** (page 207), but below this they are connected by a bridge of white matter, the **corpus callosum**. Each cerebral hemisphere is described as having frontal, parietal, temporal and occipital **lobes**, which correspond in position with the skull bones of the same name. The anterior part of the frontal lobe is called the **frontal pole**, that of the temporal lobe the **temporal pole**, while the posterior portion of the occipital lobe is called the **occipital pole**.

The surface of the hemispheres consists of nerve cells or **grey matter**, which is called the **cerebral cortex**. This is arranged in folds or convolutions thereby greatly increasing the total amount of grey matter. The convolutions are separated by **fissures**.

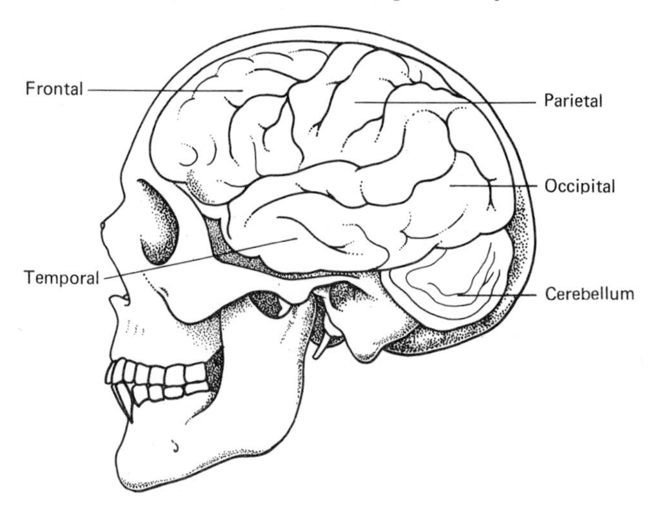

Fig. 188 Diagram of the left side of the brain, showing division into lobes and their relation to the skull.

Of the many fissures on the surface of each hemisphere, two are of special importance:

1. The **central sulcus** or **fissure of Rolando**, which passes downwards and forwards from the mid-line, separating the frontal from the parietal lobes.
2. The **lateral fissure** or **fissure of Sylvius**, which can be traced backwards and upwards from the temporal pole, separating the frontal and parietal lobes above from the temporal lobe below.

If the margins of the lateral fissure (of Sylvius) are separated,

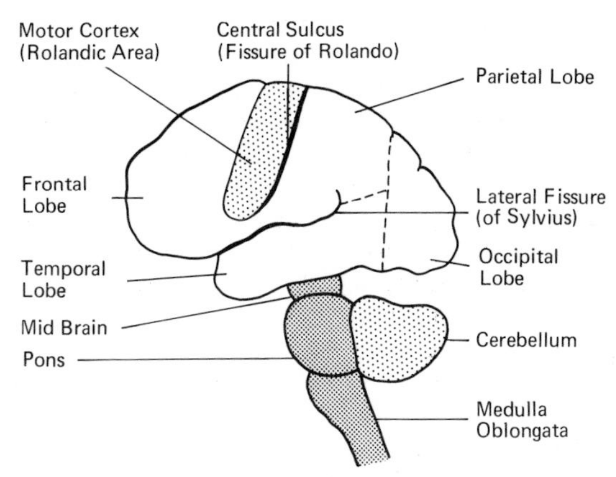

Fig. 189 Diagram of outer surface of the left cerebral hemisphere, the cerebellum and the brain stem.

a buried portion of cerebral cortex called the insula (*island of Reil*) is exposed.

The interior of each hemisphere consists mainly of a mass of white matter formed by nerve fibres passing to and from the brain and between various parts of the brain itself. Collections of grey matter found deep in the cerebral hemispheres include the basal ganglia, and hypothalamus.

The **basal ganglia** are specialized structures which are con-

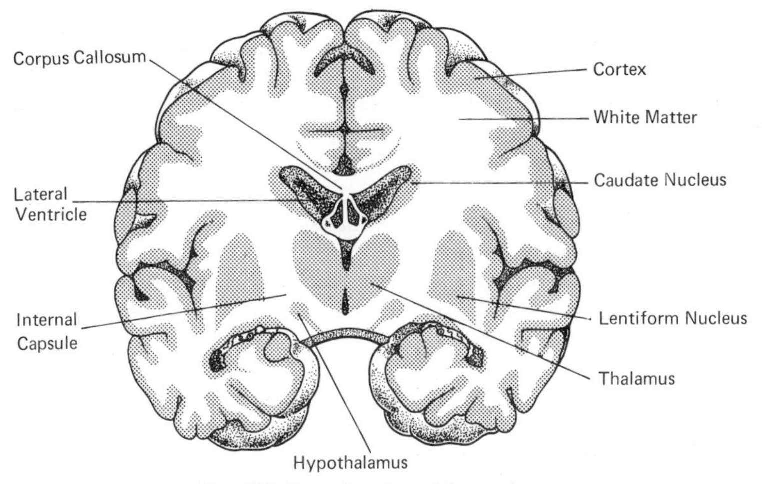

Fig. 190 Coronal section of the cerebrum.

cerned in the modulation of muscular movements, so that they are not coarse and clumsy. This is achieved by a fine balance of facilitation and inhibition. Prominent among the basal ganglia are the caudate nucleus and the lentiform nucleus which together comprise the **corpus striatum**. There are connections between the individual members of the basal ganglia and between them and the cerebral cortex, the thalamus and the reticular formation of the brain-stem.

The basal ganglia form part of a complex known as the **extra-pyramidal system**. The pyramidal system of upper motor neurones is concerned with the execution of discrete movements. The extra-pyramidal system makes these movements precise and harmonious with other muscular movements. Diseases of the pyramidal system cause spastic paralysis. Diseases of the extrapyramidal system cause involuntary movements such as **chorea** (irregular, semi-purposive, often jerky movements), **athetosis** (slow, writhing movements) and the muscular rigidity, the tremor and impoverishment of movement, of **Parkinsonism**.

The **thalamus** is a sensory relay station, with incoming fibres from the spinal cord and brain-stem and onward fibres to the cerebral cortex. 'Crude' sensation and pain may be appreciated only in the higher centres of the brain—in the sensory cortex of the parietal lobe. Lesions of the thalamus cause a peculiar hypersensitivity of the opposite side of the body.

The **hypothalamus** includes a number of structures below the thalamus, at the base of the brain. Its **functions** are:

(*a*) Manufacture of anti-diuretic hormone and oxytocin (see page 301), which are subsequently stored in the posterior lobe of the pituitary gland.

(*b*) Production of 'releasing factors' which cause rapid release of hormones from the anterior pituitary gland. Examples are GH-releasing factor, FSH-releasing factor and LH-releasing factor.

(*c*) Regulation of the action of the heart via autonomic centres in the medulla oblongata.

(*d*) Regulation of blood vessel calibre via vasomotor centres in the medulla oblongata.

(*e*) Regulation of body temperature.

(*f*) Control of appetite.

Most of the important nerve fibres passing to and from the brain lie in a relatively narrow tract of white matter situated near the basal ganglia, called the **internal capsule**.

The ventricles of the brain. The central portion of each hemi-

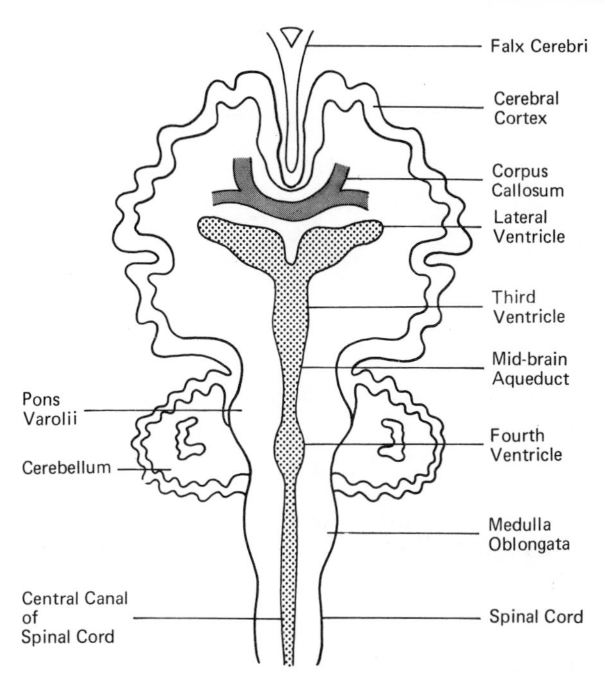

Fig. 191 Diagram of the ventricles of the brain (anterior view).

sphere is hollow and this space is called the lateral ventricle. The left and right lateral ventricles are in direct communication with each other and, below, communicate with another cavity in the mid-line called the third ventricle which in turn is connected to the fourth ventricle. The fourth ventricle constitutes the hollow portion of the pons and medulla oblongata and its roof is in contact with the under surface of the cerebellum. It will be noted that all the ventricles contain cerebrospinal fluid which is secreted by the choroid plexus in the lateral ventricles.

The cerebellum

The cerebellum or lesser brain is situated in the posterior cranial fossa of the skull and, therefore, lies below the occipital lobes of the cerebrum from which it is separated by the fold of dura mater called the **tentorium cerebelli** (page 307).

Like the cerebrum, the cerebellum consists of an outer cortex of grey matter arranged in numerous fine convolutions, and an inner

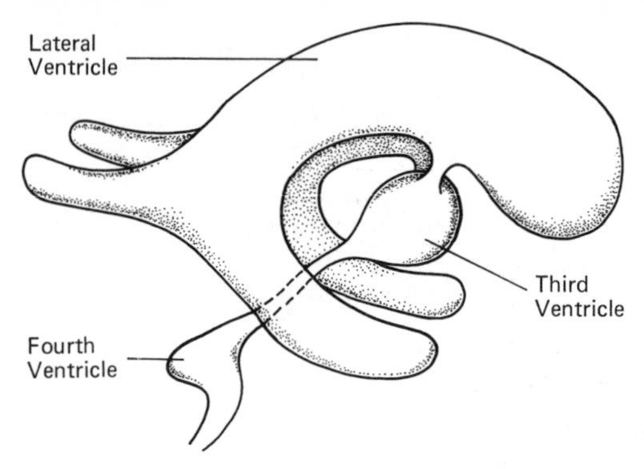

Fig. 192 Diagram of the ventricles of the brain (lateral view).

core of white matter. It has right and left lobes which are joined in the mid-line by a bridge of white matter forming part of the pons.

The brain-stem

This consists of three parts, the mid-brain, the pons and the medulla oblongata.

The mid-brain

This joins the cerebral hemispheres above to the pons below.

The **pons** consists both of grey and white matter. It acts as a bridge between the two lobes of the cerebellum, and nerve fibres pass up and down between the mid-brain above and the medulla oblongata below. The grey matter forms the nuclei of some of the cranial nerves.

The **medulla oblongata** is an important structure of grey and white matter which connects the brain with the spinal cord. It is about one inch in length and lies on the base of the skull just in front of the foramen magnum. As it passes through the foramen it becomes continuous with the spinal cord.

PHYSIOLOGY OF THE BRAIN

The brain is the control centre of the whole human body. Physiologically it may be divided into the **higher centres** which are the seat of consciousness, mind, memory and will; and the **lower centres** which control many important unconscious acts. The higher centres are situated in the cerebral hemispheres while the lower ones are found in the cerebellum and brain-stem as well as in the basal ganglia of the cerebrum.

The following mechanisms enable the brain to exercise this power of control:

1. It receives sensory or afferent impulses from all parts of the body.
2. It is able to send out motor or efferent impulses to all parts of the body.
3. There is a complicated system of connections between all parts of the brain with each other.

In other words, the white matter of the brain consists of nerve fibres passing from the body via the nerves, spinal cord and brain stem to the cerebral cortex, and nerve fibres passing from the cortex in the opposite direction to the periphery, in addition to fibres passing from one part of the brain to another. These fibres tend to run in certain well-defined tracts throughout the central nervous system.

The nature of nervous activity

Nerve cells have an active metabolism just like other cells of the body. They require oxygen and other nourishment and excrete waste products. Little is known of the actual nature of nervous activity but study of the electro-encephalogram has thrown some light on it.

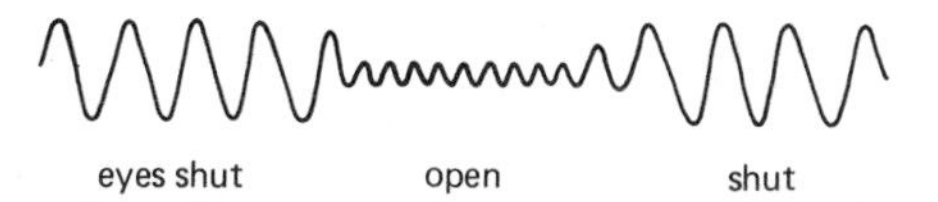

Fig. 193 Electro-encephalogram: rhythm from occipital cortex.

When special pads or electrodes are applied to the head and connected to the apparatus, rhythmic waves or oscillations can be recorded which indicate that there is a continuous discharge of energy from the neurones which normally spreads in an orderly manner. When individual parts of the brain are called into activity there is an alteration in the character of these waves.

For example, if records are taken from the region of the occipital cortex which is concerned with sight, the waves alter when the eyes are opened and shut.

In epilepsy there is an explosive irregular discharge of nervous energy and the regular waves are replaced by large irregular ones.

Physiology of sleep

'O sleep! O gentle sleep! Nature's soft nurse'. . . .
(*Henry IV,* Part II, Act III)

Sleep is a normal physiological function and a human being spends about one-third of his life in this state. However, in spite of a number of theories, no completely satisfactory and universally accepted explanation of its fundamental mechanism exists.

On the other hand, many studies and experiments have been carried out which clearly show the physiological changes which occur in the body during normal sleep.

Generally speaking, almost all physiological functions are reduced to their lowest activity. These may be summarized:

1. General metabolism and heat production are reduced.
2. The heart rate is slowed and blood pressure falls.
3. Ventilation is slowed and the thoracic type predominates over abdominal breathing.
4. Renal excretion is diminished.
5. Reflexes are reduced, e.g. the knee jerk.
6. Voluntary muscles are relaxed but the sphincter muscles remain contracted while peristalsis in the alimentary tract continues.
7. There are changes in the vaso-motor system which result in a diminished blood supply to the brain while the peripheral vessels in the skin are dilated leading to some increase in sweat secretion.

Insomnia. Prolonged loss of sleep is said to be more serious than deprivation of food and water. Although some individuals tend to exaggerate the periods of insomnia, it is an important nursing problem. Both psychological (e.g. grief and anxiety) and physical discomfort and pain may be contributary causes; they can usually be dealt with by administering suitable sedatives and analgesics.

AVERAGE OF SLEEP REQUIRED

Age	*Hours*
Infants and young children	10–12
Adolescents and young adults	7–9
Middle age	6–8
Old age	9–12

Normal adult sleep is deepest during the first two hours and later becomes lighter.

Localization of cerebral function

Certain areas of the brain are responsible for definite functions.

1. The area situated to the frontal lobe immediately in front of the central sulcus (fissure of Rolando) is called the **motor cortex**. From the cells of this area, voluntary motor impulses arise and are transmitted to the various parts of the body via the spinal cord and peripheral nerves (see motor path, page 332).

   The various groups of muscles in the body are represented with the head, arms and upper part of the trunk towards the lower end of the area, just above the commencement of the

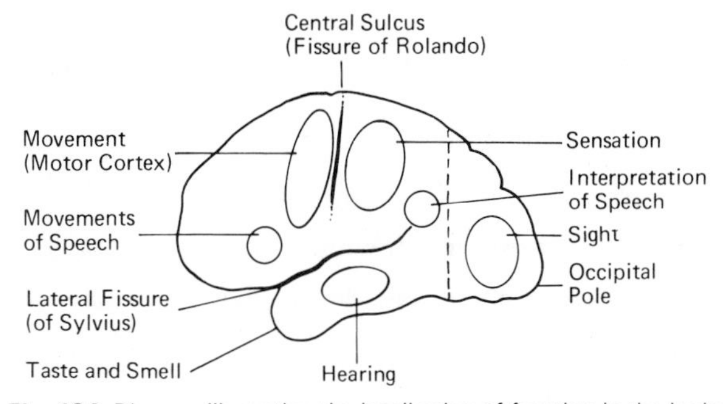

Fig. 194 Diagram illustrating the localization of function in the brain.

   lateral fissure (of Sylvius), while the feet and lower limbs are controlled by the grey matter at the upper end of the motor cortex near the superior fissure separating the two hemispheres.

   (NOTE. The right cerebral hemisphere controls the left side of the body and vice versa.)

2. **The sensory area**. General sensation is represented in the parietal lobe of the brain behind the central sulcus (fissure of Rolando).
3. **The visual area**. The special sense of sight is controlled by the grey matter in the occipital lobe.
4. **The auditory area**. The special sense of hearing is located in the temporal lobe.
5. The sense of **smell** and **taste** is located in the medial surface of the temporal lobe.

6. Special centres also exist for speech, reading and writing, which are situated around the lateral fissure (of Sylvius).

A lesion of the cerebral cortex causing inability to understand or express meanings in speech is called *aphasia*. If this only involves some difficulty rather than complete failure the term *dysphasia* is used. Difficulty in articulation (not involving understanding) is due to some nerve damage to the muscles used in speech production and is called *dysarthria*.

Reference to Fig. 194 show the representation of these areas on the surface of the brain. In addition, there are wide areas of cortex, the function of which is not clear, and these are referred to as *silent areas*.

Functions of cerebrum. These may be summarized as:

1. To receive all sensory stimuli and to convey most of them to consciousness.
2. To initiate all voluntary movement.
3. To correlate and retain all impulses received, thereby forming the basis of memory.
4. To formulate and associate ideas giving rise to intelligence.
5. To exercise unconscious control over many functions of the body.
6. To exercise control over the lower parts of the brain.

The symptoms of cerebral disease depend upon which parts of the brain are involved. For instance, a lesion (e.g. thrombosis) in the internal capsule or the motor cortex may cause spastic hemiplegia. Damage to the visual pathway will cause a loss of part of the visual field.

Functions of cerebellum. The activity of the cerebellum is unconscious and not under the control of the will and its functions may be summarized:

1. It helps to maintain muscular tone.
2. It coordinates muscular movement.
3. It helps to maintain balance and equilibrium. It is able to do this because it receives impulses from the semi-circular canals, joints and muscles.

Disease of the cerebellum, therefore, results in erratic and incoordinated movements called ataxia which may be associated with shaking or tremor and a tendency to fall to one side when standing or walking.

Functions of brain stem. The mid-brain conveys impulses to and from the cerebral hemispheres.

The pons conveys impulses to and from the cerebrum and also to and from the cerebellum.

Further, the brain-stem contains collections of grey matter which are the nuclei or points of origin of the cranial nerves.

The medulla oblongata connects the other parts of the brain with the spinal cord. It is especially important because in addition to the nerve fibres passing up and down in its substance and the nuclei of many of the cranial nerves, it contains also collections of grey matter known as the **vital centres**.

The most important of these are:

1. The respiratory centre which controls the rate and depth of respiration (see page 270).
2. The vasomotor centres which control the calibre of the blood vessels.
3. The cardiac centre which influences the rate of the heart (see page 162).
4. Also, special centres such as the swallowing centre, the vomiting centre, centres for the movements of the stomach and the secretion of saliva and gastric juice.

These centres are similar to the reflex centres in the spinal cord. They receive afferent impulses from the periphery and send out motor impulses in response to these stimuli. They may also be influenced by the higher centres of the brain.

With so many important vital structures situated in the small space of the medulla oblongata, together with the fact that all the impulses to and from the brain pass through it, it is clear that any disease, injury or pressure on the medulla are very serious and often fatal.

Centres in the brain stem are sometimes affected in poliomyelitis and other diseases of the nervous system, causing, in particular, difficulty in breathing and swallowing which may necessitate tracheostomy (Bulbar paralysis).

The cranial nerves

There are twelve pairs of cranial nerves which arise from the brain and brain stem. Some of them are sensory, bringing impulses to the brain; others are motor, carrying impulses from the brain to the periphery, while a few are mixed and contain both motor and sensory fibres.

Ist. The **olfactory nerve** is the nerve of smell and, therefore, a sensory nerve. Its fibres arise in the mucous membrane of the nose and join the olfactory bulb which lies on the cribriform plate of the ethmoid bone. From this the nerve passes backwards to reach the olfactory area of the brain situated in the under surface of the temporal lobes where the sense of smell is closely associated with that of taste.

IInd. The **optic nerve** is the nerve of sight and is sensory in character. Its fibres commence in the retina of the eye and, passing backwards through the cavity of the orbit, enter the skull. On the under surface of the brain the inner fibres of the optic nerves cross forming the optic chiasma, from which they pass backwards in the interior of the brain to reach the occipital portion of the cerebral cortex (see Fig. 221).

Number and name	*Type*	*Function and distribution*
I Olfactory	sensory	mucous membrane of nose (smell)
II Optic	sensory	retina of eye (sight)
III Oculomotor	motor	eye muscles see page 353
IV Trochlear	motor	eye muscle (superior oblique)
V Trigeminal	mixed	sensory to forehead and face and motor to muscles of mastication
VI Abducens	motor	eye muscle (lateral rectus)
VII Facial	motor	muscles of facial expression
VIII Auditory	sensory	(*a*) hearing from cochlea (*b*) equilibrium from semicircular canals
IX Glosso-pharyngeal	sensory	sensory from tongue (taste) and pharynx
X Vagus	mixed	to pharynx, larynx, trachea, bronchi, lungs, heart, stomach, oesophagus and intestine (upper part)
XI Accessory	motor	to sternomastoid and trapezius and muscles of pharynx and larynx
XII Hypoglossal	motor	to muscles of tongue

IIIrd, IVth and VIth. The **oculomotor**, **trochlear** and **abducens nerves** are the motor nerves of the eyeball, and supply the muscles situated in the orbit which move the eye (page 353). They arise from the brain stem.

Vth. The **trigeminal nerve**. This is the largest cranial nerve and is the sensory nerve of the face and anterior part of the scalp. It also has a small motor portion which supplies the muscles of mastication.

It has three main divisions:

1. The **ophthalmic branch** which supplies the forehead and

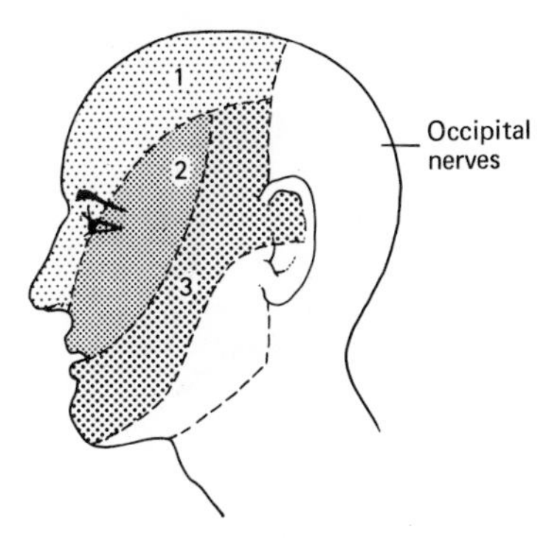

Fig. 195 Distribution of the branches of the trigeminal nerve. 1. Ophthalmic branch. 2. Maxillary branch. 3. Mandibular branch.

front half of the scalp, also the front of the eye (conjunctiva).

2. The **maxillary branch** supplying the area of the cheek and upper jaw, including the teeth.
3. The **mandibular branch** distributed to the area of the lower jaw and its teeth.

Before dividing into its three branches it passes through the important **trigeminal** (Gasserian) **ganglion** which lies on the floor of the middle cranial fossa of the skull.

VIIth. The **facial nerve** (motor) arises from the brain stem and supplies the muscles of facial expression. It also conveys fibres from the parasympathetic system to the salivary glands and sensory fibres of taste from the anterior two-thirds of the tongue.

VIIIth. The **auditory nerve** is a sensory nerve and actually consists of two parts—the true auditory nerve of hearing which passes from the cochlea, and the vestibular nerve from the semi-circular canals which is concerned with equilibrium and balance (page 348).

IXth. The **glossopharyngeal nerve** is mainly a sensory nerve sending fibres to the pharynx and the posterior third of the tongue (taste) page 342.

Xth. The **vagus nerve** is also mixed and is distributed to the pharynx, larynx, trachea, bronchi, lungs, heart, oesophagus, stomach and upper part of the intestine. In other words, it supplies mainly the viscera contained in the thorax and upper part of the abdominal cavity. The vagus arises from the medulla oblongata, passes through a foramen in the base of the skull in company with the jugular vein and continues downwards in the neck together with this vein and the carotid artery. It enters the thorax where it is closely related

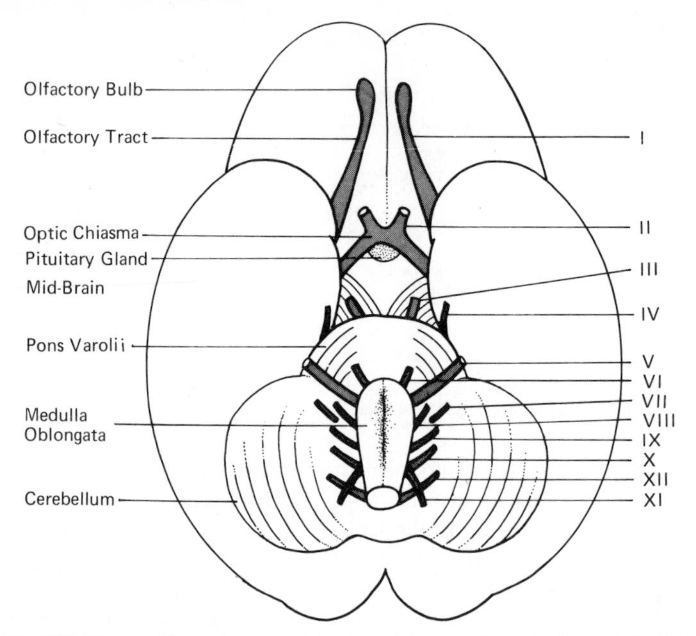

Fig. 196 Diagram illustrating the origin of cranial nerves from the base of the brain.

to the oesophagus and, with it, passes through the diaphragm to the upper abdominal organs.

XIth. The **accessory nerve** consists of two parts, one of which joins the vagus and supplies the muscles of the pharynx and larynx, the other supplies the sternomastoid and trapezius muscles in the neck.

XIIth. The **hypoglossal nerve** is a motor nerve which supplies the muscles of the tongue.

PRACTICAL CONSIDERATIONS. Injury or disease of the optic nerve will cause blindness in the affected eye.

Lesions of the IIIrd, IVth and VIth nerves will result in inability to perform various eye movements and in squints.

The Vth or trigeminal nerve may be the site of severe neuralgia called tic doloreux, characterized by pain in the face, which is sometimes treated by injecting alcohol into the trigeminal (Gasserian) ganglion.

A lesion of the VIIth or facial nerve produces loss of movement of one side of the face so that it has a flattened, ironed-out appearance, with drooping of the angle of the mouth and inability to smile on that side.

Auditory or VIIIth nerve disease produces deafness, noises in the ear and sometimes giddiness.

Surgical section of that portion of the vagus nerve supplying the stomach may be performed in the treatment of peptic ulcer. Thereby the secretion of hydrochloric acid is diminished.

The spinal cord

The spinal cord is a part of the central nervous system which lies in the vertebral canal. It commences above at the level of the foramen magnum where it is continuous with the medulla oblongata. In the adult, it is about 45 cm (18 in) long and ends at the level of the first lumbar vertebra; it therefore does not completely fill the lower part of the vertebral canal. In infancy, the cord extends to a lower level than the first lumbar vertebra. Surrounding the spinal cord are the meninges. The dura mater and arachnoid mater form a sac which encloses the cord and extends downwards to the lowest part of the vertebral canal, while the pia mater closely covers the surface of the cord in the same way as it adheres to the surface of the brain. Cerebrospinal fluid fills the subarachnoid space between the arachnoid mater and pia mater.

The spinal cord is an elongated oval structure, which is increased in circumference in two areas, viz. the **cervical** and **lumbar enlargements**. From these enlargements the important nerves to the arms and legs arise (see Fig. 199).

It ends in a fibrous strand, the **filum terminale**, which continues downward and is attached to the coccyx.

The spinal nerves arise from the sides of the spinal cord by two roots, anterior and posterior, which unite as they leave the vertebral column to form the spinal nerve trunk.

There are, in all, thirty-one pairs of spinal nerves corresponding to the segments of the vertebral column:

viz. 8 cervical,
12 thoracic or dorsal,
5 lumbar,
5 sacral,
1 coccygeal.

The nerves from the lumbar, sacral and coccygeal regions occupy the space in the lower part of the vertebral canal below the termination of the spinal cord, before passing between their corresponding vertebrae. This mass of nerves is called the **cauda equina** (like a horse's tail).

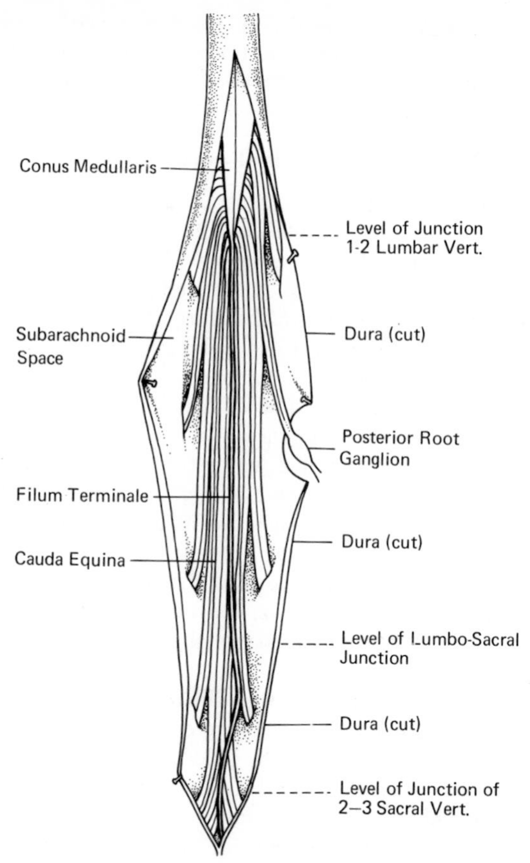

Fig. 197 Lower end of the dura and arachnoid laid open to show the cauda equina.

Transverse section of spinal cord

Study of the cross-section of the spinal cord is of importance. It has been noted that the grey matter of the brain is situated on the surface and the white matter in the interior. In the case of the spinal cord this arrangement is reversed and the white matter of the cord is on the surface while the grey matter is placed in an H-shaped manner in its interior.

It will be seen (Fig. 198) that the cord is oval in shape and that the H-shaped grey matter divides the white matter into anterior, lateral and posterior columns. The anterior columns of each side are separated by a well-marked anterior fissure. In the centre of the spinal cord is a minute canal (the **central canal**) which is continuous above with the fourth ventricle in the medulla oblongata.

Grey matter of spinal cord. The H-shaped grey matter is described as having anterior and posterior horns. The anterior horns contain a number of special cells which are concerned with motor functions and from them arise the nerve fibres which form the anterior roots of the spinal nerves.

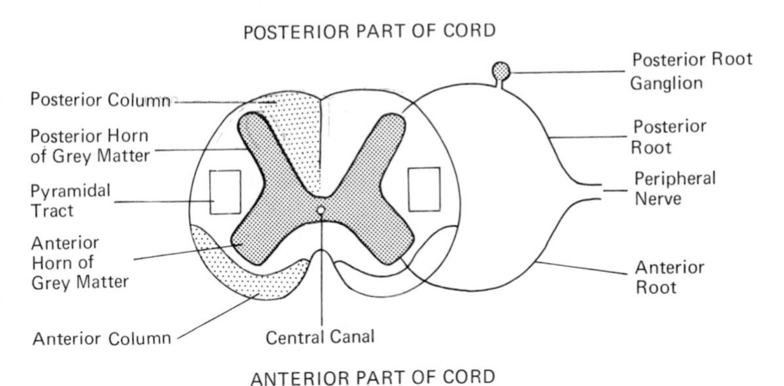

Fig. 198 Diagram illustrating a cross-section of the spinal cord and showing the relative positions of the grey matter and various tracts of white matter.

The posterior or sensory root of the spinal nerve enters the posterior horn of grey matter, the cells of which are sensory in function.

White matter of spinal cord. It has been seen that the H-shaped grey matter divides the white matter into three main columns, anterior, posterior and lateral.

The white matter of the spinal cord consists of sensory fibres running upwards to the brain and motor fibres passing downwards from the brain for distribution to the periphery. The motor fibres are situated in the lateral columns of white matter in a special tract, the **pyramidal tract.**

The sensory fibres entering from the posterior root pass to the anterior and posterior columns. (The senses of pain and temperature pass upwards in the anterior columns, while the senses of touch and position reach the brain via the posterior columns.)

Injury or disease resulting in a complete transection of the lower part of the spinal cord leads to paralysis of both lower limbs (paraplegia). If the lesion is higher up all four limbs may be affected (quadriplegia). These conditions present many nursing and social problems, e.g. care of the bladder, prevention of bedsores, and provision of a suitable invalid carriage.

The spinal nerves

It has been seen that there are thirty-one pairs of spinal nerves coming off the spinal cord at intervals and that they are classified according to the regions from which they arise. Further, that each spinal nerve consists of anterior and posterior **roots** which unite to form the main nerve **trunk** as it leaves the vertebral column. On the posterior nerve root of each spinal nerve is a collection of nerve cells called the **posterior root ganglion.**

It has also been pointed out that all sensory nerve fibres reach the spinal cord by the posterior root and all motor nerve fibres leave by the anterior root. It follows that the peripheral nerve trunk is a mixed nerve containing both motor and sensory fibres.

The individual nerve trunks arising from certain regions of the

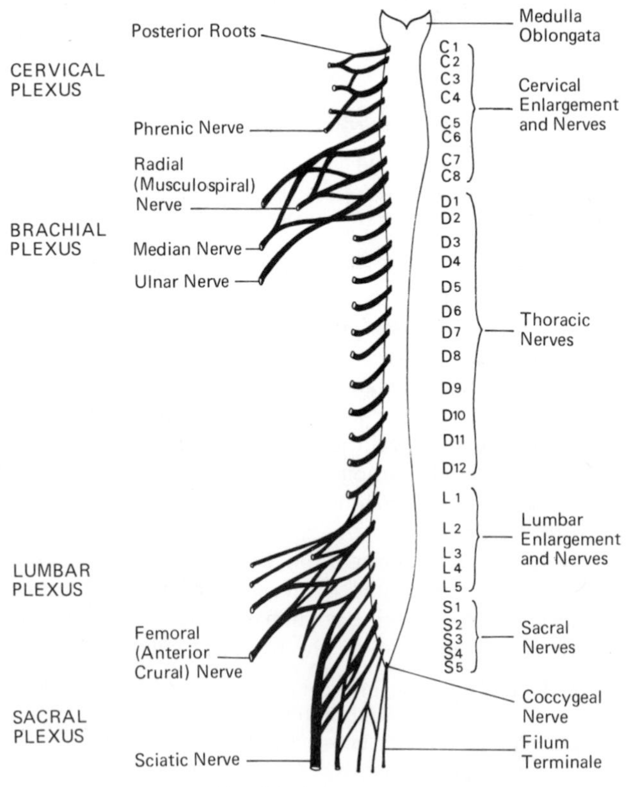

Fig. 199 Spinal nerves.

spinal cord join up together to form what is called a **plexus** from which they emerge rearranged as the individual peripheral nerves. Two plexuses are formed by the cervical nerves and one by the lumbar and sacral nerves.

Cervical plexus

The upper four cervical nerve trunks unite to form a plexus lying deeply in the upper part of the neck from which peripheral nerves are distributed mainly to the skin and muscles of the head and neck.

The most important branch of the cervical plexus is the **phrenic nerve** which passes down in the neck to enter the thorax, where it is closely related to the side of the pericardium. It terminates in the diaphragm of which it is the motor nerve.

Brachial plexus

This is formed mainly from the lower cervical nerve trunks and is situated deeply in the lower part of the neck behind the clavicle. From this plexus the important nerves of the arm are derived. These nerves enter the axilla where they are closely related to the axillary artery.

Among the more important branches of the brachial plexus are:

The **circumflex** (axillary) **nerve** which is related to the surgical neck of the humerus and supplies the deltoid muscle.

In this region it may be injured by a fracture of the bone or in dislocations of the shoulder joint.

The **radial** (musculospiral) **nerve**. This, after leaving the axilla, winds round the posterior aspect of the shaft of the humerus in the spiral groove and supplies branches to the skin, the triceps muscle and the extensor muscles in the back of the forearm.

It is liable to injury in the upper part of its course and when paralysed produces the characteristic deformity known as 'drop wrist', in which the patient is unable to extend the wrist or fingers. The nerve is liable to crutch pressure in the axilla.

The **ulnar nerve** is at first related to the axillary artery and then accompanies the brachial artery in the arm. It passes down behind the medial epicondyle of the humerus to reach the ulnar side of the forearm. In addition to supplying muscle in the forearm it also gives branches to the skin of the ring and little fingers and small muscles in the hand.

It is the close relationship with the medial epicondyle where the nerve is relatively superficial which has given rise to the popular name for the point of the elbow, 'funny bone'. Any severe knock in this area stimulates the ulnar nerve and produces pain together with a tingling sensation in the ring and little fingers, which clearly demonstrates the distribution of the nerve.

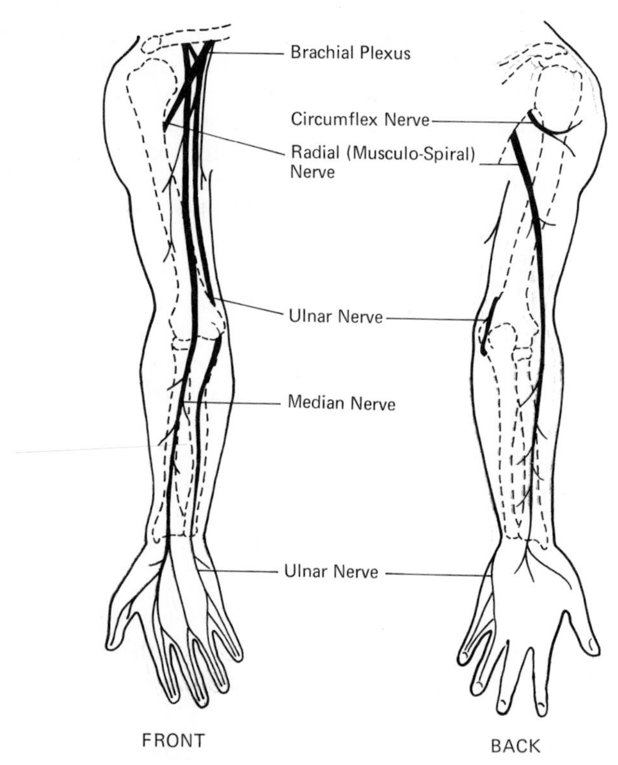

Fig. 200 The important nerves of the right upper limb.

Injury of the nerve is not uncommon and the resulting paralysis is characteristic. The hand assumes a claw-like attitude in which the terminal joints of the phalanges, especially of the ring and little fingers, are flexed.

The **median nerve** also lies close to the brachial artery in the arm. It passes down in the mid-line of the front of the forearm to the hand. It supplies muscles in the forearm and hand and gives cutaneous branches to these parts and the thumbs and first two fingers.

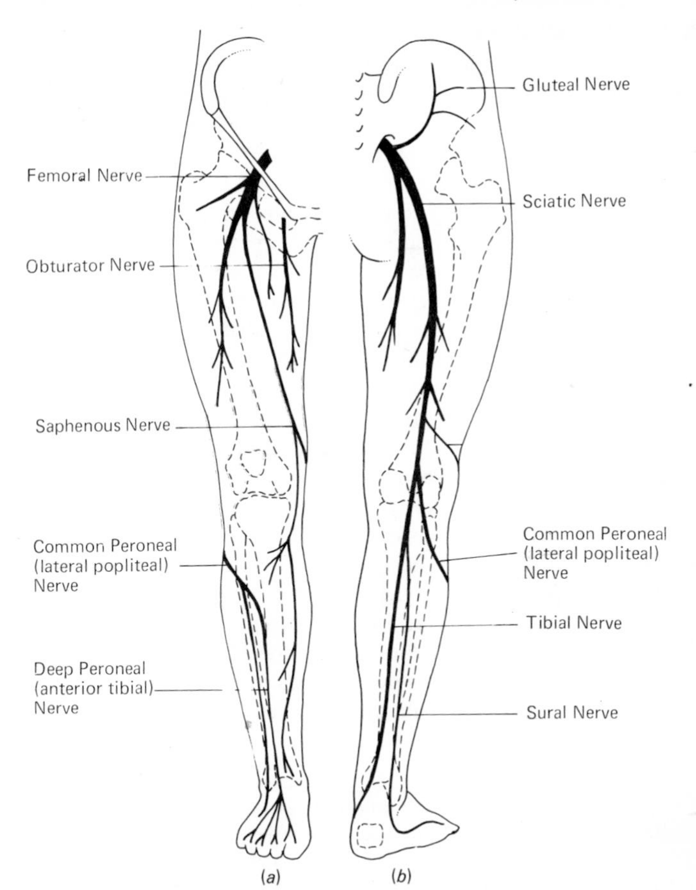

Fig. 201 The important nerves of the right lower limb. (*a*) Front of limb. (*b*) Back

Injury to the median nerve usually occurs just above the wrist and results in weakness and loss of sensation in the thumb and first two fingers. If damaged in the upper part of its course, there will also be weakness in the power of pronation.

The thoracic nerves

These are twelve in number and one passes to each intercostal space. They run forwards between the ribs, supply branches to the intercostal muscles and muscles of the anterior abdominal wall and are finally distributed to the skin covering the thorax and front of the abdomen.

The lumbosacral plexus

This large plexus is formed by the union of lumbar and sacral nerve roots. Its upper part lies mainly in the psoas muscle which forms part of the posterior wall of the abdominal cavity, and its lower part lies on the anterior aspect of the sacrum. The more important branches are:

The **femoral** (anterior crural) **nerve** which passes under the inguinal (Poupart's) ligament in company with the femoral artery and vein to enter the femoral (Scarpa's) triangle in the front of the thigh. It supplies the muscles of the front of the thigh (quadriceps) and skin on the front of the thigh and medial side of the leg.

The **obturator nerve** which leaves the pelvis through the obturator foramen in the innominate bone to supply muscles in the medial part of the thigh (adductors).

The sciatic nerve. This is the largest nerve in the body. It is formed on the anterior aspect of the sacrum and leaves the pelvis through the great sciatic notch of the innominate bone into the gluteal region. Thence it passes down the back of the thigh, the muscles of which it supplies, to the popliteal space and becomes related to the popliteal artery. Here it gives off the **lateral popliteal** (peroneal) **nerve** which reaches the front of the leg by winding round the lateral aspect of the neck of the fibula.

The main nerve continues downwards in the calf as the **posterior tibial nerve.** This supplies the muscles of the calf and ends in branches to the skin and muscles of the sole of the foot.

PRACTICAL CONSIDERATIONS. The sciatic nerve is of importance because it is a common site for painful symptoms (sciatica). The patient complains of pain in the thigh, which passes down the leg often to the ankle. The nerve is tender on pressure and pain is caused on stretching it by extending the leg when the thigh is flexed. In severe cases there may be wasting of the calf muscles. This condition is usually due to a hernia of the intervertebral disc which presses on nerve roots.

Sensation and the sensory path

Two types of sensation may be considered, special sensation and general sensation (see also Chapter 20).

Special sensations are those which can only be detected by specialized organs, i.e. smell, sight, hearing and taste. Sensory impulses received from these organs are conveyed to the brain by the appropriate cranial nerve (see page 319). **General sensations** are the feelings which are appreciated by all parts of the body. They include the superficial sensations detected by the skin and the deep sensations felt in the muscles, joints and other organs.

Superficial sensation. The important superficial sensations are appreciated by the sensory nerve endings in the skin and include:

Pain
Touch
Temperature (heat and cold)

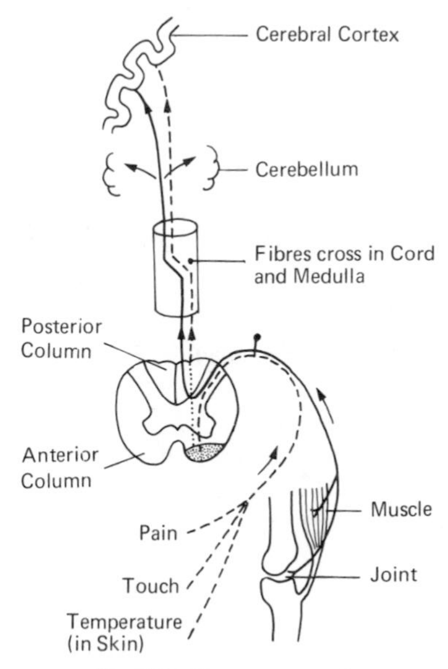

Fig. 202 The sensory path.

Deep sensation. In addition to pressure and deep pain, the most important deep sensation is that of the position of muscles and joints. We are always aware of the exact position in space of any limb or part of a limb. This may be demonstrated by closing the eyes and bringing the tip of the index finger to the end of the nose. This can only be done accurately because we know exactly where the respective parts are by muscle and joint sense, without the aid of vision.

The sensory path

Both superficial and deep sensations travel in the peripheral nerve from the skin, joint or muscle towards the spinal cord. After reaching the trunk of the nerve, the sensory fibres enter the posterior horn of grey matter in the spinal cord via the posterior nerve root. The two types of nerve fibre then take separate courses.

The fibres conveying sensations of position, pressure and light touch pass upwards in the posterior columns. Those taking the superficial sensations of pain and temperature go to the anterior columns.

Both sets of fibres cross to the opposite side of the cord either before or when they reach the medulla oblongata. Thence they are conveyed via the brain stem, thalamus and white matter of the brain to the parietal and other sensory areas of the brain. The sensory impulses of one side of the body are, therefore, like the motor impulses, dealt with by the opposite side of the brain.

Movement and the motor path

Movement may be voluntary or involuntary. Involuntary movement is considered with reflex action.

Voluntary movement commences with an impulse sent out by the pyramidal cells of the motor cortex situated in front of the central sulcus (fissure of Rolando). The axons of these cells pass through the white matter of the brain and brain stem. In the medulla oblongata they cross to the opposite side and travel down in the lateral column of the spinal cord as the pyramidal tract. At the appropriate level the fibres leave the pyramidal tract and end round the cells in the anterior horn of grey matter.

The pyramidal nerve cell in the motor cortex and its axon extend-

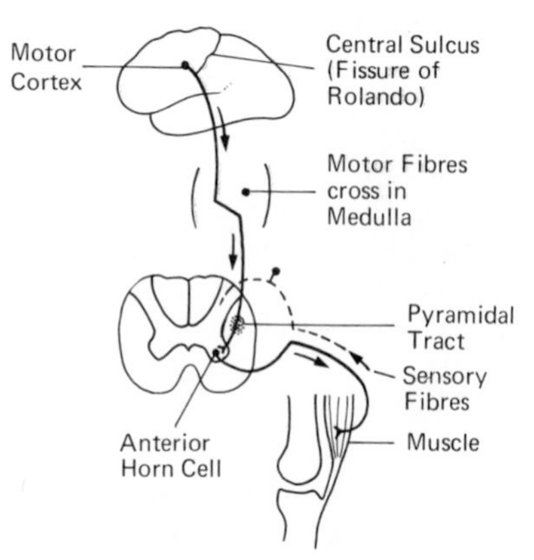

Fig. 203 The motor path.

ing as far as the anterior horn cell is called the **upper motor neurone.**

The motor impulse is then relayed through the anterior horn cell whose fibre passes via the anterior nerve root to form, with the incoming fibres of the posterior root, the main nerve trunk. The motor nerve reaches its destination in the muscle via the peripheral nerve.

The anterior horn cell and its axon passing to the muscle in the peripheral nerve is called the **lower motor neurone.**

(NOTE. The motor cortex of one side of the brain controls the muscles on the opposite side of the body.)

Reflex action

Reflex action may be defined as the automatic motor response to a sensory stimulus, and is therefore independent of the will. The structures concerned in the production of a reflex constitute 'the reflex arc'. They consist of:

1. A sense organ such as the skin or the nerve endings in a muscle, tendon or other organ.
2. An afferent or sensory nerve passing from the sense organ via the peripheral nerve and posterior nerve root to the spinal cord.
3. The spinal cord.
4. An afferent or motor nerve commencing in the anterior horn cells of the cord and passing via the peripheral nerve to the motor organ, e.g. muscle or gland.

It is clear that if this route is taken the time elapsing between the application of the sensory stimulus and the motor response will be much less than if the impulse had to pass the whole length of

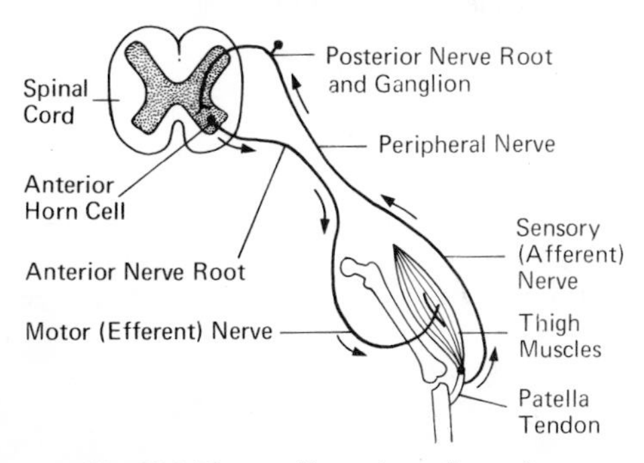

Fig. 204 Diagram illustrating reflex action.

the spinal cord to the sensory area of the brain and thence to the motor cortex, which in turn would send out a voluntary impulse down the motor path in the spinal cord before it could reach the peripheral nerve and motor end-organ.

It follows, therefore, that many reflexes are protective in character and designed to obtain the quickest possible motor response. For example, the finger is withdrawn from a hot object before we have time to think about it.

Other reflexes are concerned with automatic control of functions which do not require the supervision of consciousness, such as the secretion of gastric juice when food enters the stomach.

Common examples of reflex action include:

Withdrawal of a part of the body from any painful stimulus such as a prick or excessive heat.

The knee jerk or patellar reflex. If the patellar tendon is tapped sharply when the muscles of the leg are relaxed, the leg is jerked suddenly forward (extended) at the knee.

Stroking the sole of the foot results in the downward movement (flexion) of the big toe.

A crumb in the larynx causes coughing.

In some cases reflex action may be operated via the organs of special sense, e.g. the sight or smell of appetizing food causes a reflex flow of saliva ('makes the mouth water').

Inhibition. It has been seen that in infancy the control of the bladder and rectum is automatic or reflex in character. As age increases voluntary control over the excretions is developed. This means that the higher centres of the brain are able to control or inhibit certain reflex actions. When the controlling influence of the brain is removed by injury or disease the reflex control may again operate and, in the case of the bladder and rectum, the patient will become incontinent.

QUESTIONS

1. What do you understand by reflex action? Give three examples and mention some factors which may increase or diminish reflex action.
2. Describe the functions of (*a*) the spinal cord (*b*) a motor nerve, (*c*) a sensory nerve.
3. Describe briefly the contents of the skull.
4. Describe the spinal cord. What are its functions?
5. Describe the brain and its coverings. Give an outline of the main functions of the cerebrum.

6. Enumerate the various parts of the nervous system. Describe the cerebellum.
7. Name the various parts of the central nervous system. Give an account of the functions of the spinal cord.
8. Describe the structure of the brain and give a brief outline of its main functions.

19 Autonomic Nervous System

In addition to the central nervous system described in the previous chapter there exists in the body a second system of nerve cells and nerve fibres which have special and very important functions. The word **autonomic** means 'self-controlling'. (As well as the names given above, the terms *sympathetic* and *parasympathetic systems* are often used. In view of the fact that the working of the system is actually very complicated and only general outlines can be discussed here these terms will be applied to a part of the system rather than to the whole.)

Although the involuntary system exercises its functions indepen-

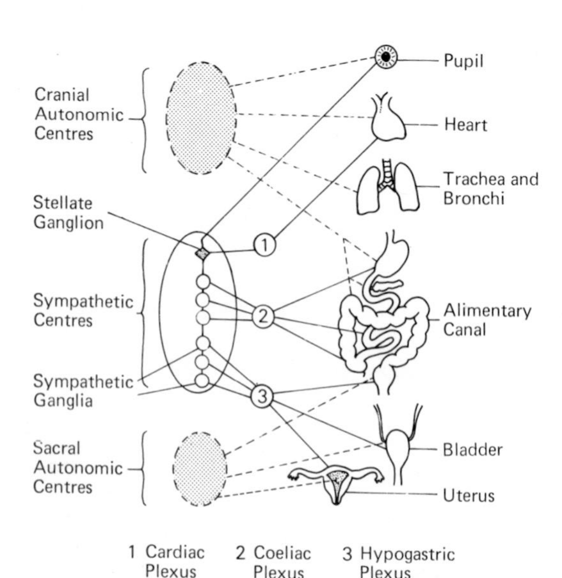

Fig. 205 Diagram illustrating distribution of involuntary nervous system.

dently of the central nervous system they are both closely associated and are able to influence each other.

Whereas the central nervous system is largely concerned with the interpretation of sensory impulses and in sending motor impulses which pass to the voluntary or striped muscle in the body, the involuntary system is concerned mainly with the control of all the involuntary, unstriped muscle in the body such as is found in the blood vessels, abdominal viscera, bladder, etc. Many glands, for example the salivary and sweat glands, also come under its influence.

When the central nervous system exercises any control over these structures it does so indirectly through its connections with the involuntary system. It has been seen, for example, that the vaso-motor centres in the medulla oblongata control the blood vessels. The actual nerves through which this is carried out are involuntary nerves specially distributed to the vessels and not the ordinary peripheral nerves.

The essential feature of the involuntary system is a series of **ganglia** or masses of nerve cells situated outside the brain and spinal cord. These ganglia may be classified into three groups which, in turn, form the main subdivisions of the involuntary system:

1. The cranial autonomic system.
2. The sympathetic system proper.
3. The sacral autonomic system.

For convenience, the cranial autonomic and sacral autonomic portions may be grouped together and called the **parasympathetic** system.

It will be seen later that, in many instances, the involuntary system sends two sets of impulses of directly opposite character to the same organ. Thus there are involuntary fibres which increase the rate of the heart-beat and fibres which decrease the rate. In such instances the opposite impulses are conveyed by the sympathetic and parasympathetic respectively. In fact almost all the unstriped involuntary muscles have both sympathetic and parasympathetic supplies.

1. Cranial autonomic system (parasympathetic)

This consists of the ganglia and fibres which are associated with the cranial nerves. The most important are:

(*a*) Fibres which control the movements of the pupil of the eye and are associated with the IIIrd or oculo-motor nerve.

(*b*) Fibres which run with the Xth or vagus nerve to the heart, bronchi and alimentary tract.

These fibres convey impulses opposite in action to those distributed by the sympathetic system to the same organs.

2. Sympathetic system

This consists of two chains of ganglia situated one on either side of the front of the vertebral column. Each of these ganglia has connecting fibres which pass to the corresponding nerve roots of the spinal nerves. Instead of separate ganglia for each segment of the spinal cord, however, in certain segments a number are collected together to form a single large ganglion (e.g. the cervical ganglion in the neck and the stellate ganglion in the upper thoracic region).

From certain of these ganglia nerve fibres are collected into special groups of plexuses from which afferent and efferent fibres pass to the organs to which they are distributed. The most important plexuses are:

(*a*) The **cardiac plexus** which is connected via the stellate ganglion to the lower cervical and upper thoracic (dorsal) spinal nerves and supplies the heart and lungs.

(*b*) The **coeliac plexus** (solar plexus) connected with the lower thoracic (dorsal) spinal nerves. This lies in the epigastrium behind the stomach and close to the coeliac axis artery. It supplies sympathetic fibres to the stomach, intestine, suprarenal glands and other viscera in the abdominal cavity.

3. Sacral autonomic system

This supplies parasympathetic fibres to the structures in the pelvis including the rectum, bladder, uterus and other reproductive organs.

It is important to remember that reflex action can take place through the afferent and efferent fibres and the ganglia of the involuntary system just as in the central nervous system.

In mentioning the opposite actions of the sympathetic and parasympathetic systems it was indicated that both sets of fibres were conveyed to the heart. The parasympathetic fibres reaching the heart from the vagus carry impulses which tend to slow its rate (inhibitors), while those from the sympathetic (cardiac plexus) tend to increase its rate (accelerators).

It is also true to say that the structures innervated by the involuntary nervous system are very largely influenced by the secretions of ductless glands. In particular the secretion of the adrenal gland, called adrenaline, stimulates the muscle in the walls of small arteries and acts as a vasoconstrictor.

Chemical transmitters. Messages are transmitted from one neuron to another and from neurons to muscle or gland cells by chemical transmitters. The best known of these are noradrenaline and acetylcholine.

Adrenergic receptors. The sympathetic nervous system exerts its effects through special receptors which are found in certain tissues. There are two distinct types of receptor, known as the alpha (α)-receptor and the beta (β)-receptor. **α-receptors** are generally **excitatory** in function; their stimulation causes, for example, contraction of smooth muscle in the walls of blood vessels, resulting in vasoconstriction. **β-receptors** are generally **inhibitory** in function; their activation causes relaxation of smooth muscle in the walls of blood vessels and bronchi, resulting in vasodilatation and bronchodilatation. The β-receptors in cardiac muscle are exceptional in being excitatory; their stimulation causes the heart to beat more rapidly and more forcibly.

These facts have important applications in pharmacology and therapeutics where, for example, β-receptor stimulants (e.g. salbutamol) are used in asthma and β-receptor blocking agents (e.g. propranolol) are used in certain diseases of the heart.

QUESTION

What do you understand by the term *autonomic nervous system*? Give a brief account of its functions.

20 The Special Sense Organs

It has been seen that the afferent or sensory nerves reaching the central nervous system have their beginnings (although these are sometimes referred to as nerve endings) in various peripheral structures such as the skin, muscles, joints and special organs like the eye and ear.

Sensations are the conscious results of processes taking place in the brain following the arrival of impulses derived from the sensory nerves. The structures concerned in the production of a sensation are:

1. An end-organ or receptor, situated in the periphery at the terminations of the sensory nerves.
2. The afferent nerve fibres in the peripheral nerve and spinal cord.
3. The thalamus, which is a cell station relaying sensations to the cerebral cortex.
4. The sensory reception areas in the brain which are connected with various psychic areas where the impulse is interpreted and may be stored as memory.

It is important to remember that, although the brain receives and appreciates the sensation, it projects it back to the site or end-organ at which it was received and it is actually felt by the individual in the peripheral region.

The end-organs for each sense are specially adapted in structure to receive only the particular stimulus for that sense. Thus, waves of light are described as being the 'adequate stimulus' for the nerve endings in the retina of the eye. Sound waves are the adequate stimulus for the endings of the auditory nerve in the ear, and have no effect on the eye, nose or skin.

Sensation may be classified in various ways. In describing the nervous system it was sufficient to make the simple subdivisions of (*a*) special senses, i.e. sight, hearing, taste and smell, (*b*) general sensations, including all the others.

It is more scientific to classify them in the following way:

1. Sensations produced by stimuli arising outside the body, or external sensations, such as sight, hearing, taste, smell, pressure, temperature and external pain. The sense organs receiving those sensations are said to be **exteroceptive**.
2. Sensations arising from stimuli inside the body, or internal sensations. These may occur:
   (*a*) In the viscera and include hunger, thirst and internal pain. These are called **enteroceptive** organs.
   (*b*) In the muscles, joints and semicircular canals by which the sense of position is appreciated. These are called **proprioceptive** organs.

Sensation in the skin

It has already been pointed out that the sensations felt by the skin are those of pain, temperature (hot and cold) and light touch. Situated in the skin is a mosaic of minute sensory areas, which correspond with the nerve endings. Special little areas are present for each of the varieties of sensation. Further, the spots for one type of sensation may be more numerous in one area of skin than in another. Thus the tips of the fingers are more sensitive than the skin of the forearm.

The sense of smell

The mechanism of smell is dependent on:

1. The receptor or end-organ, i.e. the endings of the 1st cranial or olfactory nerve in the mucous membrane of the upper part of the nasal cavity (page 319).
2. The olfactory bulb and tract which conveys the impulses to the brain.
3. The olfactory centre situated in the medial surface of the cerebral hemisphere (page 317).

The appropriate or adequate stimulus for the sense of smell must be either in the form of gas or minute particles which are soluble in the secretions of the nasal mucous membrane. These gases or particles are conveyed to the nasal cavities by the air and their concentration in the upper part of the nasal cavities may be increased by the process of 'sniffing'.

Smell is a sense characterized by its extreme delicacy and the ease with which it becomes fatigued. This is shown by the fact that persons sitting in a closed room for some time may cease to be aware of an odour which is very apparent to someone entering from the fresh air.

The sense of smell is also diminished by inflammation or excess of secretion in the nasal mucous membrane which occurs in the common cold.

Odours are most simply classified into pleasant and unpleasant. It is important to remember that this sense is closely associated with the sense of taste and that the majority of flavours are actually appreciated by the olfactory organ. This is clearly shown by the loss of taste which accompanies a severe head cold.

The sense of taste

The true sense of taste is localized in the tongue. There are four primary tastes, viz. bitter, sweet, sour and salt. These only are appreciated by the tongue. All other flavours are appreciated by the sense of smell as already stated.

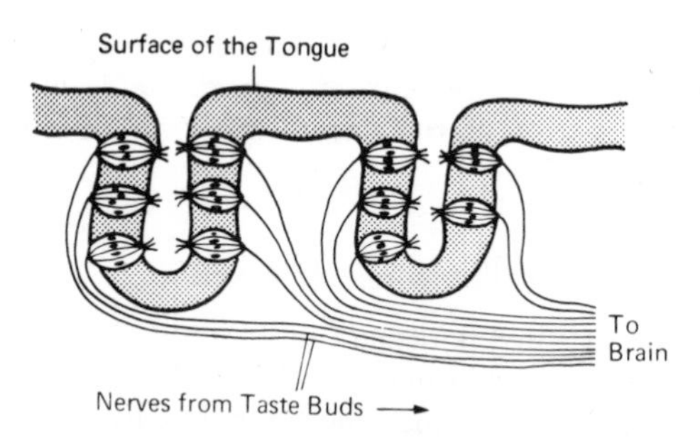

Fig. 206 Nerve supply to taste buds.

The tongue (page 192) is a very mobile organ, consisting of muscles some of which arise from the hyoid bone and the lower jaw. It is covered by mucous membrane. The roughness of its upper surface is due to numerous minute elevations called papillae. The end-organs for the sense of taste are called **taste buds,** which are situated most densely at the sides and base of the tongue. They consist of collections of special cells arranged in 'nests' around which the sensory nerve endings are placed.

The nerve fibres of the sense of taste reach the brain mainly via the facial (anterior two-thirds of tongue) and glossopharyngeal (posterior third) nerves. The sensation of pain and touch are con-

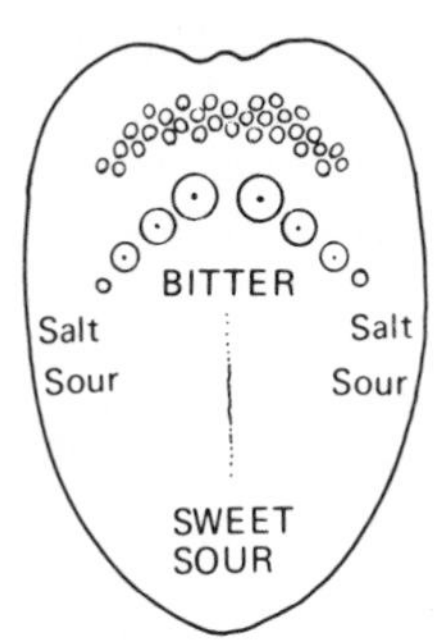

Fig. 207 Diagram illustrating the parts of the tongue in which the various sensations of taste are appreciated.

veyed by the trigeminal nerves. The motor nerve to the tongue is the XIIth cranial or hypoglossal nerve.

As in the case of smell, in order that substances may have taste they must be soluble in the watery secretion of the mouth and salivary glands.

The functions of the tongue are:

1. Motor: (*a*) speech,
   (*b*) mastication (and the act of swallowing).
2. Sensory: (*a*) taste (bitter, sweet, sour and salt),
   (*b*) touch.

The sense of hearing

The auditory apparatus consists of (1) the external ear, (2) the middle ear, (3) the internal ear, (4) the auditory nerve and centre in the brain.

THE ANATOMY OF THE EAR

External ear

This consists of:

(*a*) the auricle or pinna;
(*b*) the external auditory canal or meatus:
  (i) cartilaginous portion,
  (ii) bony portion.

The **auricle** or **pinna** is attached to the side of the head about midway between the forehead and the occiput. It consists of the concha or shell-like upper part and the lobe. The concha is formed of yellow elastic cartilage covered by skin. The lobe is soft and consists of skin covering connective and adipose tissue.

The function of the auricle is to collect sound waves and conduct them to the external auditory canal. This function is more marked in many animals than in man and in consequence their auricles are relatively larger and more mobile. Although there are several small muscles attached to the human ear, in only a few people is a limited amount of movement possible.

The **external auditory canal** or **meatus** leads from the pinna to the tympanic membrane or ear-drum. It is a tubular passage about 2·5 cm (1 in) long. Its course is not direct but shows a slight double or S-shaped bend, being directed at first medially, forwards, and slightly upwards, and then medially and slightly backwards. In order to bring the meatus into a straight line the pinna should be lifted upwards and backwards in adults, but in children owing to a slightly different course the pinna should be pulled downwards and backwards. This is of importance when it is desired to examine the tympanic membrane.

Structurally the external auditory canal consists of two parts: the outer **cartilaginous** portion, which is continuous with the cartilage of the pinna; and the inner **bony** part, the walls of which are formed by the external auditory canal of the temporal bone.

The canal is lined by skin which is continuous with that covering the pinna, but is characterized by containing special glands, the **ceruminous glands**, which secrete a yellow greasy substance called cerumen or wax. They are modified sweat glands and their secretion helps to prevent the entry into the meatus of foreign bodies, especially insects. A few hairs are present which also assist in this function.

Middle ear

The middle ear or tympanic cavity is a small irregular cavity situated in the petrous portion of the temporal bone. It contains a chain of small bones or ossicles by which the sound waves are transmitted from the tympanic membrane to the internal ear. Roughly speaking, it is a narrow oblong box having anterior, posterior, medial and lateral walls with a roof and floor.

Its walls are both bony and membranous in structure. The outer or lateral wall is formed by the tympanic membrane. Its medial wall, though mainly consisting of bone, has two openings which are covered by membrane, viz. the **fenestra ovalis** or oval window

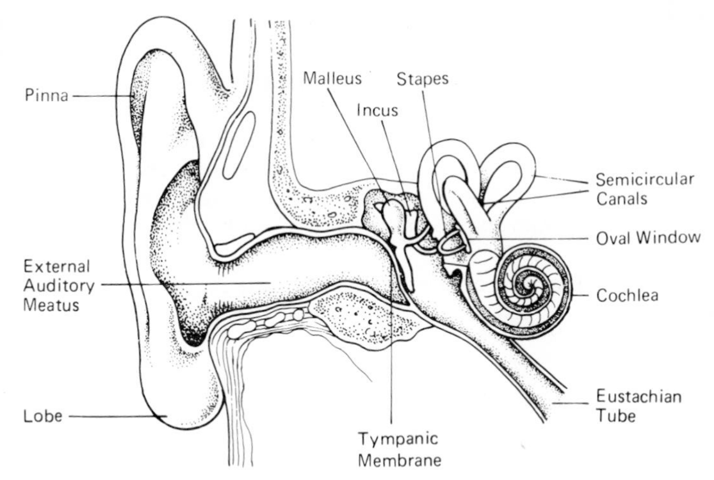

Fig. 208 The structure of the ear.

of the vestibule above, and the **fenestra rotunda** or round window of the cochlea below.

In addition to these membranous defects in its outer and inner walls, both the anterior and posterior walls have openings. Entering the middle ear in its anterior wall is the outer or lateral end of the **Eustachian** (auditory) tube which communicates with the nasopharynx. Posteriorly the middle ear communicates with the mastoid antrum and the mastoid air cells which occupy the mastoid process of the temporal bone (page 70).

The whole of the cavity of the middle ear is lined by mucous membrane which, therefore, forms the inner lining of the tympanic membrane and is continuous with the mucous lining of the Eustachian tube and with that of the mastoid antrum and cells.

The Eustachian tube is about 4 cm ($1\frac{1}{2}$ in) long and connects the nasopharynx with the middle ear. There is thus a connection between the middle ear and the outer air so that the pressure of air on each side of the tympanic membrane is equalized.

The pharyngeal opening of the Eustachian tube is normally closed by the approximation of its walls, but is opened by the action of muscles in swallowing and yawning. In catarrhal conditions of the nasopharynx, as in a severe head cold, the opening of the tube may be defective and a degree of deafness is not uncommon from this cause.

The effect of atmospheric pressure is also demonstrated in

climbing up a mountain or going to the top of a high hill rapidly in a car, or going up in an aeroplane. Until air escapes from the middle ear by the Eustachian tube the atmospheric pressure on the outer surface of the tympanic membrane, being diminished by reason of the height, is less than the pressure on its inner side. An uncomfortable sensation in the ear and slight temporary deafness is produced but disappears with the adjustment of pressure, which is often accompanied by clicking sounds and is aided by the act of swallowing.

The main function of the Eustachian tube is, therefore, to equalize the pressure in the middle ear with the atmospheric pressure outside.

The ossicles. The ossicles or small bones of the middle ear are three in number, the malleus, the incus and stapes. They stretch from the tympanic membrane to the fenestra ovalis or oval window of the vestibule.

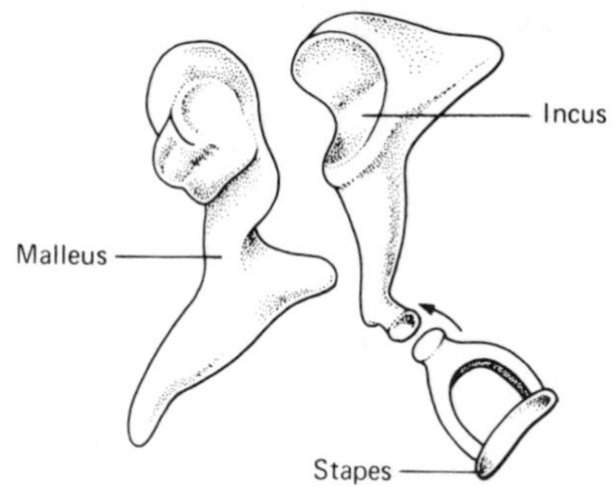

Fig. 209 The bones of the middle ear.

The **malleus** or hammer bone consists of a head which articulates with the incus and a handle which is attached to the tympanic membrane. The **incus** or **anvil** is the middle of the three bones and consists of a body and two short legs, one of which articulates with the roof of the middle ear, the other with the stapes. The **stapes** or stirrup bone is the smallest of the three. Its head articulates with the incus while its base or foot-plate is attached to the membrane covering the fenestra ovalis.

These three bones act as a series of levers transmitting the movements or vibrations of the tympanic membrane caused by sound waves impinging upon it to the membrane covering the fenestra ovalis. It will be seen later that from the fenestra ovalis the vibrations are passed on to the internal ear.

(The movements of the ossicles are controlled to some extent by two tiny muscles, the tensor tympani inserted into the handle of the malleus, and the stapedius muscle inserted into the head of the stapes.)

The **tympanic membrane** (tympanum or ear-drum) is situated at the deepest part of the external auditory canal which it separates from the middle ear. It lies obliquely so that its upper part is nearer the exterior than its lower part. In structure, its outer surface consists of epithelium continuous with the skin lining the external auditory canal, while its inner lining is mucous membrane continuous with that of the middle ear. Between these two layers is a small amount of fibrous tissue. Firmly attached to its inner wall (and passing downwards and slightly backwards from its upper edge to a point just below and behind its centre) is the handle of the malleus.

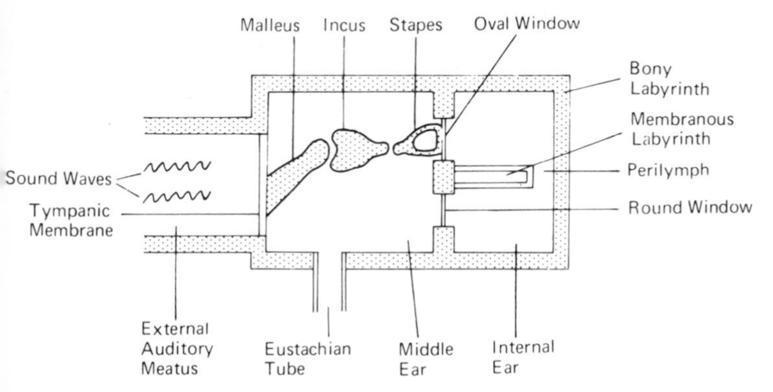

Fig. 210 Diagram illustrating the anatomy of the auditory apparatus.

It appears as an almost circular structure tightly stretched between the walls of the bony meatus except for a small area in its upper part (known as the flaccid membrane of Shrapnell).

Internal ear

The internal ear or labyrinth consists of a series of irregular cavities situated in the petrous portion of the temporal bone. These cavities constitute the bony or osseous labyrinth. Within these bony walls is a membranous structure which more or less follows the shape of the bony labyrinth.

Between the bony walls and the membranous part of the labyrinth is a clear fluid called **perilymph**, while the membranous labyrinth itself is a sac filled with a similar fluid called the **endolymph**.

The internal ear or labyrinth consists of the following parts:

1. The vestibule or entrance which communicates with—
2. The cochlea (little shell) or organ of hearing in front and—
3. The semicircular canals behind. These are concerned with equilibrium and the sense of position.

The **vestibule** is closely connected to the middle ear, from which it is separated by the membrane covering the fenestra ovalis (the oval window of the vestibule) with the attached foot-plate of the stapes.

The **cochlea** or anterior portion of the labyrinth contains the organ of hearing. In some respects it resembles the shell of a small snail and, in section, looks something like a spiral staircase with a central bony structure called the **modiolus** from which ridges project to the outer wall. The membranous portion of the cochlea is contained within these bony walls and in it is the most important part of the structure, the organ of Corti, which is the true end-organ of hearing.

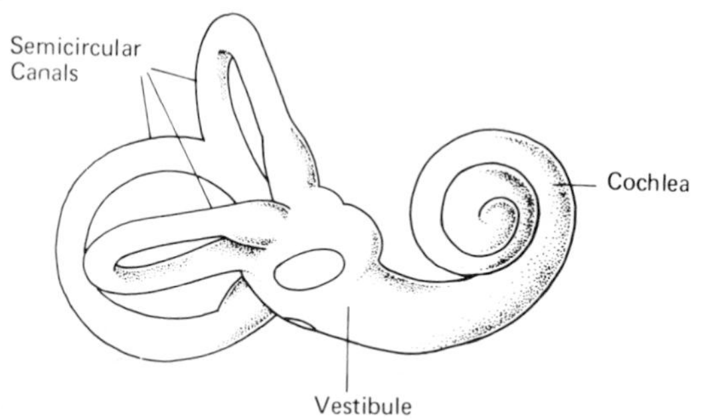

Fig. 211 The right bony labyrinth.

The **organ of Corti** consists of special epithelial cells called rods of Corti on either side of which are layers of cells (hair cells) having hair-like processes on their free surface rather like ciliated epithelium in appearance. It is around these cells that the fibres of the auditory or VIIIth cranial nerve commence. The organ of Corti, therefore, is a spiral structure contained in the membranous labyrinth, which follows the shape of the cochlea and is bathed in endolymph.

The semicircular canals. These are three canals situated in the posterior part of the bony labyrinth and set at right angles to each other. They are known as the superior, posterior and lateral canals. Within the bony walls are the membranous canals or ducts surrounded by perilymph and containing endolymph. Each canal is enlarged at one end into an **ampulla** where the special nerve fibres end around cells which have fine hair-like processes projecting from them. Movements of the head and alteration in its position cause movement of the endolymph in the semicircular canals. This movement of fluid acts as a stimulus to the nerve endings in the ampullae and the impulses are conveyed to the brain (both the cerebrum and cerebellum) by that part of the VIIIth cranial nerve known as the vestibular portion (page 321). NOTE. Lesions of this portion of the nerve cause vertigo (dizziness).

It will be remembered that the sense of equilibrium and position is also dependent on impulses received from the muscles and joints and that these receptor organs are said to be proprioceptive.

Mechanism of hearing

Sound is due to waves or vibrations in the air and has three main qualities:

1. Pitch, which depends on the frequency of the vibrations. The more rapid the frequency the higher the pitch of the note produced.
2. Intensity or loudness, which depends on the amplitude of the vibrations.
3. The quality, which is due to the combination of various vibrations. These may blend to produce harmony or music, or fail to unite giving rise to a discord or noise.

Sound waves in the air are collected by the pinna and directed along the external auditory canal to the tympanic membrane which they cause to vibrate. These vibrations are transmitted across the middle ear by the movements of the malleus, incus and stapes to the membrane covering the fenestra ovalis. The inner surface of this membrane is in contact with the perilymph in the vestibule which picks up the vibrations and, in turn, passes them on to the endolymph by means of which they reach the organ of Corti.

The stimulus thus reaching the organ of Corti is conveyed by the cochlear portion of the VIIIth cranial or auditory nerve, which leaves the petrous portion of the temporal bone by a foramen (the internal auditory meatus) to reach the brain stem. The fibres are then carried to the auditory centre of the brain situated in the temporal lobe of the opposite side.

PRACTICAL CONSIDERATIONS. The tympanic membrane may be inspected by passing a speculum into the external auditory meatus, appropriate traction being applied to the pinna in order to bring the parts of the meatus into a straight line (page 344). When normal, the membrane appears as a greyish structure with the handle of the malleus as a whitish streak passing downwards and backwards to about its centre. From the tip of the handle of the malleus a bright cone-shaped area of reflected light may be observed passing downwards and forwards to the periphery of the membrane.

When the middle ear becomes inflamed the membrane appears red and, if an abscess forms, may bulge outwards and then perforate, with the discharge of pus into the meatus.

Inflammation of the middle ear (otitis media) is common, and infection usually reaches the cavity by spreading up the Eustachian tube from the nasopharynx. There is always the added risk that in this condition the inflammation may spread backwards through the mastoid antrum to the mastoid cells (mastoiditis). The close relation of the mastoid process of the temporal bone to the lateral sinus in the posterior cranial fossa will be recalled.

The portions of the brain nearest the middle and internal ears are the temporal lobe of the cerebrum and the cerebellum. Spread of infection from the ear to within the skull may, therefore, result in meningitis or abscess in the temporal lobe or cerebellum.

SUMMARY OF THE SENSE OF HEARING

Sound waves → pinna → external auditory meatus → tympanic membrane → malleus → incus → stapes → fenestra ovalis → vestibule → cochlea (perilymph → endolymph) → organ of Corti → VIIIth cranial nerve → temporal lobe of brain (of opposite side).

ANATOMY OF THE EYE

The sense of sight—vision

The eye or organ of sight is situated in the orbital cavity of the skull and is well protected by its bony walls except on its anterior aspect.

In addition to the essential organs of the visual apparatus, namely, the eyeball, the optic nerve and the visual centres in the brain, there are certain accessory organs which are necessary for the protection and functioning of the eye. These include 1. the eyebrows, 2. the eyelids, 3. the conjunctiva, 4. the lacrimal apparatus, 5. the muscles of the eye.

Accessory organs of the eye

The eyebrows. These are formed by the skin covering the orbital process of the frontal bones, which in its natural state is plentifully supplied with short thick hairs. Their main function is protective and by their shape prevent the sweat of the brow from pouring into the eye.

The eyelids. These are two movable folds, upper and lower, which form the anterior protection for the eye. The upper is the larger and more mobile of the two and is provided with a muscle which elevates it. The eyelids are covered externally with skin and their inner lining consists of a mucous membrane, the conjunctiva. Between these layers is a dense mass of fibrous tissue called the **tarsal plate**. Into the tarsal plate of the upper lid the muscle which raises it is inserted (levator palpebrae superioris). Surrounding both lids is a circular sphincter muscle which closes them and, when fully contracted, 'screws them up' (orbicularis oculi).

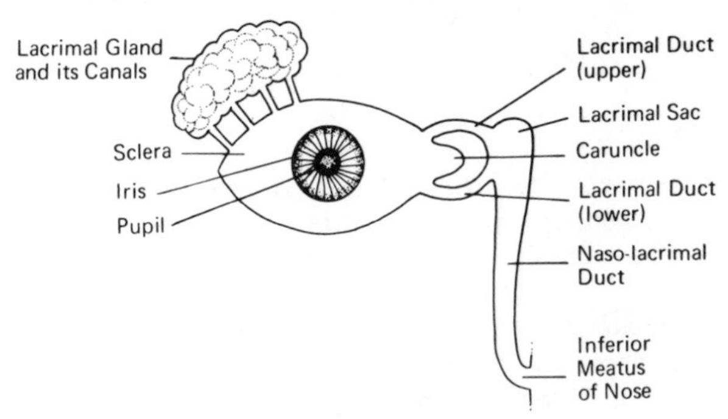

Fig. 212 The lacrimal apparatus.

The space between the two lids is called the **palpebral fissure**. Its outer angle is the **lateral canthus** and its inner angle the **medial canthus**.

The eyelids blink every few seconds. This movement keeps the front of the eye free from dust and helps to move the tears across the conjunctival sac. In addition to protecting the eyes from the entrance of foreign bodies the eyelids also prevent the entry of excessive light.

The eyelashes. A row of short thick hairs project from the free margin of each eyelid. Arranged immediately behind the eyelashes are the openings of the *Meibomian* (*tarsal*) *glands,* which are modified sebaceous glands and are sometimes the site of small cysts. Infection of the hair follicles of the eyelashes results in the common condition known as stye.

The conjunctiva. This is a delicate mucous membrane which lines the inner surface of the eyelids and is then reflected on to the outer surface of the eyeball. The space between the two layers is called the conjunctival sac.

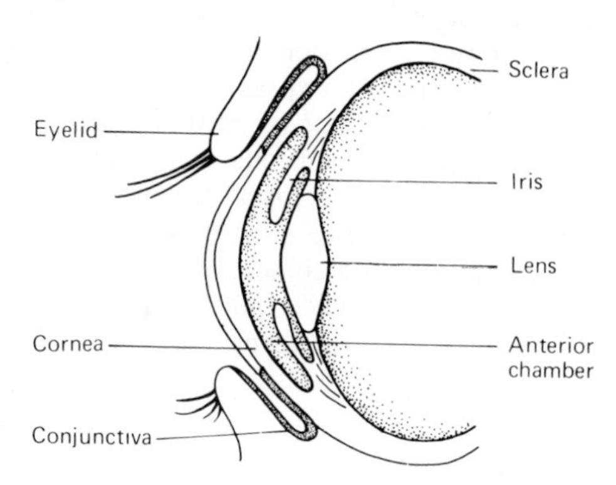

Fig. 213 Diagram of the conjunctiva.

The lacrimal apparatus. This is concerned with the formation of tears and consists of the following structures: 1. the lacrimal glands, 2. the lacrimal ducts, 3. the lacrimal sac and nasolacrimal duct.

The lacrimal glands. These are situated in the orbital cavity immediately above the outer angle of the eye (lateral canthus). Each is almond-shaped and lies in a depression in the orbital plate of the frontal bone. A number of small canals leads from it to the outer angle of the conjunctival sac.

The lacrimal ducts. If the inner end of each eyelid is carefully examined the orifice of a minute duct can be seen. From this opening the upper and lower lacrimal ducts pass inwards, one above and one below the small red elevation at the internal angle (canthus) of the eye called the *lacrimal caruncle.* These ducts enter the lacrimal sac.

The lacrimal sac and nasolacrimal duct. The lacrimal sac may be regarded as the upper expanded portion of the nasolacrimal duct which passes downwards inside the bony wall of the nasal cavity to open into the inferior meatus of the nose.

Tears are a slightly alkaline watery fluid containing a small amount of sodium chloride which give them a salty taste. Normally there is a constant secretion from the lacrimal glands just sufficient to keep the interior of the conjunctival sac moist and free from dust. By the frequent movement of the eyelids the tears pass across the front of the eye from the lateral to the medial side, where they pass through the openings of the lacrimal ducts and are drained into the nose by the nasolacrimal duct to mix with the secretions of the nose.

Having observed the minute openings of the lacrimal ducts at the medial corners of the eyelids it is clear that any excess of tears must overflow from the conjunctival sac and run down the cheek. Some of the excess, however, does pass down the ducts and accounts for the excess of watery secretion from the nose after crying which, if severe, requires the use of a handkerchief, although in these circumstances its place is frequently taken by 'sniffing'.

The secretion of tears is increased by the presence of foreign bodies and inflammation caused by bacteria or irritating vapours. Irritation of the nasal mucous membrane and very bright light provoke reflex lacrimation, while emotional states and pain also result in the flowing of tears.

The functions of the tears may be summarized:

1. Keeping the eyes moist, thereby allowing free movement of the lids.
2. Removal of dust and foreign bodies, including bacteria.
3. Acting as a mild antiseptic.
4. Expression of emotion or pain.

The muscles of the eye. Each eyeball is moved by muscles which arise from the posterior wall of the bony orbit close to the entrance of the optic nerve and are inserted into the outer fibrous coat (**sclera**) of the eye. There are four straight and two oblique muscles in addition to the muscle elevating the upper lid (levator palpebrae superioris).

Straight muscles. These are the **superior rectus, inferior rectus, medial rectus** and **lateral rectus**. Their position in relation to the eyeball is indicated by their names. The action of these muscles is not too difficult to follow if two facts are remembered. 1. If one eye alone is considered, contraction of the superior rectus turning the eye upwards will be associated with relaxation of the inferior rectus, and vice versa. The medial and lateral recti move the eye to one side or the other respectively and work together in the same way. This action is similar to the opposing action of the flexor and extensor muscles of the forearm. When one set contracts

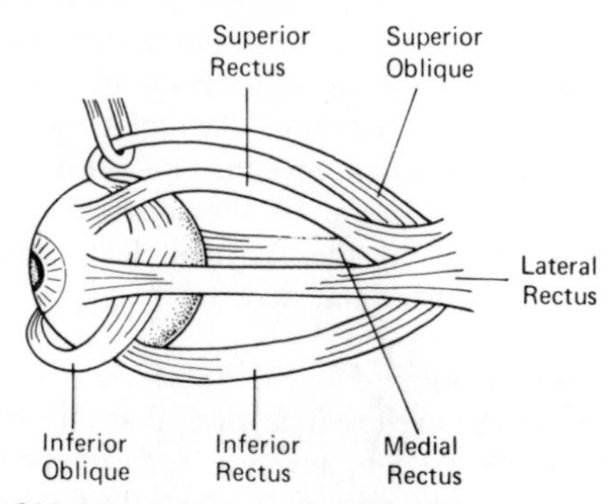

Fig. 214 Diagram of eye muscles. Left orbit (lateral aspect).

the opposite set relaxes. 2. In normal vision both eyes move together. Therefore, if the eyes are turned upwards both right and left superior recti will contract and both inferior recti will relax. On the other hand, if both eyes are turned to the right it follows that the right lateral rectus and the left medial rectus will contract while the right medial rectus and left lateral rectus will relax. The converse is true if the eyes are turned to the left. This is quite simple, and with a little though the individual can work it out for himself.

Oblique muscles. The **superior** oblique muscle is so arranged to direct the eye downwards and outwards. The **inferior** oblique muscle turns the eye upwards and outwards. The muscles of the eyes are supplied by the cranial nerves (IIIrd, IVth and VIth). The lateral rectus is supplied by the VIth or abducens nerve. The superior oblique is supplied by the IVth or trochlear nerve. All the others (levator palpebrae superioris, medial rectus, inferior rectus, inferior

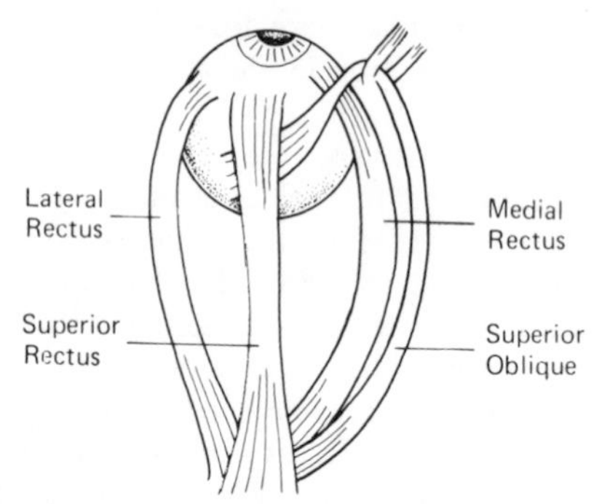

Fig. 215 Diagram of eye muscles. Left orbit (from above).

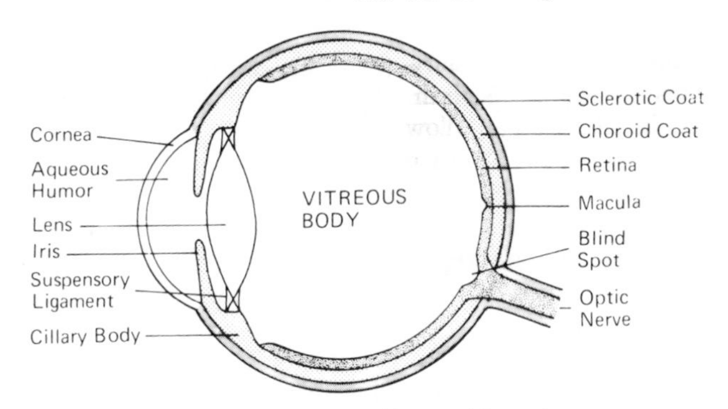

Fig. 216 Section through eyeball viewed from above.

oblique and superior rectus) are supplied by the IIIrd or oculomotor nerve. These muscles are sometimes referred to as the external or extrinsic muscles of the eye. The internal or intrinsic muscles are mentioned later (page 357).

The common condition of squint is usually due to imperfect balance between opposing muscles and may be treated by operations designed to shorten or lengthen the appropriate muscles.

The eyeball or bulb of the eye

The bulb of the eye is situated in the anterior part of the orbital cavity and is almost spherical in shape. It is surrounded by a pad of fat which occupies the rest of the orbital space.

The structure of the eye may be considered in the following way:

1. (*a*) Fibrous: (*a*) sclera, (*b*) cornea.
   (*b*) Vascular: (*a*) choroid, (*b*) ciliary body, (*c*) iris.
   (*c*) Nervous: retina.
2. The light-transmitting mechanism:
   (*a*) Aqueous humour.
   (*b*) The lens.
   (*c*) Vitreous body.

The fibrous coats. (*a*) The **sclera** (or sclerotic coat). The posterior five-sixths of the outer coat of the eyeball consists of strong, opaque fibrous tissue and is called the sclera or sclerotic coat. It is protective in function and helps to maintain the shape of the eyeball. When viewed from the front it is that portion which is referred to as 'the white of the eye' and, in this position, is covered by the conjunctiva. Posteriorly, the optic nerve passes through it to reach the retina inside the eye, and in the orbit the nerve is protected by a sheath of fibrous tissue continuous with the sclera.

(*b*) The **cornea** occupies the anterior one-sixth of the external surface of the eyeball and, being transparent, allows light to enter the interior of the eye. The cornea is sometimes described as the 'window of the eye' and its anterior surface can be seen to be slightly curved or convex. Over the cornea the conjunctiva becomes very thin and is only represented by a few layers of epithelial cells. The cornea has no blood vessels but derives its nourishment from the lymph.

The vascular coat. This is the middle layer of the eye. It contains many blood vessels and capillaries which are derived from the ophthalmic branch of the internal carotid artery and is pigmented.

(*a*) *The choroid.* This is a thin pigmented membrane, dark brown in colour, which lines the posterior compartment of the eye. It is situated between the inner surface of the sclera and the retina.

(*b*) *The ciliary body.* This is a circular structure continuous with the anterior part of the choroid which surrounds the periphery of the iris immediately behind the outer margin of the cornea where it joins the sclera (see Fig. 204). It contains muscle fibres (the ciliary muscle) and to it is attached the ligament which helps to suspend the lens in position.

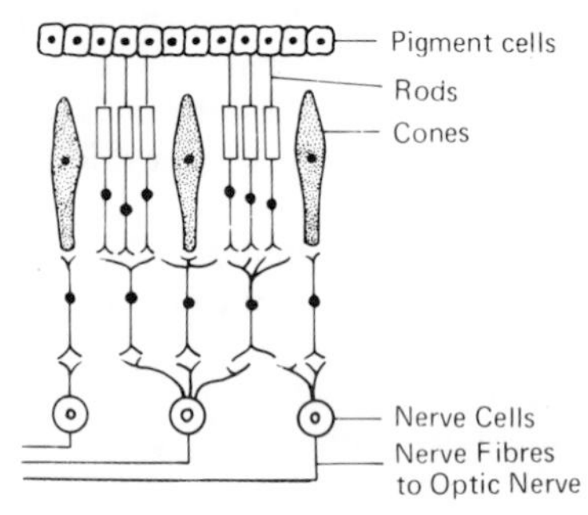

Fig. 217 Diagram illustrating the layers of the retina.

(*c*) *The iris.* This is the pigmented membrane which surrounds the pupil of the eye. It arises from the margin of the ciliary body and forms a diaphragm with a black central opening (the pupil) immediately in front of the lens. The colour of the eye is dependent on the pigment in the iris. In dark eyes the pigment is plentiful, but in blue eyes it is scanty.

The iris contains two sets of muscle fibres, one (sphincter pupillae) arranged to encircle the pupil, the other (dilator pupillae) passing in a radial direction from the outer margin of the iris to the edge of the pupil. It will be clear that the circular muscle acting as a sphincter will reduce the size of the pupil when it contracts. Contraction of the radial fibres, on the other hand, increase its size and they are, therefore, dilators. These, with the ciliary muscle, are the intrinsic muscles of the eye.

The function of the iris is to regulate the amount of light entering the posterior part of the eye. Thus, when a bright light shines on the retina the pupil contracts.

Drugs also vary the size of the pupil. Physostigmine (Eserine) causes it to contract while atropine dilates it.

The retina. The retina is the innermost coat of the eye. It lines the posterior chamber and ends anteriorly at the margin of the ciliary body. It is a delicate membrane consisting of neurones, that is, nerve cells and nerve fibres, together with a layer of special structures called **rods** and **cones**. These are situated on the outer or choroidal surface of the retina, while the nerve fibres are on the inner surface facing the chamber of the eye. The rods and cones are the actual receptors of sight and light reaching them sets up the impulses which are transmitted to the nerves.

The retina is the nervous portion of the eye and is therefore the true end-organ of vision. The fibres of the optic nerve commence in the cells of the retina and are collected together at a point just medial to the most posterior part of the eyeball where they pierce the choroid and sclera and pass backwards as the optic nerve through the orbit to the optic chiasma and brain.

The point at which the optic nerve fibres all converge contains no nerve cells and no rods and cones. It is, therefore, insensitive to light and is called the **blind spot**.

The retina is supplied with blood by a branch of the ophthalmic artery which enters the eye with the optic nerve and is called the **central artery of the retina**.

The **macula** is situated just to the lateral side of the entrance of the optic nerve, i.e. at the very centre of the posterior part of the eye. It is a small area of the retina of great sensibility on which the images seen by direct or near vision are focused. (No rods are present in this area but cones are especially numerous.)

The retina may become partially detached from the choroid, particularly in people with severe myopia (short sightedness).

The light-transmitting mechanism.

1. The aqueous humour. Situated between the cornea in front and the iris and ciliary body behind is the anterior chamber of the eye which contains a clear watery fluid called the aqueous humour.

2. The lens. This is a firm transparent structure, convex in shape, which is suspended in its capsule by a ligament attached to the ciliary body. It is placed immediately behind the iris and pupil of the eye. Its function is to focus rays of light entering the eye through the pupil on to the retina.

3. The vitreous body. This is a semi-fluid or jelly-like substance which fills the posterior four-fifths of the eye. It helps to preserve the spherical shape of the eyeball and to support the retina.

The presence of fluid and semi-fluid material in the interior of the eye maintains its shape and in doing so keeps up a constant pressure on its walls. This is referred to as the **intra-ocular tension.**

In certain conditions drainage of the fluid may be impaired and there will be a consequent increase in the intra-ocular tension, a serious effect which may disturb the nutrition of the retina and lead to blindness. This is known as **glaucoma**. It may be treated:

1. By placing drugs in the eye which cause the pupil to contract (myotics e.g. eserine or physostigmine), thereby helping to open up the canals situated at the point of attachment of the iris to the ciliary body where the excess of fluid is normally drained off into the circulation.
2. By making a small hole (trephine) through the sclera into the anterior chamber of the eye so that fluid can drain under the conjunctiva and so relieve the tension within the eyeball.

Diminution in intra-ocular tension is seen in cases of severe shock and marked fluid loss from the body. Complete loss of tension is observed after death.

The mechanism of sight

From a structural point of view the eye may be compared with a simple camera. The eyelids act as a shutter and there is an entrance window for light—the cornea; a diaphragm to regulate the amount of light entering—the iris; a lens to focus the image; a darkened interior formed by the choroid, and a light-sensitive plate which receives the image—the retina.

The optic nerve and its connections convey the details of the image to the occipital region of the cerebral cortex where they are developed into consciousness.

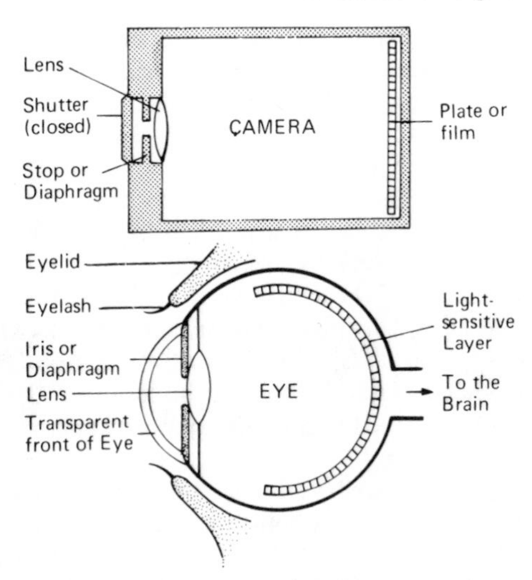

Fig. 218 The eye compared with a camera.

In order to understand the mechanism of vision it is necessary to know something about light and the action of lenses. Light consists of waves in the ether which travel faster than anything else known in the universe at the rate of 186,000 miles per second.

Some objects, such as the sun, electric light or candle, emit light rays and are self-luminous sources of light. Other objects, such as the things we normally see, merely reflect light received from other sources. If there is no source of light, complete darkness exists and no object can be seen.

Rays of light ordinarily travel in straight lines. A lens, which may be roughly defined as a curved transparent structure, has the power of bending or refracting rays of light. A lens which is thicker at the centre than at the periphery is described as convex. One which is thinner in the middle is called concave.

A convex lens has the power of bending rays of light so that they converge and meet at a point of focus behind the lens. The stronger the lens, i.e. the greater the degree of curvature of its surfaces, the nearer is the focal point. A concave lens, on the other hand, bends the light rays so that they diverge and do not focus behind the lens. The lens of the eye is convex and focuses the rays of light passing through it on to the retina.

Actually the image reaching the retina is inverted but this is turned the right way up by the visual cortex in the brain.

Accommodation. Rays of light from distant objects are for all practical purposes parallel and therefore strike the vertical axis of the lens at right angles. The eye is so adjusted that such rays are bent by the lens to focus exactly on the retina, forming a sharp image. Rays of light from a near object (say 10 inches) are divergent and strike the lens obliquely. In order that such rays may be accurately focused on the retina the lens must be made more powerful (of greater focusing power) by becoming thicker. This is accomplished by the action of the ciliary muscles. At the same time the clearness of the image is increased by cutting down the number of rays entering the eye by contraction of the iris. This process of altering the shape of the lens and contraction of the iris to produce a small pupil is called accommodation and operates every time a near object is looked at.

When an object is placed very near the eyes, in order to obtain a clearly focused picture on both retinae the eyes turn slightly inwards towards each other. This is called convergence. The extreme of convergence is illustrated by the trick of 'squinting down the end of the nose'.

It is of interest at this point to note some of the common defects of vision requiring the use of spectacles. Whereas the normal eye is practically spherical, in some people it tends to be slightly elongated and in others flattened. In other words, in the former case the distance from the lens to the retina is increased and in the latter is is decreased. It follows, therefore, that the lens will not naturally focus the image accurately on the

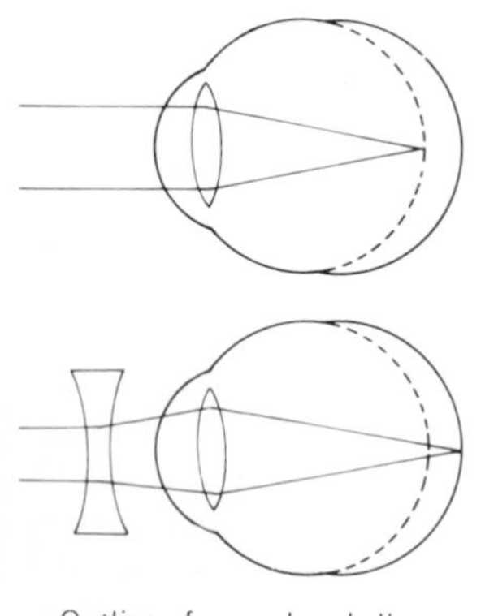

-----Outline of normal eyeball

—Outline of short-sighted eyeball

Fig. 219 Myopia (short sight) and its correction.

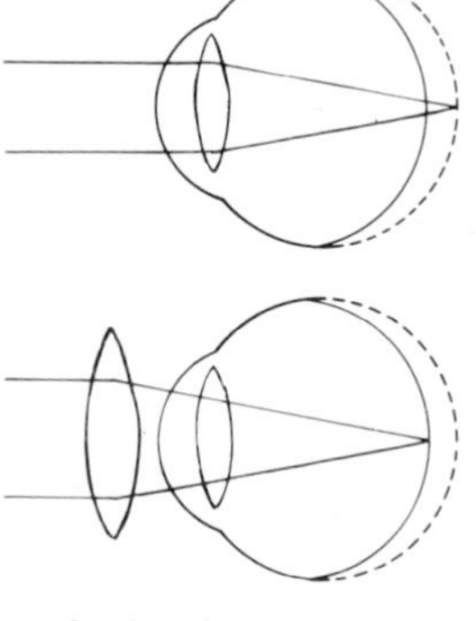

-----Outline of normal eyeball

—Outline of long-sighted eyeball

Fig. 220 Hypermetropia (long sight) and its correction.

retina in these conditions. In the former, elongated or myopic eye (short sight), the image will tend to fall in front of the retina, while in the shortened or hypermetropic eye (long sight) it will fall behind the retina. In both instances the objects seen will be blurred and out of focus.

These defects can be compensated by using additional lenses in the form of spectacles. By placing a concave glass lens in front of a myopic eye the rays of light will become divergent before reaching the lens, so that its point of focus is shifted back on to the retina. A convex lens in front of a hypermetropic eye will bring the image nearer the front of the eye by increasing the convergence of the rays so that they are focused by the retina.

(Presbyopia is a defect of accommodation which comes on with advancing age and may be caused by lack of elasticity in the lens. Astigmatism is due to the unequal curvature of the surfaces of the cornea, i.e. it may be curved more vertically than horizontally.)

Vision. It has been seen that rays of light reflected from visible objects fall on the retina and, in near vision, are focused on the macula. It is probable that light falling on the retina produces certain chemical changes there which stimulate the endings of the optic

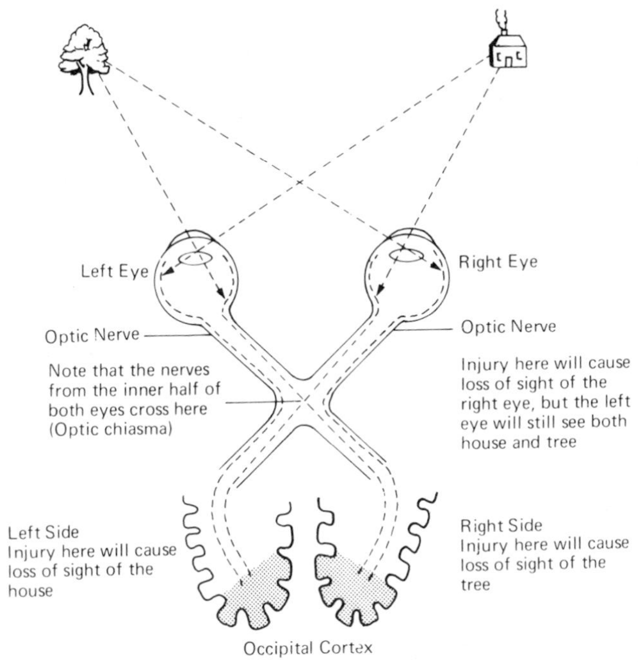

Fig. 221 Diagram of the visual fields and paths.

nerve. These stimuli are conveyed by the optic nerve to the optic chiasma where certain of the fibres cross and are then carried back to the cortex of the occipital lobe to be interpreted into consciousness.

Binocular vision. In considering the sense of sight it must be remembered that although we can see with each eye separately, normal stereoscopic vision is obtained by the simultaneous use of both eyes.

Rays of light strike the retina from all directions. Those coming from the left-hand side of the body will fall on the nasal side of the retina of the left eye and the temporal side of the retina of the right eye. These images produced by objects on the left side of the body are eventually received by the occipital lobe on the right side of the brain. This explains the crossing of certain fibres of the optic nerve in the optic chiasma and is part of the principle that one side of the body is controlled by the cerebral hemisphere of the opposite side. These facts are easily appreciated from Fig. 216.

SUMMARY OF SENSE OF SIGHT

Light waves → cornea → aqueous humour → lens → vitreous body → retina → optic nerve → optic chiasma → occipital lobe of brain.

QUESTIONS

1. Which are the 'special senses'? Give an account of the structure of the organ concerned with any one of these.
2. Give an account of the structure and function of the eye.
3. Describe the structure and function of the organs of hearing.

21 The Reproductive System

The power of reproduction is one of the essential characteristics of life. It is illustrated in its primitive form by the action of the single-celled amoeba in dividing into two. Most of the cells of the human body have the same power of division by virtue of which growth and repair are possible.

Reproduction of the species in man and the higher animals is a complicated process involving the existence of two sexes, both of which play their parts in the formation of a new individual.

The reproductive organs of the male and female differ in anatomical structure and arrangement, each being adapted to the functional activities they are required to perform.

The function of the male organs is to form **spermatozoa** or **sperms** and implant them within the female so that they can meet the ova. The female organs are adapted to form **ova** or eggs which, if fertilized by spermatozoa, remain in the cavity of the uterus. Here an embryo or fetus is formed and is retained until the second individual is capable of a separate and independent existence.

THE FEMALE ORGANS OF GENERATION

The female sex organs are situated in the pelvis and for purposes of description may be divided into:

1. Internal organs: uterus,
   ovaries,
   Fallopian (uterine) tubes,
   vagina.
2. External organs (external genitalia or vulva) mons veneris,
   labia majora and minora,
   clitoris
   hymen.
3. Secondary organs: the breasts or mammae.

Internal organs of generation

The uterus

The uterus is a hollow, pear-shaped organ situated in the pelvis between the bladder in front and the rectum behind. It has thick muscular walls and a small central cavity. In the adult it measures about 8 cm (3 in) in length, 5 cm (2 in) in width and 2·5 cm (1 in) in thickness.

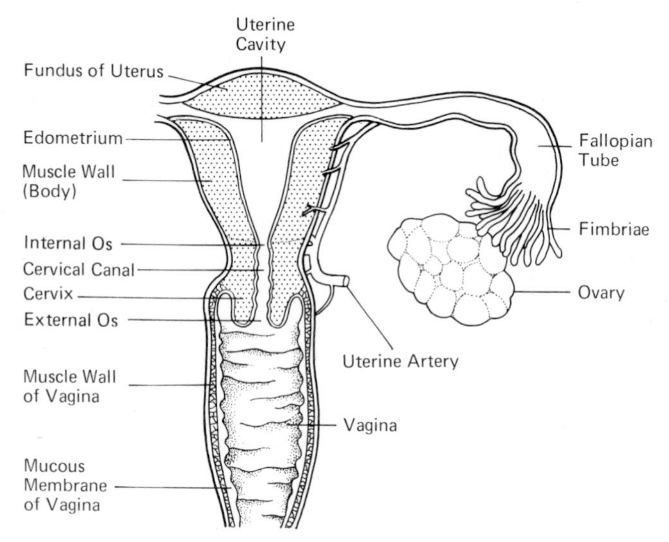

Fig. 222 Diagram of vagina, cervix, uterus, Fallopian tube and ovary.

It consists of: (*i*) the **fundus,** (*ii*) the **body** and (*iii*) the **cervix.** The fundus is the upper part of the uterus situated between the two **Fallopian tubes.** The body forms the greater part of the organ and is the portion between the fundus and the cervix. The cervix or neck is the lowest portion, part of which projects like an inverted dome into the vagina below. It is traversed by a canal opening above into the cavity of the uterus by an orifice called the **internal os,** and below into the vagina by the **external os.**

Attached to either side of the fundus of the uterus are the hollow Fallopian tubes. The cavity of the uterus has, therefore, three openings—one into each Fallopian tube and one through the external os of the cervix into the vagina.

The fundus, the body and the cervix, except for that part which projects into the vagina, are covered on their outer surface by peritoneum. The peritoneum on the anterior surface of the body of the uterus, if traced forwards, is found to be reflected on to the superior surface of the bladder. That from the posterior surface lines the lowest part of the pelvic cavity before passing on to the rectum. This space between the uterus and the rectum is called the **recto-uterine pouch** (of Douglas). (The posterior fornix of the vagina is closely related to the pouch of Douglas—see below.)

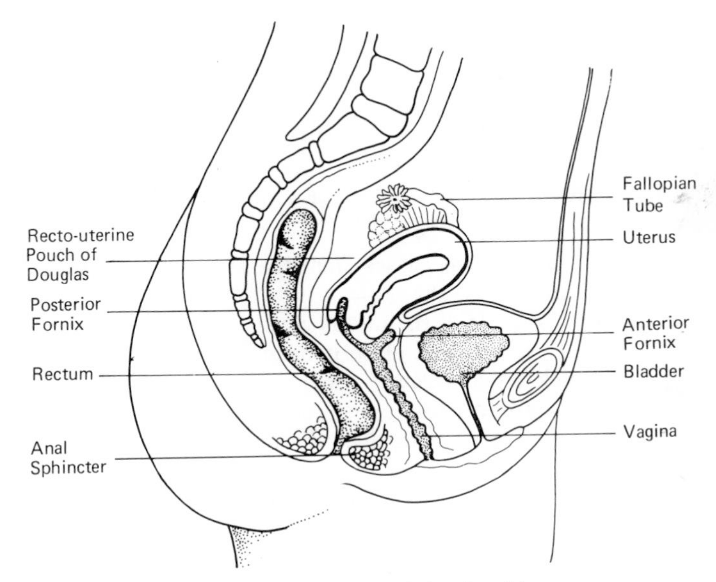

Fig. 223 Section through female pelvis.

The peritoneum passing laterally from the uterus extends to the side wall of the pelvis. It consists of two layers, the front layer being continuous with the peritoneum covering the anterior surface of the uterus and the posterior layer with that covering the posterior surface of the uterus.

This double fold of peritoneum passing from the side of the uterus to the wall of the pelvic cavity is called the **broad ligament**. Between the two layers forming its upper margin is situated the Fallopian tube. It is, therefore, rather like a piece of material draped to hang down on either side of a horizontal pole—the pole being represented by the curved Fallopian tube. Also enclosed between the layers of the broad ligament is a fibrous band, the **round liga-**

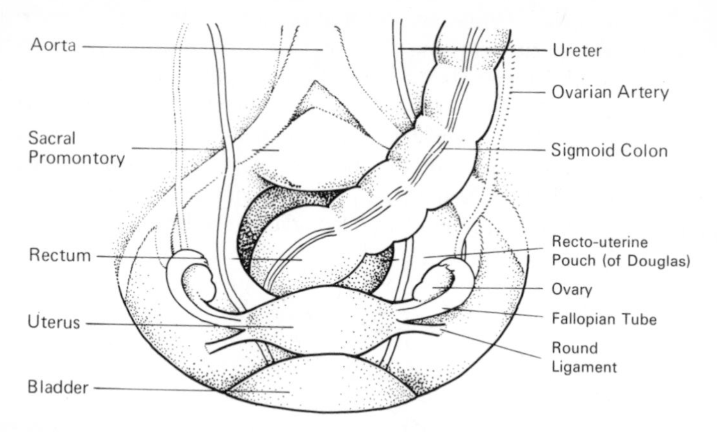

Fig. 224 The female pelvic organs seen from above and in front. All the organs except the ovaries are covered by the peritoneum.

ment of the uterus which passes from the side wall of this organ to the inguinal canal.

Structure. The walls of the uterus consist of three layers:

(*i*) The outer serous coat of peritoneum.
(*ii*) The thick middle layer consisting of involuntary, plain muscle (**myometrium**).
(*iii*) The inner mucous coat called the **endometrium**.

The uterus receives its blood supply from the uterine branch of the internal iliac artery.

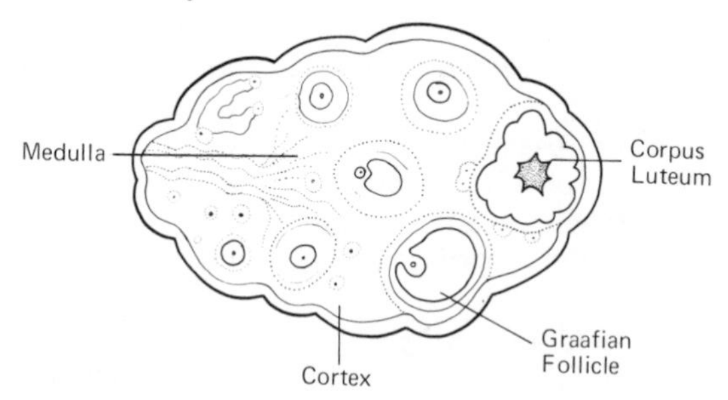

Fig. 225 Transverse section of the ovary (diagrammatic).

Functions

1. To receive the fertilized ovum and to retain and nourish the developing fetus throughout the duration of pregnancy.

2. To expel the fetus at the end of pregnancy by the contractions of its muscular walls.
3. To play a part in the phenomenon of menstruation.

The ovary

There are two ovaries, right and left, situated in the pelvis on either side of the uterus. Each is about the size of a large almond and is attached to the posterior aspect of the broad ligament of the uterus by a fold of peritoneum. It lies immediately below the Fallopian tube which forms an arch over the top of the ovary and ends just below its lateral margin.

Structure. Briefly the ovary may be described as having (*i*) a **medulla** in the centre consisting mainly of fibrous tissue or stroma, and (*ii*) a **cortex** on the surface consisting of a layer of epithelium (the **germinal epithelium**), a number of cystic spaces of various sizes—the **Graafian follicles**—which contain the ova surrounded by a little fluid, and a yellow body—the **corpus luteum**—formed after a Graafian follicle has ruptured and discharged its contained ova and fluid.

Functions

1. To produce ova.
2. The Graafian follicle secretes the hormone oestrogen (page 302).
3. The corpus luteum secretes the hormone progesterone.

The Fallopian tube (uterine tube)

The Fallopian tubes, two in number, are about 10 cm (4 in) in length and lie in the upper margin of each broad ligament of the uterus, thus being surrounded by peritoneum. The outer end of the tube is expanded and has an opening into the peritoneal cavity. This is surrounded by a number of fringe-like processes, the *fimbriae,* which lie close to the lateral part of the ovary. It has already been seen that the middle of the tube curves round the ovary like an arch.

Structure. The Fallopian tube has a muscular wall continuous with that of the uterus. Its outer surface is covered by peritoneum, while its inner lining or mucous membrane is formed of ciliated epithelium.

Function. Its function is to collect the ova discharged from the ovary in its fimbriated end, and pass them along its interior towards the cavity of the uterus by the action of its ciliated epithelium. Fertilization of the ovum by spermatozoa usually takes place in the tube.

The vagina

This is a canal with muscular walls 8–10 cm (3–4 in) long which passes in a downward and forward direction from the cervix of the uterus to its lower orifice in the vulva.

It is lined by a thin type of skin which is thrown into a number of transverse folds and is kept moist by the secretion of the mucous glands present in the cervix. This secretion is slightly acid in reaction (due to lactic acid).

External organs of generation

In order to appreciate the somewhat complicated anatomical arrangement of the external genitalia, it must be remembered that (*i*) they are closely related to the urethra, and (*ii*) their ultimate function is to permit the passage of the fully developed fetus from the cavity of the uterus at birth.

The female external genitalia are known collectively as the **vulva** and are enclosed in an area bounded in front by the mons veneris and on either side by the folds of skin known as the **labia majora** which unite posteriorly in the perineum.

The **mons veneris** (mons pubis) is a pad of fat in front of the symphysis pubis in the female. At puberty it becomes covered with pubic hair.

The labia majora. These are two rounded folds of skin which pass backwards from the mons veneris in a curve and unite at their posterior ends in the perineum, in front of the anus. At the posterior ends of each is a small gland which secretes mucus, called *Bartholin's gland.* (labium majus (singular) = large lip.)

The **labia minora** or **nymphae** are two thin folds of skin lying within the space enclosed by the labia majora. They enclose a triangular area called the **vestibule**, within which are the openings of the urethra and vagina. Anteriorly the labia minora meet to form a hood (prepuce) to the clitoris. Behind they unite in a small fold, the **fourchette**. (labium minus (singular) = smaller lip.)

Projecting into its upper part for about 1·5 cm ($\frac{1}{2}$ in) like the lower half of a sphere is the cervix of the uterus. The short recess of the vagina extending above the cervix is called the **fornix** (anterior, posterior and lateral according to the exact part of the recess referred to). The posterior fornix is closely related to the peritoneum forming the pouch of Douglas (Fig. 223).

Lying in front of the vagina are the bladder and urethra. Behind, it is separated by a pad of tissue from the rectum and anus. This tissue between the external orifice of the vagina and the anus is called the **perineal body**. The external opening of the vagina in the virgin is partially guarded by a membrane called the **hymen**.

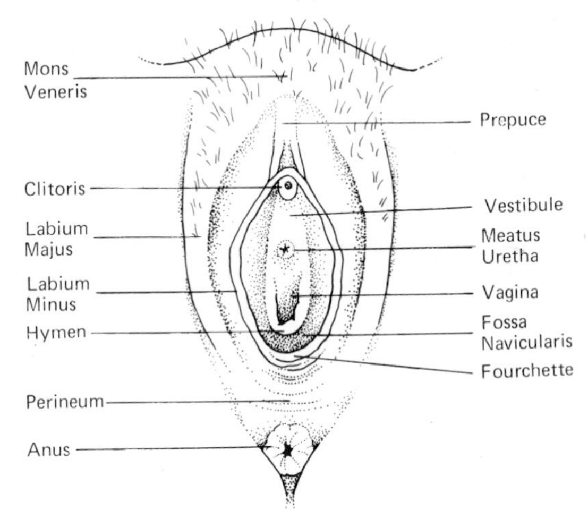

Fig. 226 The external genital organs.

The clitoris. This is a small body situated at the apex or front part of the triangle formed by the labia minora, which as they unite in front provide a hood-like covering for the organ. It is composed of erectile tissue, i.e. tissue which becomes firm and rigid when congested with blood. It is very sensitive and well supplied with nerves. It corresponds developmentally with the penis in the male.

The hymen. In the virgin this forms an almost complete membranous diaphragm across the vaginal entrance. It is perforated in its centre by a small circular opening which permits the exit of the menstrual flow. In rare instances it completely closes the vagina (imperforate hymen), but this condition is not discovered until after puberty when the menstrual fluid is unable to escape.

The **perineum** is the mass of tissue and the skin situated between the external genitals and the anus. It stretches to a remarkable extent during labour, but in spite of this elasticity is frequently torn and requires surgical repair. When this rupture appears inevitable a deliberate surgical cut may be made (perineotomy) immediately before the baby is born. This is easily sutured and minimizes the damage to the pelvic floor.

It will be observed that the female urethra is situated in the vestibule between the folds of the labia minora. It is placed behind the clitoris and in front of the vaginal opening.

The breasts or mammary glands

The two breasts are glands which are accessory to the genital system, that is, take no part in the actual process of reproduction. They are present in an undeveloped form in the female before puberty and also in the male.

The fully developed female breast, while varying considerably in size, is circular in outline and approximately hemispherical in shape. It lies on the pectoralis major muscle, extending from the second rib above to the sixth rib below and from the margin of the sternum on its medial side to the axilla on the lateral side.

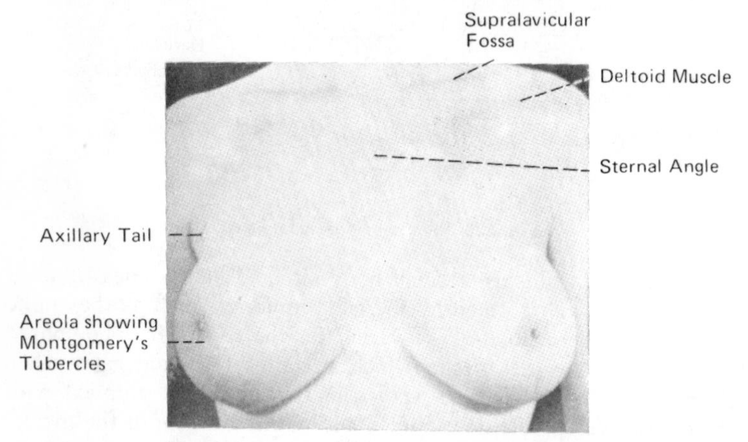

Fig. 227 The breasts of a girl aged eighteen years.

Just below the centre is a small elevation, the nipple, in which the ducts of the breast open. The nipple, in addition, contains a few plain muscle fibres which cause it to become erect when stimulated, a function which makes it an efficient teat for the infant to suck.

Structure. The gland consists of a number of lobules formed of columnar epithelium. From these lobules the **lacteriferous ducts** pass towards the nipple in a radiating direction, and close to their termination each is dilated into an ampulla. The lobules of the gland are surrounded by fat which gives the general contour of the organ.

Surrounding the nipple is a pigmented area of skin, the **areola** which contains a number of specialized sebaceous glands (the glands of Montgomery).

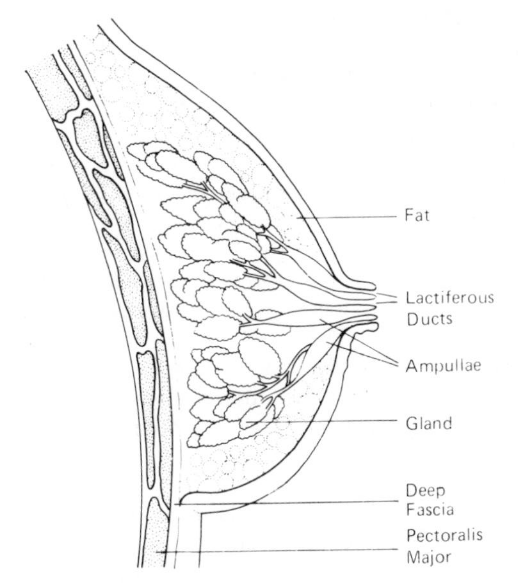

Fig. 228 Section of the breast.

The gland is only stimulated into activity during pregnancy. After conception there is a general enlargement and hardening, the veins on the surface become dilated and dark brown pigment is deposited in the areola. Towards the end of pregnancy a watery fluid called **colostrum** can be expressed from the breast. After delivery, this fluid increases in amount for a day or two and is then replaced by the secretion of milk, or lactation.

During pregnancy oestrogen and progesterone are the hormones responsible for the growth of the ducts and lobules of the breast respectively. At first these hormones are secreted by the ovary and later by the placenta. After childbirth the level of these hormones in the mother's blood falls. As a result of this fall, prolactin from the anterior lobe of the pituitary gland is able to stimulate milk secretion.

The continuation of the milk supply is largely dependent upon the continued suction of the infant or its withdrawal by mechanical means and, therefore, theoretically lactation can continue indefinitely.

It has been seen that the process of lactation is dependent upon the hormones—prolactin and oxytocin—secreted by the anterior lobe of the pituitary gland (page 301).

Physiology of the female reproductive organs. The functions of the female reproductive organs are directed to the following ends:

1. The formation of ova or ovulation.
2. The preparation of the uterus to receive the fertilized ovum.
3. The retention of the fertilized ovum within the cavity of the uterus until a mature fetus is formed, capable of leading an independent existence, i.e. pregnancy.
4. The expulsion of the mature fetus, i.e. labour or parturition.

In the young female child these processes are in abeyance. At a variable age, as a result of the activities of the ductless glands preparation for the reproductive period in a woman's life commences and is called **puberty**. The period during which reproduction is possible usually extends from the early teens until the age of forty-five to fifty, when it ends in the **menopause** (climacteric or 'change of life'), after which pregnancy does not occur.

Ovulation

The ovary contains many thousands of eggs or ova which lie dormant until the onset of puberty. Active changes then take place in the ovary which result in the periodic discharge of an ovum at intervals of a month. (NOTE. This does not correspond with the menstrual period.)

A Graafian follicle is a small cystic sac containing fluid and having the ovum attached to its wall, which comes gradually to the surface of the ovary and ruptures about two weeks after the commencement of the last period.

The ovum therefore actually passes into the peritoneal cavity but is soon caught up in the fimbriae of the Fallopian tube which closely surround the ovary. By the action of the ciliated epithelium of the Fallopian tube, the ovum is carried slowly towards the cavity of the uterus, possibly taking ten days over this stage of its journey.

The ovum is either fertilized, in which case it becomes embedded in the wall of the uterus and commences to grow into an embryo; or else it is discharged unfertilized from the uterus in the menstrual flow.

Certain changes take place in a Graafian follicle after its rupture and it becomes a solid yellowish body called the **corpus luteum**. This body goes on developing until the next menstrual period, when it gradually disappears and is replaced by fibrous tissue. If the ovum is fertilized, however, the corpus luteum persists throughout pregnancy and, it will be recalled, acts as a gland of internal secretion, producing the hormone progesterone.

Menstruation

This is a function of the uterus established at puberty (average age, 12 to 13 years) as a result of ovarian activity and consists of the periodic discharge of blood from its cavity. It occurs on the average every twenty-eight days until the menopause or climacteric is reached, and lasts for about five days. The amount of fluid, which consists of blood, mucin and epithelial cells, varies between 90–200 ml (3–7 fl. oz). Menstruation ceases during pregnancy and is often not re-established until lactation is completed.

The purpose of the monthly cycle is to prepare the mucous membrane of the uterus (endometrium) to receive a fertilized ovum. The endometrium undergoes constant changes between one menstrual period and another and these changes are made in preparation to receive the fertilized ovum. They are largely brought about by the follicle-stimulating (FSH) and the luteinizing (LH) hormones secreted by he pituitary gland, and by oestrogen and progesterone secreted by the ovary. Menstruation is really a clearing up of these changes in the endometrium when no fertilized ovum has arrived, and therefore in this sense it gives the endometrium an opportunity to make a fresh preparation.

These changes are described as the menstrual cycle and may be conveniently divided in the following way:

1. The secretory (pre-menstrual) phase, lasting for about 14 days before the period, during which the endometrium becomes thickened and congested and is in a state of preparedness to receive a fertilized ovum.
2. The menstruation period (average five days) in which some

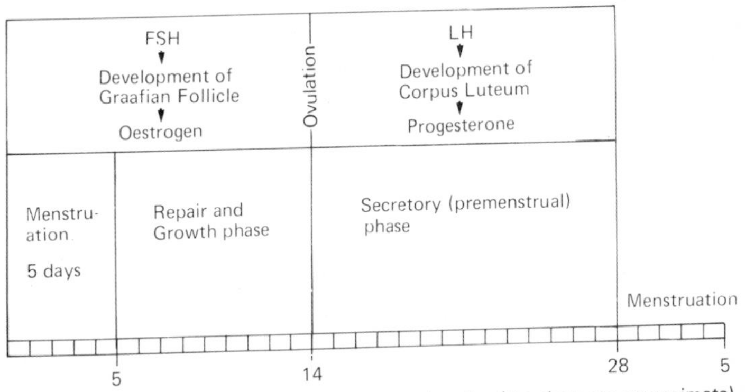

Fig. 229 Diagram demonstrating the menstrual cycles (the days are approximate).

of the epithelium of the uterine mucosa is shed and is accompanied by bleeding. In other words, no fertilized ovum has been received and the work of preparation has been useless.

3. The stage of repair begins in the third or fourth days of the menstrual cycle.
4. The growth phase (fourth to fourteenth day) before the next secretory phase.

THE MALE ORGANS OF GENERATION

It has been seen in the female that the urinary tract, although in close anatomical proximity to the genital organs, is entirely separated from them except for the opening of the urethra in the vulva. In the male, however, the urethra is shared by the urinary and genital systems.

The system is conveniently described as consisting of internal and external organs.

Internal organs

(*i*) Two testes, contained in the scrotum, which produce spermatozoa and an internal secretion (testosterone).
(*ii*) The vas deferens or seminal duct on each side.
(*iii*) The seminal vesicles and their ducts.
(*iv*) The prostate gland.

External organs. The penis, containing the urethra.

The testes

The testes are two glandular organs, oval in shape, situated in the scrotum. Each testis or testicle consists of a number of tubules lined by epithelium, the **seminiferous tubules**, between which are the **interstitial cells** responsible for the internal secretion of the gland. It is in the walls of the seminiferous tubules that the spermatozoa are formed. The seminiferous tubules unite at the upper end of the testis to form the **epididymis**. This structure is applied to the lateral surface of the testis and forms the commencement of the **seminal duct** or **vas deferens**.

In the embryo the testis is actually developed within the abdominal cavity but shortly before birth each testis migrates down the posterior abdominal wall and leaves the cavity by passing into the inguinal canal. As it does so it carries with it a coat of peritoneum which normally becomes separated from the rest of the peritoneum within the abdomen. Finally, by the time the child is born the testis

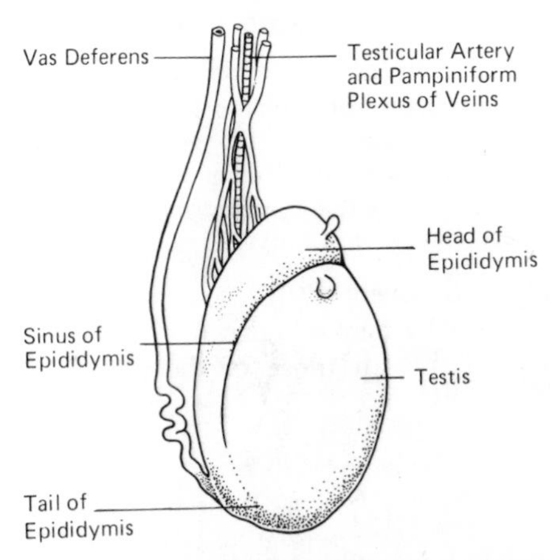

Fig. 230 The right testis and epididymis (lateral aspect).

has left the inguinal canal and has reached the scrotum. In this journey the testis has carried with it its blood vessels and duct (the vas deferens) which form the **spermatic cord**. The spermatic cord, therefore, after birth and in adult life occupies the inguinal canal. This is a very important fact in connection with the anatomy of hernia or rupture, a condition of considerable surgical importance (see page 384).

The testicle is enclosed in a sac—the **tunica vaginalis**. This

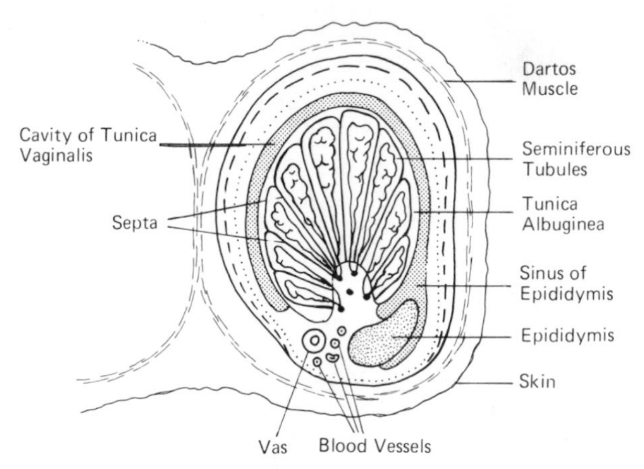

Fig. 231 Section through scrotum and right testis.

sometimes becomes distended with fluid—a condition known as hydrocele.

Occasionally one or both testes fail to leave the abdominal cavity or inguinal canal to reach the scrotum at the period of birth or shortly afterwards. In some of these cases the descent is completed naturally by the age of puberty. In instances of doubt, endocrine treatment may be successful; in others surgery may be necessary.

The **scrotum** is a pouch or bag consisting of skin and having some muscle fibres (the dartos muscle) in its walls. Its skin and tissues are continuous with those of the perineum and groin. It contains the two testes.

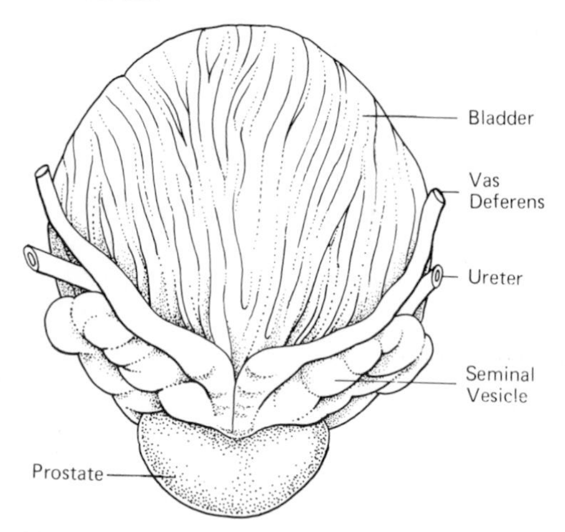

Fig. 232 The structures at the base of the bladder (from behind).

The vas deferens or seminal duct

The testis is a gland having an internal secretion formed by the interstitial cells which enters the blood, and an 'external' secretion (spermatozoa) from the seminiferous tubules which leaves the gland by passing through the epididymis. The vas deferens commences in the epididymis at the lower pole of the testis and leaves the scrotum in the spermatic cord. It passes upwards through the inguinal canal and, after entering the abdominal cavity but remaining outside the peritoneum, it turns inwards to join the seminal vesicle of its own side which is situated at the back of the bladder.

The **seminal vesicles** are two in number, right and left. Each is a reservoir for the external secretion of the testis which is con-

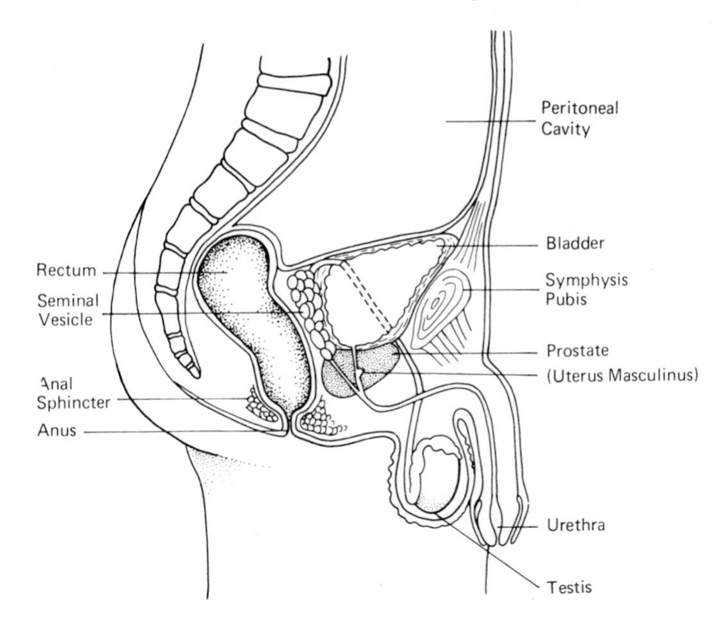

Fig. 233 Section through male pelvis.

veyed by the vas deferens. A small duct, the ejaculatory duct, passes from the seminal vesicle, between the lobes of the prostate gland, to enter the first part of the urethra.

The prostate gland

This is about the size of a chestnut and is situated below the base of the bladder. It is anterior to the rectum and, if enlarged, can be felt by a finger inserted into the rectum. It surrounds the first part of the urethra into which its secretion, the prostatic fluid, passes.

It is of considerable surgical importance because enlargement of the gland, which is common in elderly men, tends to cause obstruction to the outlet of urine from the bladder.

The penis

This structure is made up of two **corpora cavernosa** and the **corpus spongiosum** enclosed in skin. The corpus spongiosum is situated on the under surface and is traversed by the last part of the urethra (see page 281).

The end of the organ is slightly enlarged to form the **glans penis** which is covered by a fold of skin—the **prepuce**. The opening in the prepuce is situated immediately in front of the external opening of the urethra (the urinary meatus). If this fold of skin covers the gland too tightly a condition known as phimosis is caused which results in difficulty in micturition. This is not uncommon in young children and may sometimes require circumcision.

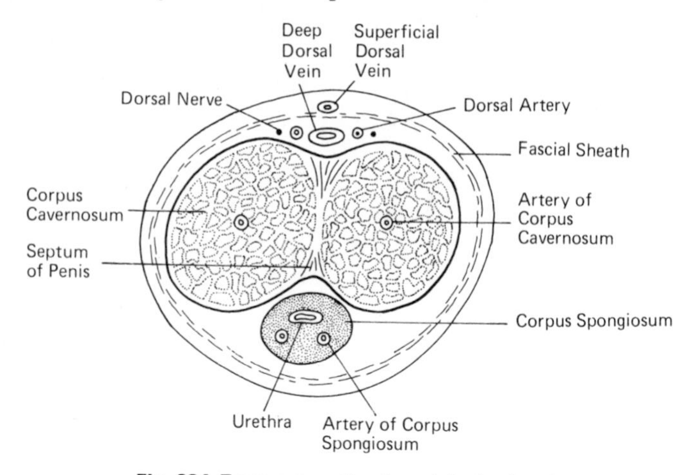

Fig. 234 Transverse section through body of penis.

The penis consists of erectile tissue similar in structure to that of the clitoris (page 369).

Functions of the male organs of generation. The functions of the testis and prostate gland are to secrete the seminal fluid or semen which contains the spermatozoa, formed by the epithelial

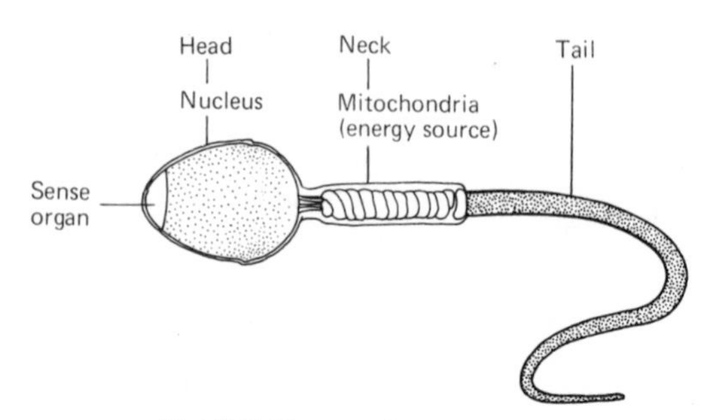

Fig. 235 Diagram of spermatozoon.

cells of the seminiferous tubules. Each spermatozoon is an actively motile body consisting of a head and tail and its function is to fertilize the ovum.

The main function of the hormone testosterone is to control the changes which take place at puberty, such as the growth of pubic and body hair and deepening of the voice. It subsequently influences the male sexual appetite (see page 302).

QUESTIONS

1. Describe either the female or the male reproductive organs.
2. Describe the contents *either* of the male *or* female pelvis.

Appendix

The scalp

The scalp forms the soft parts covering the vault of the skull or cranium. It consists of:

1. **Skin** containing a very large number of hairs in their follicles and many sebaceous glands.
2. **Superficial fascia** consisting of a mesh of fibrous tissue enclosing small lobules of fat.
3. **Deep fascia**, a very strong layer, sometimes called the **epicranial aponeurosis**, which is attached in front to the frontalis muscle and behind to the occipital bone.
4. The following *muscles* (see Fig. 80, page 112):
   (*a*) The frontalis.
   (*b*) The occipitalis.
   (*c*) The temporal muscle on each side.
   (*d*) The small muscles attached to the ears.
5. Superficial or cutaneous **nerves**.
6. Numerous **arteries** which supply the scalp plentifully with blood:

e.g. (*a*) the superficial temporal
(*b*) the posterior auricular
(*c*) the occipital
} branches of the external carotid

Most of the *veins* drain into the external jugular but a few pass directly through the skull bones from the venous sinuses in the interior of the cranium (sometimes a source of danger in infections of the scalp).

7. The deepest layer of the scalp is the periosteum covering the cranial bones, here called the **pericranium**.

The triangles of the neck

The neck looked at from the side (see Fig. 80, page 112) is roughly quadrilateral in shape, bounded above by the lower jaw, below by the clavicle, in front by the mid-line and behind by the trapezius muscle. It will be seen that the sternomastoid muscle divides this area into two (anterior and posterior) triangles. The **anterior triangle** contains, in its upper part, the external carotid artery and its important main branches, various nerves, part of the pharynx and larynx and some lymphatic glands. The lower part of the anterior triangle is covered by muscles and also contains the trachea and thyroid gland.

The **posterior triangle**, situated between the anterior border of the trapezius and the posterior border of the sternomastoid, contains the occipital artery (page 138), various cutaneous nerves and some lymph glands. Situated deep in its lower part just above the clavicle are parts of the subclavian artery and vein, and the brachial plexus.

The axilla

The axilla is the space or hollow between the upper part of the arm and side of the chest. When the arm is by the side, the space is pyramidal or cone-shaped, having an apex directed upwards towards the root of the neck and situated close to the coracoid process of the scapula.

The anterior wall is formed by the pectoralis major and minor muscles.

The posterior wall is formed by the subscapularis and latissimus dorsi muscles.

The medial wall is formed by the upper four or five ribs, the intercostal muscles and the serratus anterior.

The lateral wall consists of the shaft of the humerus and the attached muscles.

The floor or base consists of skin which contains hair and many sweat glands and a thick fascia—the axillary fascia.

The axilla contains:

(*a*) The axillary artery and vein.

(*b*) The important nerves of the arm, including the radial (musculospiral) ulnar and median, which are arranged round the axillary vessels.

(*c*) Lymphatic glands which receive lymph from: (*i*) the arm, and (*ii*) the breast (especially important in cancer of the breast).

The cubital fossa

The cubital fossa is a small space situated in front of the elbow joint at the bend of the elbow. Superficially are placed the median basilic and median cephalic veins, more deeply the brachial artery divides into the radial and ulnar branches. To the lateral side of

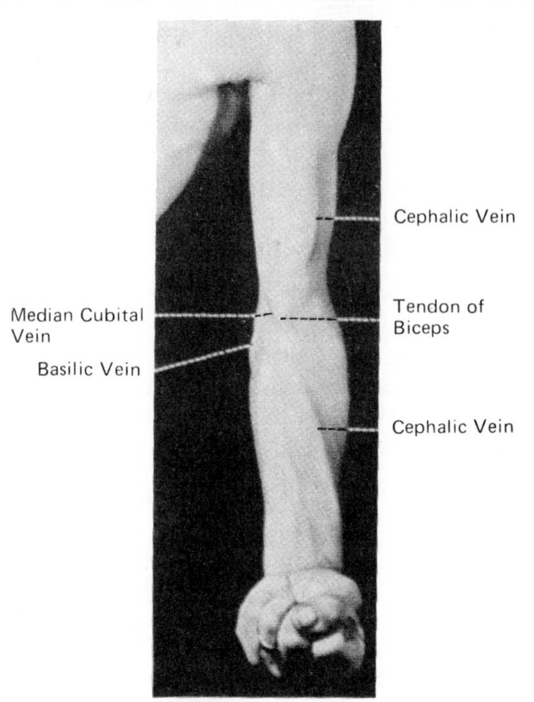

Fig. 236 The cubital fossa.

the brachial artery the tendon of the biceps muscle passes to be inserted into the tubercle of the radius. The radial (musculospiral) nerve also lies to the lateral side. On the medial side is the median nerve. The boundaries of the space are formed by various muscles (see Fig. 81, page 113).

The palm of the hand

The palm of the hand is of considerable practical importance on account of its liability to injury and infection. It may be conveniently considered in layers.

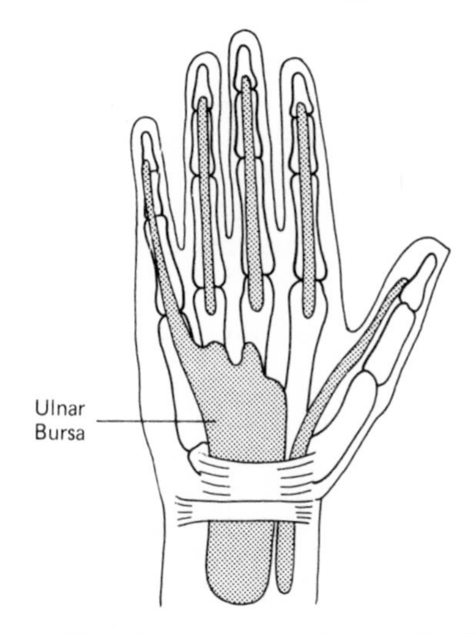

Fig. 237 Diagram illustrating arrangement of tendon sheaths in the palm.

1. The **skin** and superficial fascia.

2. The deep or **palmar** fascia is a strong fibrous layer consisting of a central and two lateral portions, the latter being spread over (*a*) the thenar eminence formed by the short muscles of the thumb and (*b*) the hypothenar eminence formed by the short muscles of the little finger.

3. (*a*) The **superficial palmar arch** which is formed by the continuation of the ulnar artery into the hand joining with a branch from the radial artery on the radial side.

(*b*) The median and ulnar **nerves** and their branches to the fingers.

(*c*) The **tendons** of the flexor muscles which are enclosed in synovial sheaths, the arrangement of which is shown in Fig. 237. These are very important in connection with infections of the fingers and hand. It is clear that infection of the tendon sheath of the little finger can spread and involve the main synovial sac in the palm which is called the **ulnar bursa**. The tendon sheaths of the other fingers extend upwards only as far as the heads of the metacarpal bones.

4. The deepest part of the palm is formed by the metacarpal **bones,** the short muscles of the fingers (interossei) on which lies the deep **palmar arch** formed by the continuation of the radial artery joining a deep branch of the ulnar artery on the ulnar side.

The umbilicus

The umbilicus or navel is a puckered depression in the skin of the abdominal wall, situated in the mid-line rather nearer the symphysis pubis than the xiphoid cartilage. It is actually a scar representing the site of the attachment of the umbilical cord which connects the fetus to the placenta and contains blood vessels.

The inguinal canal

The inguinal canal is a passage, 4 cm ($1\frac{1}{2}$ in), in the lower part of the abdominal wall. It passes obliquely downwards forwards and medially and contains part of the spermatic cord in the male and the round ligament of the uterus in the female. It is situated immediately above the medial half of the inguinal (Poupart's) ligament. The lateral and deeper end is called the **deep inguinal** (internal abdominal) **ring**. The superficial end nearer to the midline and just above the spine of the pubis is called the **superficial inguinal** (external abdominal) **ring**. It has—

(*a*) a floor formed mainly by the inguinal ligament;
(*b*) an anterior wall consisting of skin, fascia and the aponeurosis of the external oblique muscle;
(*c*) a posterior wall formed by peritoneum covered by fascia and the tendinous insertion of the internal oblique and transversus muscle, called the conjoined tendon.

It is through this canal that the testis travels in its descent from the abdominal cavity to the scrotum, carrying with it the spermatic cord (page 375 and Fig. 86).

Sometimes a rupture or hernia develops in this area. An indirect inguinal hernia consists of a protrusion of the peritoneum through the deep inguinal ring which may extend along the inguinal canal and pass through the superficial inguinal ring to reach the scrotum. This protrusion or peritoneal sac may contain intestine, omentum or some other abdominal viscus. A direct inguinal hernia does not pass through the deep inguinal ring. It bulges directly through the posterior wall of the inguinal canal. It emerges through the superficial inguinal ring but does not usually descend into the scrotum.

A femoral hernia passes down through the femoral ring, which is medial to the femoral vein and behind the inguinal ligament, and emerges at the saphenous opening.

The femoral (Scarpa's) triangle

This is a triangular hollow in the upper part of the thigh. Its base is above and is formed by the inguinal ligament. Its lateral boundary is the sartorius muscle, its medial boundary is the adductor longus muscle. The psoas, iliacus and pectineus muscles form the floor.

It contains the femoral artery and vein and the femoral nerve which pass vertically downwards from the middle of the inguinal ligament to its apex. Placed superficially are lymph glands and the saphenous opening through which the long saphenous vein passes to reach the femoral vein (see Figs 111 and 113).

The adductor (Hunter's) canal

After leaving the femoral (Scarpa's) triangle, the femoral artery enters Hunter's canal. This is a channel situated deeply in the muscles on the lower third of the medial side of the thigh.

The sartorius muscle forms the roof, the adductor longus the posterior wall and the vastus medialis the lateral wall.

The femoral artery and vein leave the lower end of the canal by passing into the popliteal space (see Fig. 111, page 144).

The popliteal space

The popliteal space is situated behind the knee joint. Its anterior wall or floor is formed from above downwards by—

1. the popliteal surface of the femur;
2. the posterior ligaments of the knee joint;
3. the upper part of the head of the tibia and the popliteus muscle which covers it.

The space is diamond-shaped and has the following boundaries:

1. Upper and medial: the semi-membranosus and semi-tendinosus muscles.
   Upper and lateral: the biceps muscle and tendon.
2. Lower and medial: the medial head of gastrocnemius.
   Lower and lateral: the lateral head of gastrocnemius.

Passing vertically downwards through the space are the popliteal artery and vein and the medial popliteal nerve.

The sole of the foot

This corresponds to the palm of the hand and is noted for the strong plantar fascia which extends throughout its length. It has been seen

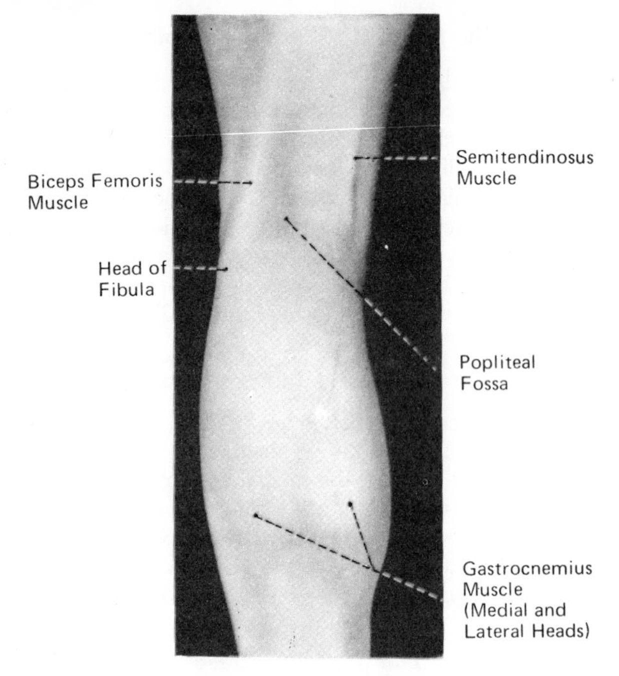

Fig. 238 Popliteal fossa.

that the foot is arched both longitudinally and transversely (page 61).

The sole of the foot contains the flexor tendons of the toes, short muscles passing to the big and little toes, nerves and the arteries which form the plantar arch.

Summary of the anatomical and physiological changes in pregnancy

Anatomical

1. Gradual enlargement of the uterus, with softening of the cervix.
2. Gradual enlargement of the breasts with increased pigmentation around the nipples.
3. Stretching of the skin of the abdominal wall. Pink striations, which after delivery may have a silvery appearance, sometimes develop (*striae gravidarum*). Increased pigmentation may be

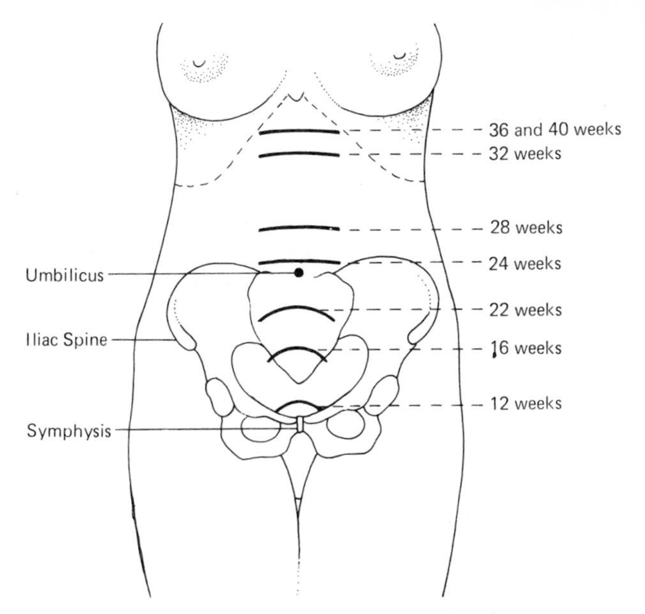

Fig. 239 Height of fundus of uterus during weeks of pregnancy.

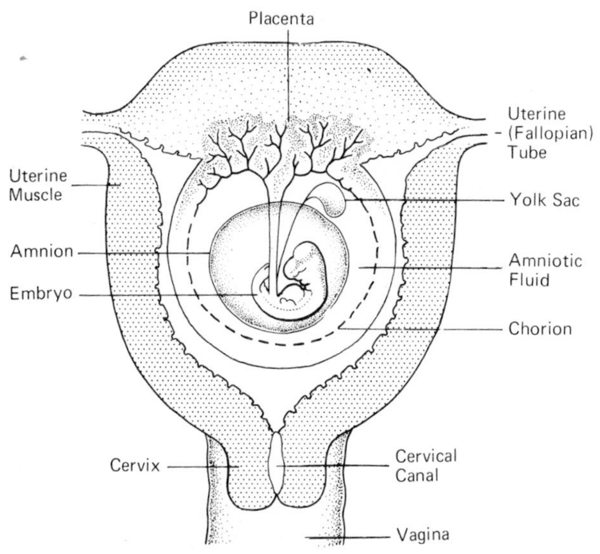

Fig. 240 Diagram of uterus and fetus in early pregnancy.

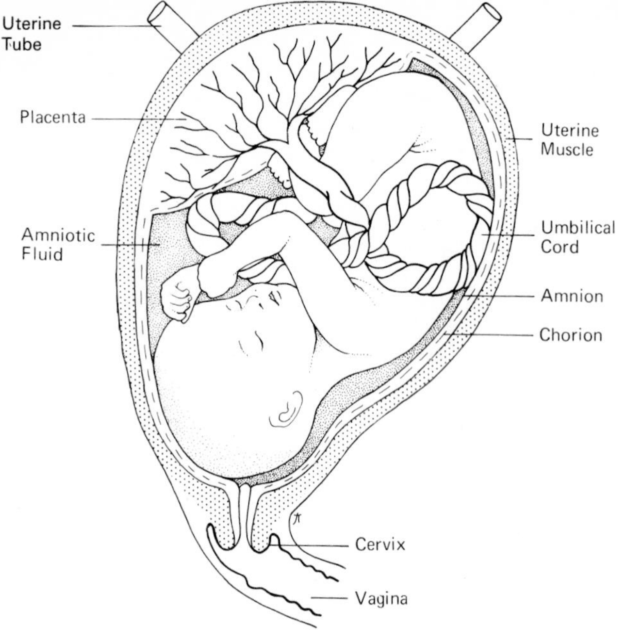

Fig. 241 Diagram illustrating full-term pregnancy.

noticed. A dark line (*linea nigra*) may be seen extending from the umbilicus to the pubis.

4. The enlargement of the uterus causes the diaphragm to rise, so that towards the end of pregnancy thoracic respiration may predominate.
5. The pelvic joints become looser.

Physiological

1. In the early weeks of pregnancy the corpus luteum in the ovary grows progressively larger and secretes increasing quantities of oestrogens and progesterone. These hormones are essential for the growth of the uterus and the maintenance of pregnancy. The corpus luteum shrinks and becomes inactive by the fourth month of pregnancy and the placenta takes over the production of oestrogens and progesterone. The hormonal activity of the corpus luteum depends on the stimulating effect of luteinizing hormone (LH), which is produced at first by

the anterior pituitary gland and later by the chorion (when it is known as chorionic gonadotrophin).

2. Important metabolic changes take place. There is usually some gain in body weight irrespective of the increase in the size of the fetus. The infant obtains its nutritional requirements from the mother's food, but if this is inadequate it will obtain its nourishment at the expense of the tissues of the mother, who will lose weight and may suffer in health as a result.

   The fetus lays up a large store of iron in its liver. This is important because for the first six or more months of its life it will be living on milk which contains no iron. Iron is essential for the formation of haemoglobin and any deficiency will lead to anaemia.
3. The calcium and phosphorus requirements of the fetus are considerable and are necessary for the formation of bone. Unless sufficient quantities are present in the maternal diet they will be taken from the mother's own reserves. This may account for the increased tendency to dental caries during pregnancy and indicates the importance of proper dental care during this period.
4. It follows that careful attention must be given to the mother's diet during pregnancy. She should have adequate supplies of carbohydrate, protein and fat such as are obtained in a normal diet supplemented by extra milk, vitamins, calcium and iron.
5. In the early weeks of pregnancy the corpus luteum in the ovary enlarges and secretes increasing quantities of oestrogens and progesterone which are responsible for the growth of the uterus and the breasts (see page 302).

   By the fourth month of pregnancy, the corpus luteum becomes inactive and the placenta produces these hormones.

   The hormone-secreting activity of the corpus luteum is stimulated by the luteinizing hormone (LH) produced by the anterior pituitary gland (page 299).
6. The plasma volume rises by 40–50 per cent during pregnancy. The extra plasma dilutes the blood and the concentration of haemoglobin consequently falls. This must not be interpreted as anaemia which, however, does occur if the dietary intake of iron and folate is inadequate for the combined needs of the mother and fetus.
7. The work-load on the heart is increased by the increased demand for oxygen (oxygen consumption rises by 20 per cent

in pregnancy), the increased blood volume and the increased size of the vascular bed, i.e. the placenta.

Ageing

Ageing is not something that happens suddenly after 40, 50, 60 or some other determinate number of years. Old age is, with certain exceptions, the culmination of many years of slowly declining function and degenerating structure. People do not all age at the same rate, and in clinical medicine biological age is more important than chronological age.

It is particularly important for doctors, nurses and others to know what changes ageing causes, as these must be taken into account in the treatment of disease.

Normal elderly people have a physiology that is adequate for resting conditions but, compared to young people, they show slower adjustment to environmental change and a diminished reserve capacity. All physiological functions decline with age.

Changes due to ageing

Skin. The skin loses its elasticity.

Eyes. The lens of the eye also loses its elasticity, with the result that it is unable to accommodate to a shape sufficiently convex to focus near objects. The 'long sight' resulting from this ageing of the lens is known as presbyopia. Sometimes the lens becomes opaque (cataract).

Stature. Body stature diminishes, mainly as a result of diminution in the thickness of intervertebral discs.

Body weight. The usual pattern is a slow increase in body weight up to the age of 60 years and then a decline. Although body fat tends to increase with age, the lean body mass (composed mostly of muscle and bone) decreases.

Muscles. Muscle power decreases with age.

Heart. Myocardial function diminishes and cardiac output falls gradually with age. Myocardial reserve is diminished and cardiac failure is more easily induced (e.g. by over-zealous intravenous therapy).

Arteries. These generally lose their elasticity and the blood pressure, especially the systolic pressure, increases.

Kidneys. The number of nephrons (glomeruli and tubules) decreases and the kidneys become smaller. Therefore, renal function diminishes.

Lungs. Chest movement diminishes because of increasing rigidity of the thoracic cage, and vital capacity is reduced. Ageing also results in less efficient diffusion and tissue utilization of oxygen. It is not surprising that elderly people more easily become short of breath than do young people.

Alimentary tract. Changes with ageing include atrophy, diminished enzyme secretion, and reduced gastrointestinal activity. As a result of these changes, the digestion of food takes longer and its absorption may be impaired. The reduced gastrointestinal movement may lead to constipation.

Endocrine glands. The secretion of sex hormones declines sharply in women (at the menopause) and progressively in men. Thyroxine secretion and hence basal metabolic rate (BMR) gradually diminishes with age. Glucose tolerance also diminishes.

Autonomic nervous system. Vasomotor control is often impaired in elderly people. Diminished blood supply to the brain may cause unsteadiness, faintness, and falling after sudden assumption of the erect posture. Nurses have to remember this when getting elderly patients out of bed, and may have to proceed very slowly.

Brain. There is a daily loss from the brain of about 1,000 nerve cells but fortunately there is a large reserve and many elderly people never show obvious senile mental changes.

Temperature regulation. This may be impaired and infections, particularly pneumonia, in old people commonly produce no pyrexia. The elderly tend to be liable to hypothermia because of impairment of their mechanism for heat production and preservation. Some people feel the cold more as they age whilst others are uncomplaining in a cold home environment which their younger relatives, accustomed to efficient modern heating, would not tolerate.

Thirst. The feeling of thirst may be diminished in elderly people and help to cause dehydration, particularly when they become ill.

Pain. Pain is sometimes less keenly appreciated by elderly people. A myocardial infarction or 'acute abdomen' may be painless (*silent*).

Resistance to infection. There is increased susceptibility to infection in old age because of a diminished antibody response and a reduced immunoglobulin concentration.

Relevance of ageing to clinical medicine

Nurses, physiotherapists and other professional people will realize that, because of impaired vasomotor control, elderly people often cannot be got out of their beds hurriedly without making them feel faint and unsteady.

Doctors avoid misguided attempts at treatment of a raised blood pressure by relating the degree of elevation to the patient's age. For example a blood pressure of 200/100 mm Hg in a patient aged 70 years would not call for any treatment. Indeed, effective treatment would be likely to induce disabling postural hypotension.

Drug treatment of any condition may be fraught with special problems in old age. The absorption, metabolism and excretion of a drug may all be slowed. Therefore it is often necessary to reduce the standard dosage of some drugs. This is a factor which accounts for the increased incidence of digitalis intoxication in elderly patients.

Miscellaneous Short Questions

Write brief notes on:

Cartilage
Periosteum
Bone marrow
Epiphysis
Plain muscle
Mucous membrane

Clavicle
Carpus
Great trochanter
Tarsus
Astragalus

Shoulder joint
Hip joint
Knee joint

Tendon
Diaphragm
Endocardium
Aorta
Mitral valve
Carotid artery
Blood supply of brain
Circle of Willis
Superior vena cava
Inferior vena cava

Blood pressure
The pulse

Red blood cells
Haemoglobin
White blood cells
Blood platelets
Plasma
Serum

Lymph
Lymph glands
Thoracic duct
Spleen

Enzyme
Salivary glands
The tongue
Pharynx
Oesophagus
Peristalsis
Large intestine
Vermiform appendix
Gastric juice

Gall bladder
Bile

Write brief notes on:

Calorie
Vitamins

Sweat glands
Nails

Air Sinuses
Larynx
Eustachian tube
Vocal cords
Trachea
Pleura
Carbon dioxide

Nephron

Thyroid gland
Parathormone

Suprarenal gland
Pituitary gland
Meninges
Cerebrospinal fluid
Medulla
Cerebellum
Vagus nerve

Tympanum
Middle ear
Ossicles
Iris
Retina
Tears

Fallopian tube
Ovary
Mammary gland.

Weights and Measures

1 ounce (oz) = 437·5 grains (gr) = 28·35 grams (G)
1 pound (lb) = 16 ounces = 7000 grains = 0·45 kilogram (kg)

1 yard = 36 inches (in) = 0·91 metres (m)
1 foot = 30·48 centimetres (cm)
1 inch = 2·54 centimetres

1 gram (G) = 15·43 grains
1 kilogram = 2·2 pounds = 35·27 ounces

1 litre = 1·75 pints = 61 cubic inches = 0·22 gallon

1 metre = 1·09 yards = 39·37 inches = 3·28 feet

To convert:
grains to grams multiply by 0·065
grams to grains multiply by 15·0
grams to ounces multiply by 0·03
ounces to grams multiply by 28·0
minims to millilitres multiply by 0·06
pints to litres multiply by 0·57

Fahrenheit to Centrigrade, subtract 32 and multiply by $\frac{5}{9}$
Centigrade to Fahrenheit, multiply by $\frac{9}{5}$ and add 32

Some approximate equivalents used in prescribing:

1 milligram (mg)	= $\frac{1}{60}$ grain
1 gram	= 15 grains
30 grams	= 1 ounce
1 kilogram	= $2\frac{1}{4}$ pounds
6·5 kilograms	= 1 stone (14 pounds)
1 millilitre (ml)	= 15 minims
30 millilitres	= 1 fluid ounce
600 millilitres	= 1 pint
1 litre	= 35 fluid ounces = 1·75 pints

Index

Abdomen, boundaries of, 37; contents of, 37; regions of, 203
Abdominal ring, 384
Absorption of carbohydrate, 221, 236; of fat, 222, 238; of protein, 221, 239
Accommodation, 360
Acetabular labrum, 98
Acetabulum (innominate bone), 54
Achilles, tendon of, 61, 124
Acid, ascorbic, 247; hydrochloric, 216; lactic, 28; uric, 283
Acidosis, 284
Acromegaly, 299
ACTH, 295
Addison's disease, 297
Adenoids, 262
Adrenaline, 297
Adrenergic receptors, 339
Afferent impulse, 306
Ageing, 390
Agglutinins, 175
Air sacs, 264
Alcohol, 252
Alimentary canal, scheme of, 189
Alkali reserve, 284
Alveoli, 264
Amoeba, 12
Ampulla of Vater, 210, 227
Amylase, 232
Anabolism, 233
Anaemia (pernicious), 167
Anatomy, definition of, 2
Aneurine, 246
Angle, epigastric, 88; of ribs, 87
Antecubital fossa, 382
Antrum of Highmore, 72
Anus, 214
Aorta, 135
Aphasia, 318
Aponeurosis, 109
Appendix, vermiform, 212
Aqueous humour, 358
Arachnoid mater, 308
Arch, palmar, 141, 383; plantar, 143, 386
Arteries, 135 *et seq*; hepatic, 225; retinal, 357
Arteriole, 127
Articulation, definition of, 90
Astragalus, 61
Athetosis, 312
Atlas, 81
Atom, 3
Atrium (heart), 129
Auditory (Eustachian) tube, 262, 345
Auricle (ear), 343
Auricle (heart), 129
Autonomic nervous system, 336
Axilla, 381
Axis, 81
Axon, 30

Basal ganglia, 311; metabolism, 235
Beri-beri, 247
Bile, duct, 227; pigments, 228; salts, 229; secretion of, 228

Bilirubin, 228
Biliverdin, 228
Bladder, 280
Blind spot, 357
Blood, 164; clotting of, 171; functions of, 174; groups, 174; plasma, 171; platelets, 169; pressure, 157; reaction of, 284; sugar, 285; transfusion, 174; vessels, nervous control of, 162
Blushing, 162
Bolus, 201
Bone, composition of, 21; development of, 22; general structure, 21; growth of, 23
Bones, long, 40; sesamoid, 41; short, 40
Bowman's capsule, 278
Brain, anatomy of, 309; physiology of, 314; stem, 314; functions of, 318
Bread, 251
Breast, 370
Bronchus, 264
Bursa, 109

Caecum, 211
Calcaneum, 61
Calciferol, 247
Calcification, 23
Calcitonin, 293
Calorie, 234
Calyx, 279
Canal, central of spinal cord, 325; Haversian, 22; Hunter's, 143, 385; inguinal, 384; semicircular, 348
Canthus of eye, 351
Capillary, 127
Capitate bone (carpus), 53
Capitulum (of humerus), 50
Carbohydrates, 221, 236
Cardia (of stomach), 207
Cardiac catheterization, 153
Carotene, 246
Carotid sheath, 137
Carpus, bones of, 53
Cartilage, 20; arytenoid, 263; costal, 82; cricoid, 263; semilunar (of knee-joint), 100; thyroid, 262; triangular (of wrist), 95
Caruncle, lacrimal, 352
Castle, intrinsic factor, 167, 217
Cauda equina, 323
Cavity, glenoid (of scapula), 46, 92; nasal, 260
Cell, 8; nerve, 30
Centre, heat-regulating, 257; respiratory, 270, 319; vasomotor, 319
Cerebellum, 313; functions of, 318
Cerebrospinal fluid, 308
Cerebrum, 309; function of, 318
Cerumen, 344
Cheese, 251
Cholesterol, 228
Chordae tendineae, 134
Choroid, 356, plexus, 308
Chromosome, 9
Chyle, 182
Chyme, 217
Cilia, 16
Ciliary body, 356
Circle of Willis, 139
Circumvallate papillae, 192
Cisterna chyli, 180
Clavicle, 47
Climacteric, 372
Clitoris, 369
Coccyx, 83
Cochlea, 348
Coffee, 251
Collagen, 19
Colloid (thyroid), 291
Colon, 212
Colostrum, 371
Combustion, 11
Compound (chemical), 2
Concha, 344
Condiments, 252

Condyle, definition of, 41; of femur, 58; of humerus, 49; of mandible, 76; of occipital bone, 69
Cones, 357
Confluence of sinuses, 152
Conjunctiva, 352
Consonants, 273
Constipation, 219
Cornea, 356
Coronoid process (of mandible), 76; of ulna, 52
Corpus callosum, 310
Corpus cavernosum, 377
Corpus luteum, 372
Corpus spongiosum, 377
Corpuscles, red, 164
Corti, organ of, 348
Corticotrophin (ACTH), 295
Cortisone, 295
Cranial nerves, 319
Cranium, 62
Crest, definition of, 41
Cretinism, 293
Cribriform plate (of ethmoid), 71
Crus (of diaphragm), 116
Cubital fossa, 382
Cuboid bone (of foot), 62
Cuneiform bone of carpus, 53; of foot, 62
Cyanocobalamin, 167, 247

De-amination, 229, 283
Defaecation, 219
Deglutition, 201
Dendrite, 32
Dens, 81
Dentine, 194
Deoxycortone, 295
Dermis, 254
Diabetes, 304
Dialysis (peritoneal), 206
Diaphragm, 116
Diaphysis, 23
Diarrhoea, 220
Diastole, 156
Diet, normal, 245
Digestive system, 187
Di-saccharides, 236
Distal, definition of, 33
DNA, 9
Douglas, pouch of, 204, 365
Droplet infection, 269
Duct, bile, 227; cystic, 227, ejaculatory, 377; hepatic, 227; lacrimal, 352; lactiferous, 370; pancreatic, 231; right lymphatic, 182; Stenson's 197; thoracic, 180; Wharton's, 197
Ductless glands, 289
Duodenum, 209; digestion in, 217
Dura mater, 307
Dysarthria, 318

Ear, anatomy of, 343; drum, 347; internal, 347; middle, 344
Efferent impulse, 306
Eggs, 251
Electro-encephalogram, 315
Element (chemical), 2
Enamel, 195
Endocardium, 134
Endocrine system, 289
Endolymph, 347
Endometrium, 366
Energy requirements, 234
Enterokinase, 218
Enzymes, 187
Eosinophile, 169
Epicondyle, of humerus, 49
Epidermis, 254
Epididymis, 374
Epigastric angle, 88
Epiglottis, 263
Epiphysis, 24
Epithelium, germinal, 367; types of, 15
Erepsin, 218
Ergosterol, 247
Erythrocytes, 164
Erythrocyte maturing factor, 166
Ethmoid bone, 71
Eustachian tube, 262, 345

Eversion (of foot), 103
Expiration, 266
Extrinsic factor, 167, 217
Eye, muscles, 353
Eyebrows, 351
Eyelash, 351
Eyelids, 351

Facet, definition of, 41
Faeces, 220
Fallopian tube, 367
Falx cerebri, 152, 307
Fascia, 108; lumbar, 120; palmar 109, 383; plantar, 109
Fat, 222, 229, 238
Fatigue of muscle, 28
Fauces, 191
Femoral triangle, 142, 385
Femur, 56
Fenestra, ovalis, 344; rotunda, 345
Fibres, nerve, 30
Fibrin, 171
Fibrinogen, 170
Fibula, 60
Filiform papillae, 192
Filum terminale, 323
Fimbria, 367
Fissure, palpebral, 351; of Rolando, 310; of Sylvius, 310
Flat foot, 61
Fluid balance, 229, 244
Folic acid, 168
Fontanelles, 76
Foodstuff, absorption of, 229; calorie value of, 60; digestion of, 250
Foot, bones of, 60; joints of, 103; sole of, 385
Foramen, definition of, 41; magnum, 65; of Magendie, 309; obturator, 56
Fossa cubital, 382; anterior cranial, 66; coronoid (of humerus), 50; definition of, 42; infraspinous (of scapula), 45; middle cranial, 66; olecranon (of humerus), 50; subscapular, 44; supraspinous (of scapula), 45
Fourchette, 368
Fracture, greenstick, 25; Colles', 53
Frontal bone, 68
Fungiform papillae, 192

Gall-bladder, 227
Gamma globulin, 170
Ganglion, Gasserian, 321; posterior root, 326; stellate, 338; sympathetic, 337
Gastric juice, 215
Gastrin, 215
Gene, 9
Gladiolus (of sternum), 85
Gland, Bartholin's, 368; ceruminous, 255, 344; ductless, 289; lacrimal, 352; lymphatic, 182; Meibomian, 351; Montgomery, 370; pineal, 303; pituitary, 298; sebaceous, 255; sex, 301; sublingual, 197; submandibular (submaxillary), 197; sweat, 256
Glans penis, 378
Gluten, 251
Glaucoma, 358
Glycogen, 229, 237
Goitre, 291
Goose-flesh, 258
Graafian follicle, 367
Graves' disease, 291
Grey matter, 30; of brain, 310; of spinal cord, 325
Gristle, 20
Groove, bicipital (of humerus, 48; musculo-spiral (of humerus), 48

Haem, 165
Haemoglobin, 164
Hair, 256
Hamate bone (carpus), 53
Haversian canal, 22

Hearing, mechanism of, 349
Heart, 129; beat, 155; muscle, 28; sounds, 157
Heat, production, loss, 257
Henle, loop of, 278
Heparin, 172, 230
Hepatic flexure (of colon), 213
Hernia (inguinal), 384; diaphragmatic, 200
Hiccough, 268
Highmore, antrum of, 72
Hirudin, 172
His, bundle of, 157
Hormone, 289
Humerus, 48
Hunter's canal, 143 385
Hydrocele, 376
Hymen, 369
Hyoid bone, 263
Hypermetropia, 361
Hypophysis, 298
Hypothalamus, 312
Hypothenar eminence, 116

Ileum, 210; digestion in, 217
Ilium, 54
Immunity, 178
Incus, 346
Inferior vena cava, 145
Ingestion, 187
Inguinal canal, 384
Immunoglobulin, 170
Innominate bone, 54
Insertion of muscles, 109
Inspiration, 266
Insulin, 231, 303
Interferon, 14
Internal capsule, 312
Intervertebral disc, 77
Intestine, large, 211; absorption from, 222; functions of, 218; small, 209; absorption from, 221
Intrinsic factor, 167, 217
Inversion (of foot), 103
Invertase, 218
Involuntary nervous system, 336
Iodine, 241, 291
Ions, 5
Iris, 356
Iron, storage, 213, 241
Ischium, 56
Islets of Langerhans, 231, 303

Jaundice, 229
Jaw, lower, 75; upper 72
Jejunum, 210; digestion in, 217
Joints, 90 *et seq.*;
Joule, 234

Keratin, 15, 254
Kidney, 276; functions of, 283
Knee-cap, 58; joint, 99
Kyphosis, 78

Labia majora, minora, 368
Lacrimal bone, 73; duct, 352; gland, 352
Lacteal, 182, 222
Lacunae, 22
Laminae (of vertebrae), 79
Langerhans, islets of, 231, 303
Larynx, 262
Lens, 358
Leucocyte, 168
Ligament, broad (uterus), 365; cruciate, 99; ilio-femoral, 98; inguinal, 119; lateral (of knee joint), 99; peritoneal, 205; Poupart's, 119; round (uterus), 365
Ligamenta flava, 19, 84
Ligaments of vertebral column, 83
Ligamentum teres (of hip joint), 57, 98
Line, ilio-pectineal, 56
Linea alba, 119
Linea aspera (of femur), 58
Lipase, 222, 232
Liver, 223; blood supply, 225; functions of, 228
Loop of Henle, 278

Lordosis, 78
Lower motor neuron, 333
Lumbago, 120
Lumbar puncture, 309
Lunate bone (carpus), 53
Lungs, 265
Lymphatic system, 179
Lymphocyte, 169

Macula, 357
Magendie, foramen of, 309
Malar bone, 73
Malleolus, lateral (of fibula), 60; medial (of tibia), 60
Malleus, 346
Malpighian body, 278
Maltase, 218
Mammae, 370
Mandible, 75
Manubrium (of sternum), 86
Marrow, bone, 22
Mastication, 187, 196
Mastoid process (of temporal bone), 70
Maxilla, 72
Maxillary air sinus, 72
Meatus, external auditory, 69; internal auditory, 70
Medulla oblongata, 314
Megakaryocyte, 169
Megaloblast, 166
Membrane, interosseous, of forearm, 50, of leg, 102; mucous, 32; of Shrapnell, 347; serous, 32; synovial, 32; tympanic, 347
Menaphthone, 249
Meninges, 307
Meniscus (of knee joint), 100
Menopause, 372
Menstruation, 373
Mesentery, 204
Metabolism, 233; basal, 235; calcium, 240; carbohydrate, 236; fat, 238; potassium and sodium, 241; protein, 239; salts, 240
Metacarpal bones, 53
Metatarsal bones, 62
Micturition, 287
Mid-brain, 314
Milk, 250
Mitosis, 12
Modiolus, 348
Mono-saccharides, 236
Mons veneris, 368
Motor cortex, 317; path, 332
Mouth, 191; absorption from, 220; functions of, 196
Mucus, 32
Muscle, composition of, 28; properties of, 28
Muscles, 108 *et seq.*; of eye, 353; papillary, 134
Myocardium, 133
Myopia, 361
Myxoedema, 291

Nails, 255
Nasal cavity, 260
Naso-pharynx, 262
Navicular bone (foot), 61
Neck, anatomical (humerus), 48; femur, 57; surgical (humerus), 48; triangles of, 380
Nerves (cranial), 319; (spinal), 326; (vasoconstrictor), 162
Nervous system, 306
Neural canal, 80
Neuroglia, 30
Neuro-muscular mechanism, 26
Neurone, 30
Nicotinic acid, 247
Node, atrio-ventricular, 157; sino-atrial, 157
Normoblast, 166
Notch, sciatic, 54; trochlear (ulna), 52

Occipital bone, 69
Ocular tension, 358

Odontoid process (of axis), 81
Oesophagus, 200
Oestrogen, 301
Omentum, 204
Opposition, 96
Orbit, bones of, 76
Organ of Corti, 348
Organs of sensation, 340
Origin of muscles, 109
Os, calcis, 61; magnum (of carpus), 53
Osmotic pressure, 243
Ossicles, 346
Ossification, 23
Osteitis fibrosa, 293
Osteoblasts, 23
Otitis media, 350
Oval window, 344
Ovary, 367; internal secretions, 301
Oxidation, 11
Oxygen (administration), 272
Oxytocin, 301

Palate bone, 73; hard, soft, 191
Palm of hand, 382
Palmar fascia, 383
Palpebral fissure, 351
Pancreas, 230, 303
Papillae, of tongue, 192
Parasympathetic system, 336
Parathyroid gland, 293
Parietal bone, 68
Parkinsonism, 312
Patella, 58
Pedicles (of vertebrae), 79
Pellagra, 247
Pelvic cavity, boundaries of, 37; contents of, 37
Pelvis of kidney, 279
Penis, 377
Pepsin, 215
Pericardium, 132
Perichondrium, 20
Pericranium, 380
Perilymph, 347
Perineum, 369
Periosteum, 22
Peristalsis, 201
Peritoneum, 203
Petrous portion (temporal bone), 70
Peyer's patches, 211
pH, 6
Phagocyte, 168
Phalanges, hand, 53; foot, 62
Pharynx, 199
Physiology, definition of, 2
Pia mater, 308
Pineal gland, 303
Pinna, 343
Pisiform bone, 53
Pitocin, 301
Pituitary gland, 298
Pituitrin, 7
Plasma, 170
Platelets (blood), 169
Pleura, 266
Plexus, 327; brachial, 327; cardiac, 338; cervical, 327; choroid, 308; coeliac, 338; lumbo-sacral, 330; solar, 338
Plicae circulares, 210
Poly-saccharides, 237
Pons, 314
Popliteal space, 385
Portal circulation, 147; fissure, 224
Pouch of Douglas, 204, 365
Prednisolone, 295
Pregnancy, anatomical changes in, 386
Prepuce, 378
Presbyopia, 361
Pressure, blood, 157; osmotic, 243; pulse, 159; venous, 160
Pressure points, 144
Proerythrocyte, 166
Progesterone, 301
Prolactin, 299
Pronation, 50, 94
Prostate gland, 377
Protein, 9, 221, 239
Prothrombin, 171
Protoplasm, 9

Proximal, definition of, 33
Pseudopodium, 12
Ptyalin, 197
Puberty, 372
Pubis, 55
Pulmonary circulation, 154
Pulse, 160
Purpura, 170
Pyloric antrum, 208
Pyramidal tract, 325

Radius, 50
Receptaculum chyli, 180
Rectum, 214
Rectus sheath, 119
Red cells, 164; development of, 165
Reflex action, 333
Reflex, gastro-colic, 218
Rennin, 216
Reproductive system, 363
Respiration, physiology of, 268; regulation of, 270
Respiratory movements, 266; system, 259
Resting juice, 215
Reticulo-endothelial system, 185
Retina, 357
Rhesus factor, 176
Rib, 86
Riboflavine, 247
Rickets, 25, 249
Rigor mortis, 29
Rods, 357
Rolando, fissure of, 310
Roughage, 218, 249
Round window, 345

Sacrum, 82
Salivary glands, 197
Salts, excretion of, 283; metabolism of, 240
Saphenous opening, 146, 385
Scalp, 380
Scaphoid bone of foot, 61; wrist, 53
Scapula, 44
Scarpa's triangle, 142, 385
Sciatica, 330
Sclera, 355
Scoliosis, 78
Scrotum, 376
Scurvy, 247
Sebum, 255
Sella turcica (sphenoid), 71
Semicircular canal, 348
Semilunar bone (of carpus), 53
Seminal vesicle, 376
Sensation, enteroceptive, 341; exteroceptive, 341
Sensory path, 330
Serum, 170
Shock, 162
Sigh, 268
Sight, mechanism of, 358
Sigmoid colon, 213
Simmond's disease, 301
Sinus, cavernous, 152; lateral, 70, 152; sigmoid, 152; superior longitudinal, 152; transverse, 70, 152
Skeleton, general description, 42
Skin, 254
Skull, bones of, 62
Sleep, 316
Smell, sense of, 341
Sodium glycocholate, 229; taurocholate, 229
Sole of foot, 385
Solutions, isotonic, etc., 243
Spectacles, 363
Speech, 273
Spermatic cord, 375
Spermatozoa, 378
Sphenoid bone, 71
sphygmomanometer, 158
Spinal cord, 323
Spine, anterior superior (of ilium), 54; as a whole, 78; movements of, 84; posterior superior (of ilium), 54
Spleen, 184
Splenic flexure, 213
Squama, of temporal bone, 69

Stapes, 346
Stercobilin, 228
Sternum, 85
Steroids, 295
Stomach, 206; absorption from, 221; functions of, 215; movements of, 216
Subarachnoid space, 308
Succus entericus, 218
Sulcus, central, 310
Superior vena cava, 145, 153
Supination, 50, 94
Suprarenal gland, 293
Suture, coronal, 65; lambdoid, 65; sagittal, 65
Swallowing, 201
Sweat, 256
Sylvius, fissure of, 310
Sympathetic system, 338
Symphysis, of mandible, 75; pubis, 55, 97
Synapse, 31
Systole, 156

Talus, 61
Tarsus, 61
Taste buds, 192; sense of, 342
Tea, 251
Tears, 352
Teeth, 192
Temperature of body, 257
Temporal bone, 69
Tendon of Achilles, 61
Tendons, 109
Tentorium cerebelli, 307, 313
Testis, 374; internal secretion, 302
Testosterone, 302
Tetany, 293
Thalamus, 312
Thenar eminence, 116
Thiamine, 246
Thoracic duct, 180
Thorax, bones of, 85; boundaries of, 35; contents of, 35
Threshold value, 286
Thrombocyte, 169
Thrombogen, 173
Thrombokinase, 173
Thymus gland, 303
Thyroid gland, 290
Thyrotoxicosis, 291
Thyroxine, 291
Tibia, 58
Tissue, adipose, 19; areolar, 19; connective, 18; fibrous, 19; fluid, 179; lymphoid, 19; muscular, 25; nervous, 29
Tone, muscular, 28
Tongue, 192, 342; functions of, 196
Tonsil, 185
Torcular Herophili, 152
Trachea, 263
Trapezium bone (carpus), 53
Trapezoid bone (carpus), 53
Triangle, femoral or Scarpa's, 385
Trigone of bladder, 280
Trypsin, 232
Triquetral bone (carpus), 53
Trochanter, definition of, 42; 8great), 57; (small), 57
Trochlea (humerus), 49; notch (ulna), 52
Tube, auditory, 345; Eustachian, 345; Fallopian, 367
Tubercle, adductor (of femur), 58; definition of, 42; of ribs, 81
Tuberosity, bicipital (radius), 50; definition of, 42; deltoid (humerus), 48; of humerus (great and small), 48; of ischium, 56; of tibia, 59
Tubule, renal, 278; seminiferous, 374
Tunica vaginalis, 375
Turbinate bones, 72
Tympanic membrane, 347

Ulna, 51
Ulnar bursa, 383
Umbilicus, 384
Upper motor neuron, 333

Urea, 283; formation of, 229
Ureter, 279
Urethra, 281
Urinary system, 275
Urine, composition of, 281; formation of, 286; retention of, 288
Uterus, 364
Uvula, 191

Vagina, 368
Valve, ileo-caecal, 211
Valves of heart, 129; of veins, 130
Valvulae conniventes, 210
Vas deferens, 374
Vaso-motor centre, 162, 319
Vasopressin, 301
Veins, 145 *et seq.*
Ventricle of brain, 312; of heart, 129
Venule, 127
Vertebrae, 78 *et seq.*
Vertebral column, 77; ligaments of, 83
Vestibule, of ear, 348
Villi, 210
Vision, 350; defects of, 360
Vital capacity, 269
Vital centres, 319
Vitamin K, 173, 249
Vitamins, 245 *et seq.*
Vitreous body, 358
Vocal cord, 263
Voice production, 272
Vomer, 73
Vowels, 273
Vulva, 368

Waste products, excretion, 234
Water, 242, 283
Wax, ear, 344
Whispering, 273
White cells, 168; matter, 31, of brain, 312, of spinal cord, 325
Willis (circle of), 139
Wormian bones, 65
Wrist, bones of, 53

Xerophthalmia, 246
Xiphoid (of sternum), 86

Yawn, 268

Zygoma, 73